Springer Texts in Social Sciences

This textbook series delivers high-quality instructional content for graduates and advanced graduates in the social sciences. It comprises self-contained edited or authored books with comprehensive international coverage that are suitable for class as well as for individual self-study and professional development. The series covers core concepts, key methodological approaches, and important issues within key social science disciplines and also across these disciplines. All texts are authored by established experts in their fields and offer a solid methodological background, accompanied by pedagogical materials to serve students, such as practical examples, exercises, case studies etc. Textbooks published in this series are aimed at graduate and advanced graduate students, but are also valuable to early career and established researchers as important resources for their education, knowledge and teaching.
The books in this series may come under, but are not limited to, these fields:

- Sociology
- Anthropology
- Population studies
- Migration studies
- Quality of life and wellbeing research

Andrei P. Kirilenko

Practical Data Mining with AI for Social Scientists

Springer

Andrei P. Kirilenko
Department of Tourism, Hospitality &
Event Management
University of Florida
Gainesville, FL, USA

ISSN 2730-6135 ISSN 2730-6143 (electronic)
Springer Texts in Social Sciences
ISBN 978-3-031-89688-0 ISBN 978-3-031-89689-7 (eBook)
https://doi.org/10.1007/978-3-031-89689-7

This Springer imprint is published by the registered company Springer Nature Switzerland AG
The registered company address is: Gewerbestrasse 11, 6330 Cham, Switzerland

To my wife and daughters

Preface

The world is running on data. Every day, we leave multiple digital footprints: making phone calls, buying groceries, walking in view of security cameras and webcams, wearing fitness trackers, searching online, sending emails, and streaming videos. We also post our thoughts, feelings, observations, insecurities, and plans on various social media platforms, acting as a huge, globally distributed, noisy network of human sensors of nature and society. Most of the time, we are unaware of the data trail we leave. Other times, we intend our data for specific audiences—for instance, Facebook friends. Less frequently, we target the highest possible data visibility (think of Internet influencers). Whatever the awareness or intention, our digital footprints are repackaged, resold, analyzed, and returned to us in the form of customized products and services, but also target advertisement and online content designed to shape our opinions, behaviors, and decisions.

Some of those changes are positive, others clearly negative, which is always true during a transformational change. From my point of view as a scientist, the abundance, diversity, and global coverage of newly available data create enormous possibilities for achieving a better understanding of societal processes. Yet, there is a challenge. Methodologically, new types of data—large in volume, noisy, unstructured, and full of errors, require analytical tools that differ from those that are traditionally taught in statistics classes targeting majors other than statistics and mathematics. These tools are loosely referred to as *data mining*.

In many aspects, data mining is similar to this "traditional" statistics. Both disciplines focus on extracting insights from data. Both require preprocessing steps such as cleaning, normalizing, and transforming raw data. Both share common techniques like clustering, classification, and hypothesis testing. Yet, there are crucial differences in scope, approach, data complexity, and primary goals of analysis. For instance, statistics focuses on hypothesis testing and making inferences about populations based on sample data. In fact, I occasionally face harsh criticism from paper reviewers claiming that research that is not hypothesis-driven is "unscientific". As opposed to that, data mining focuses on uncovering hidden patterns and relationships in large datasets, often without predefined hypotheses. In another example, an experience of working with small data samples leads to an unhealthy obsession with p-values (compulsive hypothesis testing of all results) even with analysis of datasets of thousands and millions of records.

Fortunately, it has never been easier to get your feet wet with data mining! The AI revolution has gifted us with intelligent online assistants, instrumental in translating instructions in human language into the computer code of data mining tools. To take advantage of this possibility, however, an analyst needs to know *what to ask* to achieve the intended outcomes. Hence this book.

This book aims to teach how to use AI assistants for data mining effectively. To achieve this goal, we need to know the methods and learn how these methods can be applied in practice. This is why each chapter of the book is structured in the following way. First, the theoretical part of the chapter introduces methods and discusses data considerations, results, and limitations. Then, the practical part demonstrates how the methods can be applied to the analysis of real-world data: social media posts, university basketball surveys, animosity towards the US effect on tourism, and so on. These datasets were intentionally selected to represent data typical for social scientists.

The first chapter is introductory and demonstrates how to work with an AI assistant to develop Python code for a simple analysis. The bulk of the book content is divided into several broad themes. The first theme deals with data: preprocessing of raw numerical data (▶ Chap. 2) and textual data (▶ Chap. 11), scraping raw data from websites (▶ Chap. 9), and data storage (▶ Chap. 15). The second theme is analysis of numerical data with association rules (▶ Chap. 3), decision trees (▶ Chap. 4), cluster analysis (▶ Chaps. 5 and 6), and classification (▶ Chap. 7). The third theme primarily deals with textual data and includes sentiment analysis (▶ Chap. 10), topic modeling (▶ Chapter 12), large language models (LLMs, ▶ Chap. 13), and social network analysis (▶ Chap. 14). Finally, cross-chapter topics are results validation and evaluation (▶ Chap. 8), ethics, and explainable AI (▶ Chap. 16).

The practical part of the chapters uses the Anaconda Cloud platform. Students will make a free account and learn how to use the platform in ▶ Chap. 1 lab. The platform selection was influenced by an integrated AI assistant, which simplifies code development. Yet, one may find using other AI assistants to aid in code development is easier. For instance, some labs use OpenAI's ChatGPT; another helpful option is Google's Gemini. Similarly, you may find it advantageous to use any other platforms providing tools for writing, testing, and managing Python code; my favorites are Jupiter Lab and PyCharm.

By no means is this book a complete compendium of data mining methods. The intention is to provide an introduction to a spectrum of tools, empowering the reader to continue learning novel tools to explore patterns in real-world data.

Andrei P.Kirilenko
Gainesville, Florida, USA

Competing Interests The author has no competing interests to declare that are relevant to the content of this manuscript.

Contents

1 **Introduction to Data Mining. CRISP-DM Process** ... 1
- 1.1 **Introduction** ... 2
- 1.2 **Digital Social Sciences** ... 3
- 1.3 **Introduction to CRISP-DM Process** ... 4
- 1.4 **Phases of CRISP-DM Process** ... 5
- 1.4.1 Business Understanding ... 5
- 1.4.2 Data Understanding ... 6
- 1.4.3 Data Preparation ... 6
- 1.4.4 Modeling ... 7
- 1.4.5 Evaluation ... 7
- 1.4.6 Deployment ... 7
- 1.5 **Conclusion** ... 8
- 1.6 **Review Questions** ... 8
- 1.7 **Lab** ... 8
- **Additional Reading** ... 21

2 **Data Pre-processing** ... 23
- 2.1 **Data Complexity** ... 24
- 2.2 **Data Type and Level of Measurement** ... 25
- 2.3 **Sources of Data Errors and Inconsistency** ... 28
- 2.4 **Data Cleaning** ... 30
- 2.4.1 Duplicates ... 30
- 2.4.2 Missing Data ... 30
- 2.4.3 Cleaning Text Data ... 31
- 2.4.4 Data Transformation ... 33
- 2.4.5 Detection and Cleaning Outliers ... 34
- 2.4.6 Inconsistent Data ... 36
- 2.4.7 Feature Engineering ... 36
- 2.4.8 Group Imbalances ... 37
- 2.5 **Conclusion** ... 38
- 2.6 **Review Questions** ... 38
- 2.7 **Lab** ... 39
- **Additional Reading** ... 53

3 **Introduction to Data Mining Methods. Association Rules** ... 55
- 3.1 **Statistics and Data Mining** ... 56
- 3.2 **Informal Definition of Association Rules** ... 57
- 3.3 **Metrics of Rule Interestingness** ... 59
- 3.3.1 Association Rules Formalized ... 59
- 3.3.2 Association Rules' Support and Confidence ... 59
- 3.3.3 Lift and Other Important Metrics of Interestingness ... 60
- 3.4 **Approaches to Mine Association Rules** ... 63

3.5 **Data Considerations** 64
3.6 **Conclusion** 65
3.7 **Review Questions** 66
3.8 **Lab** 66
Additional Reading 77

4 **Decision Trees** 79
4.1 **Introduction** 80
4.2 **Components of Decision Trees** 80
4.3 **Splitting Decisions** 82
4.3.1 Gini Impurity 83
4.3.2 Entropy and Information Gain 84
4.3.3 Splitting Decisions for Continuous Attribute Variables 85
4.4 **Pruning the Trees** 86
4.5 **Decision Trees with Continuous Target Variable** 87
4.6 **Method comparison** 90
4.7 **From Decision Trees to a Random Forest** 91
4.8 **Review Questions** 91
4.9 **Lab** 92
Additional Reading 109

5 **Clustering Techniques: K-means and DBSCAN** 111
5.1 **Basics of Cluster Analysis** 112
5.2 **Partitional Clustering: K-means** 114
5.3 **Evaluating the Clusters** 115
5.3.1 Euclidean Distance: Sum of Squared Errors 115
5.3.2 Manhattan Distance 117
5.3.3 Cosine Distance 118
5.4 **How Many Clusters to Select?** 119
5.4.1 Elbow Method 119
5.4.2 Silhouette Method 121
5.5 **Data Considerations and Limitations** 123
5.5.1 Levels of Measurement 123
5.5.2 Limitations 124
5.6 **Density-Based Partitioning: DBSCAN** 125
5.6.1 The Algorithm 125
5.6.2 DBSCAN Data Considerations and Limitations 127
5.7 **Review Questions** 128
5.8 **Lab** 128
Additional Reading 150

6 **Hierarchical Clustering** 151
6.1 **Introduction to Hierarchical Clustering** 152
6.2 **Agglomerative Clustering Mechanism** 154
6.2.1 Cluster Distance 155

6.2.2 Selecting Cluster Similarity Measure ... 156
6.2.3 Flattening Dendrograms ... 157
6.3 **Feature Selection** ... 159
6.4 **Methodological Considerations** ... 161
6.5 **Review Questions** ... 162
6.6 **Lab** ... 163
Additional Readings ... 181

7 **Predictive Analytics and Supervised Learning. Classification** ... 183
7.1 **Supervised Versus Unsupervised Learning** ... 184
7.2 **Probabilistic Approach to Classification** ... 185
7.2.1 Introduction to Conditional Probability ... 185
7.2.2 Naïve Bayes Classifier ... 188
7.2.3 Advantages of the Naïve Bayes Approach ... 191
7.2.4 Limitations ... 192
7.2.5 Naïve Bayes Variants ... 192
7.3 **Decision Trees and Random Forest** ... 193
7.4 **Support Vector Machines** ... 195
7.4.1 Artificial Neural Networks ... 196
7.4.2 Conclusion ... 198
7.5 **Review Questions** ... 198
7.6 **Lab** ... 199
Additional Reading ... 214

8 **Validation and Evaluation Methods** ... 215
8.1 **Introduction** ... 216
8.2 **Model Performance Evaluation Metrics** ... 217
8.2.1 Confusion Matrix ... 218
8.2.2 Common Measures of Model Performance ... 219
8.3 **Performance Metrics for Multi-class Classifiers** ... 221
8.3.1 Accuracy, Precision, Sensitivity, Specificity, and F1 ... 221
8.3.2 Accounting for By-Chance Agreement: Cohen's Kappa ... 222
8.4 **Performance Evaluation Procedures** ... 224
8.4.1 Holdout Method ... 224
8.4.2 Cross-Validation ... 225
8.4.3 Bootstrapping Procedure ... 226
8.5 **Comparing Model Performance at Different Classification Thresholds** ... 227
8.5.1 Cost-Sensitive Learning ... 227
8.5.2 Gain and Lift Charts ... 228
8.5.3 Response Operator Curve and Area Under Curve ... 230
8.5.4 Comparison of Multiple Models ... 231
8.6 **Conclusion** ... 231
8.7 **Review Questions** ... 232
8.8 **Lab** ... 233
Additional Readings ... 255

9 **Web Data Scraping** 257
9.1 **Introduction** 258
9.2 **API Scraping** 259
9.2.1 API Example 259
9.2.2 JSON Format 262
9.2.3 Authentication and Authorization 263
9.2.4 GET and POST HTTP Request Methods 266
9.2.5 API Limitations 267
9.3 **Web Scraping** 268
9.3.1 Web Page Structure 268
9.3.2 Scraping Basics 269
9.3.3 Scraping Multiple Related Elements 271
9.3.4 Navigating the Web Page Structure 272
9.4 **Scraping Tools** 274
9.4.1 No-Code Web Scraping Tools 274
9.4.2 Building Your Own Scraper 274
9.5 **Ethical and Legal Considerations** 275
9.6 **Conclusion** 277
9.7 **Review Questions** 277
9.8 **Lab** 278
Additional Reading 296

10 **Sentiment and Emotion Analysis** 297
10.1 **Introduction** 298
10.2 **Sentiment Analysis** 299
10.3 **Emotion Detection** 301
10.4 **Lexicon-Based Methods** 303
10.5 **Machine Learning Sentiment Analysis** 305
10.6 **Popular Solutions** 307
10.7 **Conclusion** 308
10.8 **Review Questions** 309
10.9 **Lab** 310
Additional Reading 322

11 **Text Mining Essentials** 323
11.1 **Introduction** 324
11.2 **Text Pre-processing** 325
11.2.1 Text Cleaning 325
11.2.2 Spelling Correction 325
11.2.3 Standardization 326
11.2.4 POS Tagging 327
11.2.5 Word Tokenization 328
11.2.6 Embedding 330
11.2.7 Visualization of the Word Space 331
11.3 **Text to Numbers** 332
11.4 **Measuring Documents' Similarity** 335

11.5 **Moving Further: Grouping the Texts** 336
11.6 **Conclusion** 338
11.7 **Review Questions** 338
11.8 **Lab** 339
Addtional Reading 358

12 **Topic Modeling: Latent Dirichlet Allocation** 359
12.1 **Introduction** 360
12.2 **Latent Dirichlet Allocation Idea** 361
12.3 **LDA Details** 362
12.3.1 Assumptions 362
12.3.2 Parameters 363
12.3.3 Interpreting Results 364
12.3.4 Finding the Optimal Number of Topics 367
12.4 **Implementation** 369
12.4.1 Best Practices 369
12.4.2 LDA Software 370
12.5 **Conclusion: Why LDA?** 371
12.6 **Review Questions** 372
12.7 **Lab** 372
Additional Reading 387

13 **Text Analysis with Large Language Models (LLMs)** 389
13.1 **Introduction** 390
13.2 **Language Models** 393
13.3 **Applications of LLMs for Text Analysis** 394
13.3.1 Sentiment Analysis with BERT 395
13.3.2 Text Classification and Summarization with BART 396
13.3.3 Topic Modeling with BERTopic 398
13.4 **Text Analysis with ChatGPT** 399
13.4.1 Examples: Sentiment Analysis and Topic Modeling 399
13.4.2 Prompt Engineering 401
13.5 **Conclusion** 403
13.6 **Abbreviations** 404
13.7 **Review Questions** 405
13.8 **Lab** 406
Additional Reading 431

14 **Introduction to Social Network Analysis** 433
14.1 **Introduction** 434
14.2 **Terminology and Theoretical Foundations** 436
14.2.1 Key Concepts 436
14.2.2 Bipartite Networks 437
14.2.3 Network Projection 438
14.3 **Measures** 440
14.3.1 Node Degree 440

14.3.2 Degree Distribution 440
14.3.3 Centrality Measures 441
14.4 **Network-Level Metrics** 442
14.4.1 Density 442
14.4.2 Homophily 443
14.4.3 Modularity and Community Detection 444
14.5 **Data Format** 445
14.6 **Software** 448
14.7 **Conclusion** 448
14.8 **Review Questions** 449
14.9 **Lab** 450
Additional Reading 465

15 **Understanding Data Storage and Databases** 467
15.1 **Introduction to Data Storage** 468
15.2 **Understanding Databases** 469
15.3 **Basic Database Operations** 470
15.4 **Merging the Tables** 471
15.4.1 The Keys 471
15.4.2 Table Join 472
15.4.3 Key Matching 473
15.5 **Data Security and Privacy** 475
15.6 **Review Questions** 476
15.7 **Lab** 477
Additional Reading 497

16 **Ethics and Explainable AI** 499
16.1 **Privacy Concerns and Data Security** 500
16.2 **Explainable AI** 502
16.2.1 Explainability and Interpretability 502
16.2.2 Algorithms 503
16.3 **Trust in AI** 507
16.4 **Review Questions** 509
16.5 **Lab** 509
Additional Reading 520

1

Introduction to Data Mining. CRISP-DM Process

A. P. Kirilenko, *Practical Data Mining with AI for Social Scientists*,
Springer Texts in Social Sciences, https://doi.org/10.1007/978-3-031-89689-7_1

1.1 Introduction

Why the sidebar?

The sidebar provides definitions of the most important terms, summarizes the most essential information from the text and provides additional illustrations

Welcome to the first chapter of our exploration into the world of data mining in the context of social science. This chapter introduces the concept of data mining and gives examples of its application in the field. Finally, we will introduce the main steps of a data mining project, as defined by the Cross-Industry Standard Process for Data Mining (CRISP-DM).

Data has become a major business resource across all industries in the past two decades (◘ Fig. 1.1). Not only has the sheer amount of data increased, but the nature of data itself has changed. Instead of well-formatted, numerical data neatly organized into tables, the vast majority of today's data is represented by unstructured texts, pictures, and videos. With the increasing availability of data, there is a growing need for techniques to extract useful information and knowledge from this data. This is where data mining comes into play.

Data mining

Data mining is the process of discovering patterns, relationships, and insights within large data sets that can guide decision-making.

Data mining aims to extract hidden knowledge from data while discovering interesting patterns and important relationships, summarizing information, and generating new knowledge to help inform decision-making. For example, data mining can summarize a hotel's customer reviews, understand traveling patterns among the points of

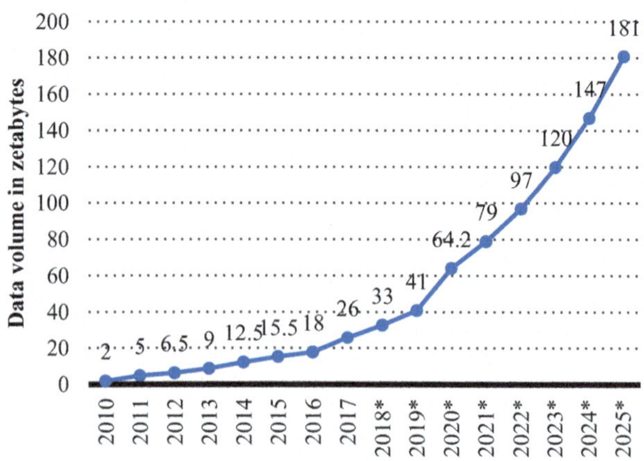

◘ **Fig. 1.1** The volume of data/information created, captured, copied, and consumed worldwide from 2010 to 2020, with forecasts from 2021 to 2025 (in zettabytes). For comparison, one zettabyte is equivalent to 100 million times the amount of text in the Library of Congress. *Source* statista.com

interest, improve museum services, and enhance tourism business performance.

By the end of this chapter, you will have a solid understanding of the concept of data mining, how it can be used in practical applications, and how to plan a data mining project following the CRISP-DM steps. This knowledge will serve as a foundation for the more advanced topics in the subsequent chapters of this book.

1.2 Digital Social Sciences

The fast progress in computer science is affecting all other research fields, including the social sciences. Humanity generates a vast amount of data on people's behavior from various sources, such as newspapers, online booking sites, social media, customer reviews, and travel blogs. A new discipline of digital social science employs Big Data, data mining methods, and Artificial Intelligence to study social phenomena. This is an interdisciplinary field that integrates traditional social sciences, such as sociology, political science, and economics, with big data analysis, machine learning, and computer modeling, to understand the complexities of modern society.

The data mining field is not new. Its basic ideas and methods have been developed since the mid-twentieth century. However, the potential of data mining has only been fully understood in the current epoch of big data. Today, data mining is used in various industries, from healthcare to finance, retail to manufacturing, entertainment to tourism and hospitality. The latter field will serve as the basis for many examples in this book, but the covered methods can be applied to many other traditional social science fields dealing with interdisciplinary data coming from the industry, surveys, interviews, social media, and expert elaborations.

Data mining in business

Data mining can be a powerful tool to drive business success. By analyzing customer data, businesses can understand customer behavior and preferences, enabling them to provide personalized services and offers and stay ahead of the competition.

Digital social science is not only theoretical but also a very practical field. Many examples in this book will be from tourism and hospitality research. In this field, data mining in tourism and hospitality can be used to understand customer behavior, preferences, and trends. For instance, by analyzing customer reviews and feedback, businesses can identify areas of improvement and work towards enhancing their services. Similarly, by understanding customer preferences, businesses can tailor their services to meet the specific needs of their customers, thereby improving customer satisfaction and loyalty. By analyzing historical data, businesses can forecast future trends and make strategic decisions accordingly. This could involve decisions related to pricing, marketing strategies, or even expansion plans.

Data mining can also be used for market segmentation. Businesses can identify customer segments by analyzing customer data and developing targeted marketing strategies for each segment. This not only improves the effectiveness of marketing efforts but also enhances the overall customer experience. Furthermore, data mining can help manage risks. For instance, by analyzing transaction data, businesses can identify unusual patterns and take necessary actions to prevent potential fraud.

Overall, data mining brings benefits to theoretical studies and applied research projects. It can help scientists to improve their theoretical understanding of people's behavioral patterns and decision-making processes. It can also help businesses make informed decisions, improve their services, understand customer behavior, identify trends and patterns, predict future outcomes, and gain a competitive edge. However, data mining methods are not to be applied randomly. They require a clear understanding of the business objectives, a well-defined data mining process, and skilled data scientists to interpret the results and translate them into actionable insights. The next section of the book covers this process in detail.

CRISP-DM

The Cross-Industry Standard Process for Data Mining (CRISP-DM) is a standardized process for conducting data mining projects. The process is divided into six phases. Each phase contributes to the overall outcome.

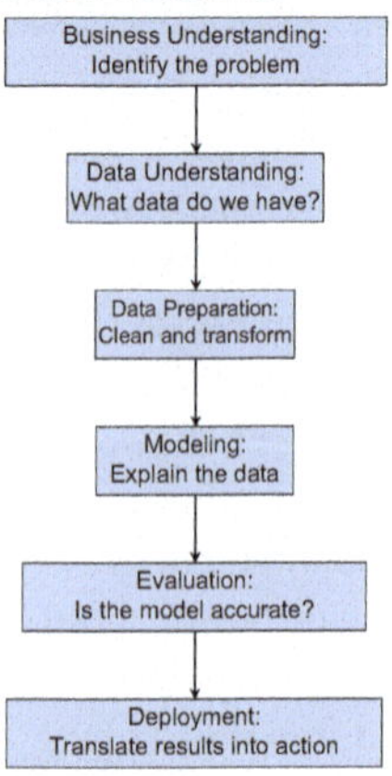

1.3 Introduction to CRISP-DM Process

CRISP-DM is a widely accepted methodology that provides a structured approach to planning and executing data mining projects. It was developed by a consortium of companies aiming to bring a level of standardization to the data mining process, making it more repeatable and reliable. Note the cyclical nature of the CRISP-DM process, resulting in continuous improvement and refinement of results as new data and insights become available.

The CRISP-DM process consists of six phases:

1. Business Understanding: This initial phase focuses on understanding the project objectives and requirements from a business perspective. It involves identifying the

business problem, determining the resources available, and defining the criteria for a successful outcome.

2. Data Understanding: This phase involves collecting, describing, exploring, and verifying data quality. It includes data acquisition, data exploration, and data quality checks.
3. Data Preparation: this phase involves cleaning, transforming, and enriching the data for modeling.
4. Modeling: This phase involves selecting and applying various modeling techniques and calibrating their parameters to optimal values. It includes tasks such as selecting modeling techniques, generating test designs, building models, and assessing model results.
5. Evaluation: This phase involves evaluating the model to ensure it adequately meets the business objectives. It includes tasks such as evaluating results, reviewing the process, and determining next steps.
6. Deployment: This final phase involves deploying the model into the operational system, creating a strategy for the deployment, and monitoring the model over time. It includes tasks such as planning deployment, monitoring, maintenance, and producing a final report.

Each CRISP-DM phase is not strictly sequential; it often occurs in parallel, is repeated, and reiterated as new data, expert insights, or business requirements emerge. The process is flexible and adaptable, making it suitable for a wide range of data mining projects. In the following sections, we will discuss each phase of the CRISP-DM process in depth.

1.4 Phases of CRISP-DM Process

In this section, we will discuss the details of each phase of the Cross-Industry Standard Process for Data Mining (CRISP-DM), including the tasks involved in each phase, the expected outcomes, and how each phase contributes to the overall data mining project.

1.4.1 Business Understanding

The first phase of the CRISP-DM process is business understanding, which means securing a clear understanding of the data mining project's goal, requirements, and deliverables. For example, a data mining project in the hospitality industry may target establishing criteria for selecting social media influencers partnering with a sustainable tourism destination. The outcome of this step is the project plan, which is developed in collaboration with field experts and representatives of the business. If the project's goal is to create generalizable knowledge, this phase involves a review of current literature to establish an understanding of the problem.

1.4.2 Data Understanding

Metadata

Metadata is data describing other data. Securing metadata is critical for error-proof data mining.

The second phase of the CRISP-DM process is data understanding. This phase involves collecting, describing, exploring, and verifying the quality of data. A social media influencer project may involve collecting Instagram posts of the perspective influencer partners together with followers' comments as well as their relevant statistics: number of followers, engagement rate, and others. Understanding the data's structure, quality, units, and limitations is essential. This phase may involve visual inspection of the data, descriptive statistics, and exploratory data analysis. Another critical piece of information is how, when, where, and by whom the data was collected. The outcome of this phase is a report describing the data quality and initial insights about the data. A brief and concise description of the collected dataset is called metadata. Metadata is data describing other data.

1.4.3 Data Preparation

Data cleaning

Data cleaning (or cleansing) is finding and correcting/removing inaccurate, incomplete, or irrelevant parts of the data. For example, in processing social media data we may want to remove advertising and spam comments.

The third phase of the CRISP-DM process is data preparation. This phase involves cleaning, transforming, and cross-referencing the data in preparation for analysis. Real-world data is noisy and inconsistent and contains outliers, errors, and missing values. For example, in a project dealing with tourists' communication in the hurricane evacuation zone, the data will likely be missing for those stranded in the hurricane landfall area. Social media data typically includes spam and advertisements.

Data cleaning can include handling missing data, removing outliers, and resolving inconsistencies in the data. Data transformation can involve scaling, aggregation, or converting variables into forms suitable for mining. The outcome of this phase is a final data set ready for modeling.

1.4.4 Modeling

The fourth phase of the CRISP-DM process is modeling. This book mainly deals with this step. During the modeling phase, multiple methods will be tested to find interesting patterns and relationships in cleaned data. The final model selection will be determined by the problem we target to solve and by the data itself. The outcome of this phase is the most suitable model (or a suite of models) of data analysis that successfully deals with the original business problem.

1.4.5 Evaluation

The fifth phase of the CRISP-DM process is model evaluation. During this step, the model performance is rigorously evaluated with a suite of indicators showing how well the selected model fits the data. Two important parts of model evaluation are model reliability and robustness. A reliable model returns the correct results even when exposed to new data that is not used in model training. A robust model successfully deals with variations in data quality, noise, and collection techniques. A reliable model is trustworthy. A robust model can handle uncertainty and variability in data. Both robustness and reliability are necessary for a model to return valid results consistently.

Methods vs. models

The method is the technique (for example, data cleaning) or approach (for example, regression analysis) used for data analysis. The model is data representation. For example, a linear regression relating hotel room sales with review star ratings is a model.

1.4.6 Deployment

The final phase of the CRISP-DM process is deployment. For a researcher, during this step, the model is applied to analyze and interpret the data to answer research questions. For business applications, this phase involves model deployment and monitoring. With time, the model may age, requiring an update as the data changes. The outcome of this phase is a completed data mining project that provides the expected results.

1.5 Conclusion

In digital social sciences, data mining can be a powerful tool for gaining insights into people's behavior. In this introductory chapter, we have learned that data mining is a process of extracting valuable information from large datasets, which can be used to make informed decisions and improve services. The types of data that can be mined are diverse, and the benefits of data mining are substantial, including enhanced customer understanding and improved business performance.

We have also introduced CRISP-DM, a system of steps for conducting data mining projects. In the following chapters, we will develop a deeper understanding of the phases of the CRISP-DM process and provide a practical review of data mining methods.

1.6 Review Questions

1. Define data mining and explain its relevance in your industry of choice.
2. What data sources can be mined in your industry—for example, hotel management? What are the potential benefits of investigating large volumes of data?
3. Describe the Cross-Industry Standard Process for Data Mining (CRISP-DM). Why is it important to follow an established process in data analysis?
4. Explain the six phases of the CRISP-DM process and discuss the tasks involved in each phase.

1.7 Lab

Objectives

1. Introduce the Anaconda Cloud programming environment
2. Analyze Titanic passenger survival data with the Decision Tree method.

Part 1. Anaconda Cloud

Welcome to your first programming lab! In this course, we will be using Python. Python is a high-level, versatile programming language known for its simplicity and readability, making it an excellent choice for beginners and experienced programmers. It's also an "interpreted" language, meaning that you don't need to complete your entire program to see results; you can execute your code as you write it, correcting the errors on the way.

Before diving into coding, note the difference between Python and the programming environment. Python is the programming language that you use **to write** the code. The programming environment is **where you write and execute** the code. For example, when you are writing an essay, English is the language you use to write the essay, and Word is the "programming environment" ("writing environment"). Just like Word, a programming environment provides tools and resources you use to write, test, and debug your Python code. This includes text editors, integrated development environments (IDEs), and libraries and frameworks that extend Python's functionality.

We will be using the Anaconda Cloud environment for Python programming. Anaconda is a distribution of Python that comes bundled with a wide range of popular libraries and tools commonly used in scientific computing, data analysis, and machine learning. Anaconda simplifies the setup process and ensures everyone has access to the same tools and libraries, creating a consistent learning environment.

Click the link to load the Anaconda Cloud programming environment (on Windows computer, Ctrl-click):

- https://tinyurl.com/dataMiningBookLab1

Many of the following steps are relevant only when running the anaconda environment for the first time.

1. You may be asked to register or to log in:

2. Then, open the code in Anaconda Cloud:

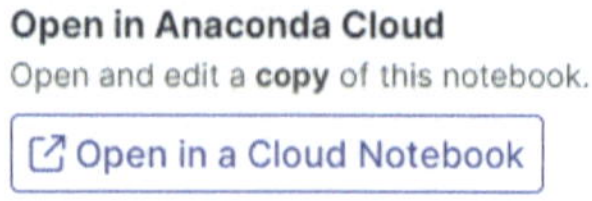

At the end of the process, the server will start up. It may take a minute or two. Just wait.

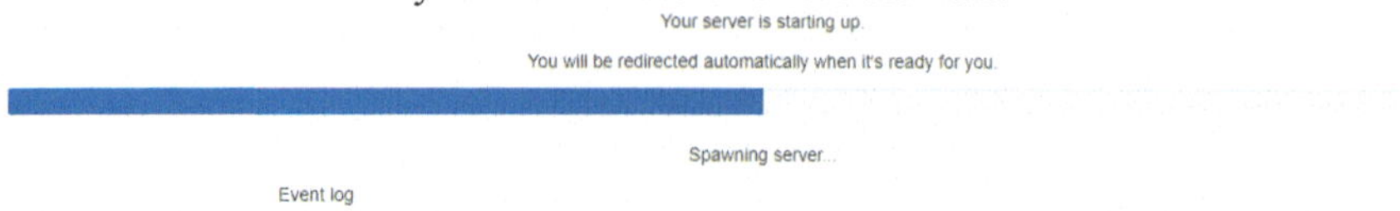

3. You may also be asked to select the kernel. "Anaconda kernel" refers to the computational engine used to run the code. Select the preferred kernel.

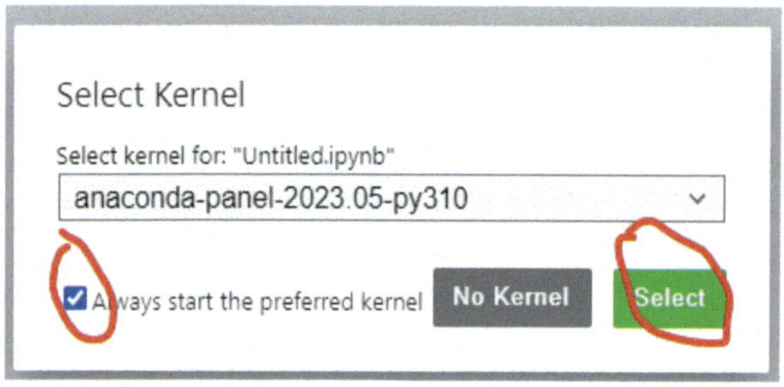

Agree to use the Anaconda Assistant if asked:

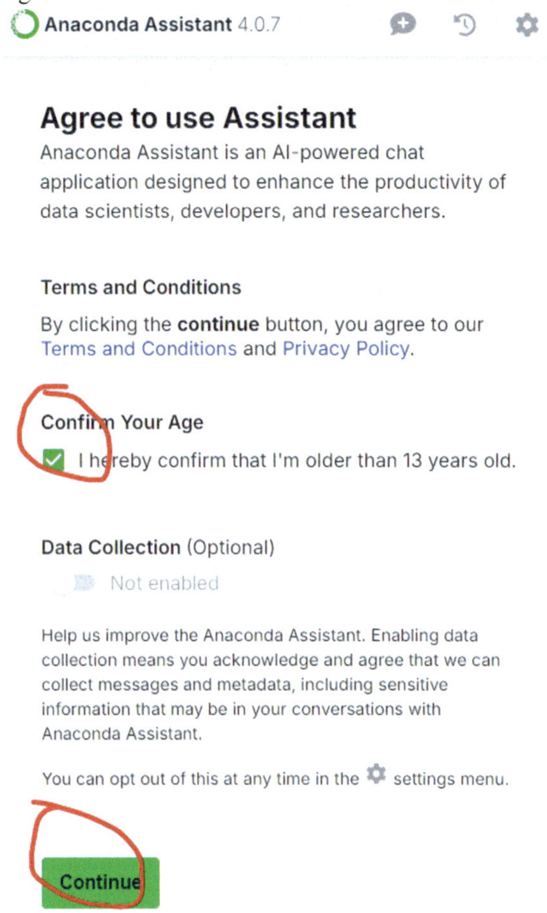

4. Finally, agree to load the code from this course if asked:

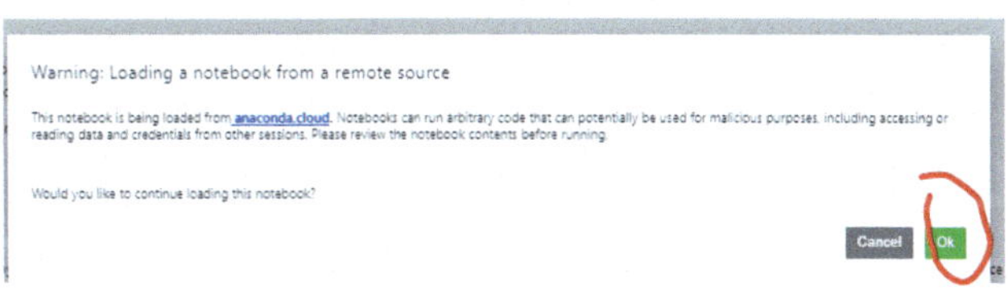

Your Programming Environment is Ready

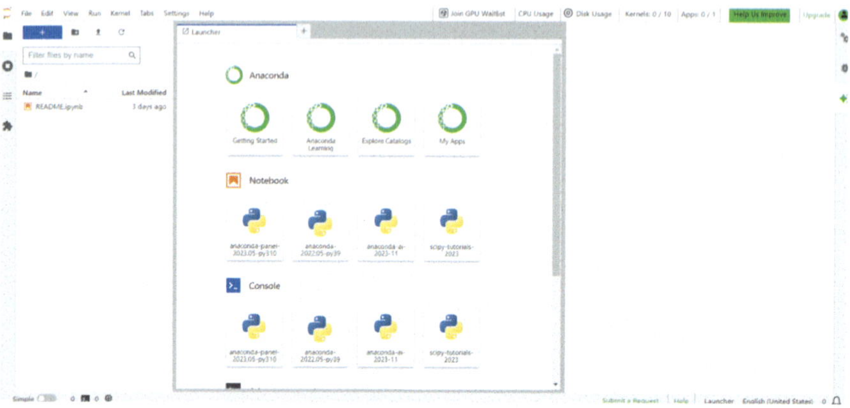

Download the Code of the Lab

Click the following link and download the lab notebook on your computer.

► https://drive.google.com/uc?export=download&id=1LyKymwxo6TkL3GztijctMiHSM5FnJr87

Drag and drop the downloaded notebook file to the left panel of Anaconda (where the Readme.ipynb file is). That will upload the notebook to the Anaconda cloud.

Double-click the notebook you just uploaded. It will open in the central panel. Your programming environment should look like the following (file names may vary).

Note: If you run Anaconda for the first time, you will be asked permission to use Anaconda Assistant. Give the permission.

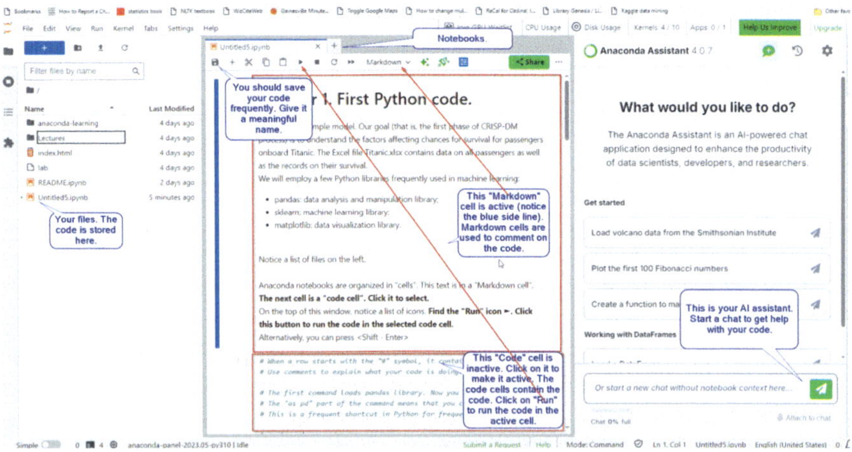

Part 2. Python Coding

Read instructions in the notebook, run active cells, and complete the lab.

Ultimately, your results will look like the following (text boxes are my explanations). Notice that your picture may differ.

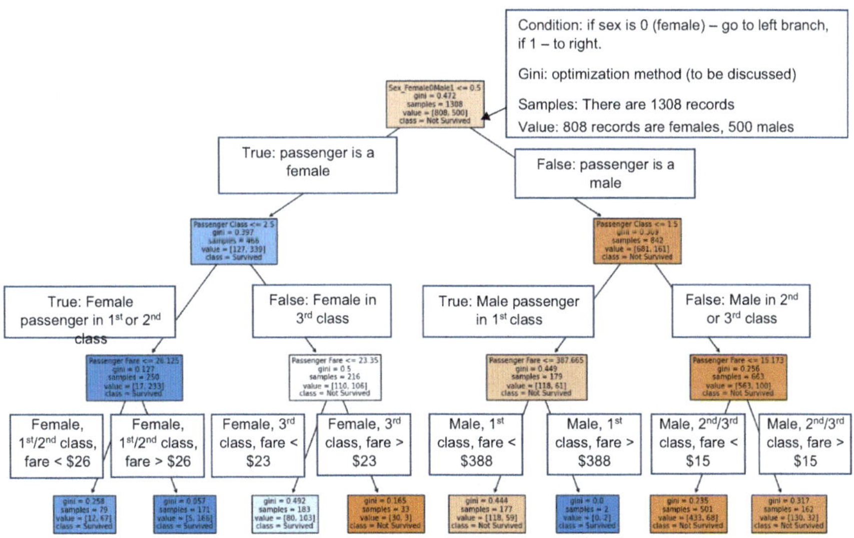

Study Picture Explanations

In a decision tree model, each internal node (a colored box) represents a decision based on the value of a particular feature. For example, the top node is split on sex variable value. Sex was the most critical condition for survival.

Based on the split condition, the decision is directed down one tree branch. **If the condition is true for a particular instance, it follows the left branch; otherwise, it follows the right branch**. Cabin class was the second most important condition for survival, and the fare was the third most important condition.

Each subsequent internal node in the tree applies a similar process, further splitting the data based on different features and conditions until reaching a leaf node. The leaf node represents the final decision or prediction made by the model for that instance.

Part 3. Using Anaconda AI Assistant

Now, we will write the same code using the AI assistant.

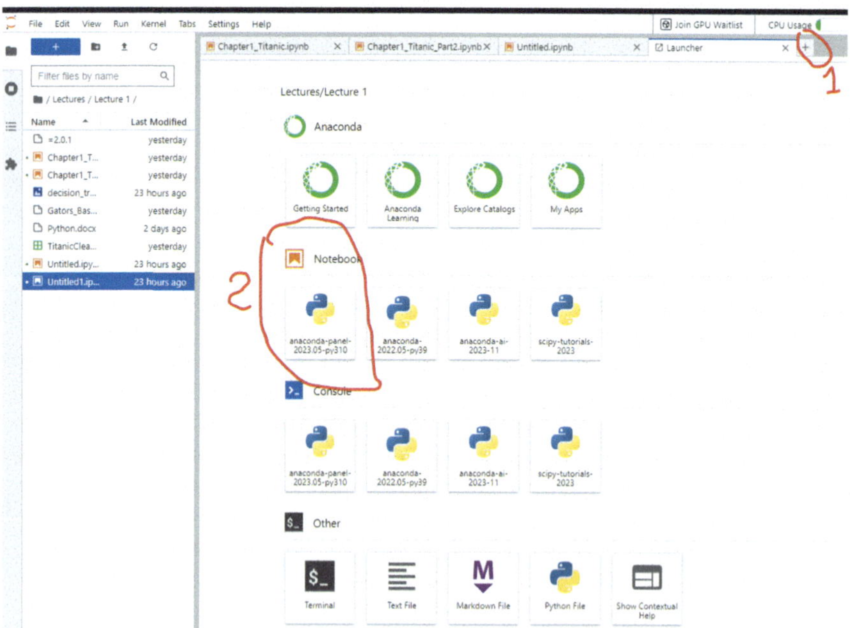

Start a new empty notebook.

Let us write the code with the same functionality using AI assistance. We will use the Anaconda Assistant (window on the left). Alternatively, you can use ChatGPT or many other large language models.

Step 1. Read Titanic Data

Once again, let us start with reading the data. This time, we will show how to read Excel data. Enter the following text to the Anaconda Assistant window on the low right and start the chat:

Read data from "▶ https://github.com/andykir/DataMiningLectures/raw/main/TitanicCleaned.xlsx".

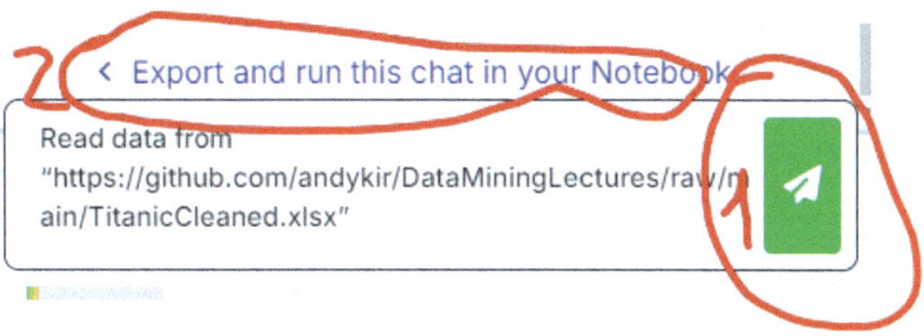

You should see the assistant writing the code with explanations. Then click "Export and run this chat in your Notebook."

Notes

1. Occasionally, the Assistant stops working. Wait 10–30 s. and repeat the request. Alternatively, after several unsuccessful attempts, you can reload the web page. You can also use ChatGPT or another generative AI instead of the Assistant; outcomes are similar, if not better, but you need to copy-paste the code between the windows.
2. Running the code will occasionally return an error or warning. You will know immediately: when you run the cell, the output cell will be red. Most warnings relate to upcoming changes in a Python library and may be ignored. When the code returns an error, the Assistant normally asks if you want to correct the error. Type "Yes"—the code will be corrected.

3. Sometimes, the error is due to a library missing. In that case, the Assistant will tell you to install the missing library. Just type the code into a notebook cell and run the cell. For example, if a library "*missing_library*" is missing, type the following:
   !pip install *missing_library*

The missing library will be installed in your workspace. Note that you have to restart the kernel to complete the update:

4. You can change the code Assistant puts in the cells as you like.
5. There are many ways to write the code; the code Assistant writes for you will differ from what it writes for me. Don't be confused.
6. SAVE your notebook! Just click on the Save icon. Give it a relevant name.

If everything works ok, you will see the assistant's code automatically copied and executed in your open notebook:

User

Read data from "https://github.com/andykir/DataMiningLectures/raw/main/TitanicCleaned.xlsx"

Assistant

To read data from the provided URL, we can use the `pandas` library in Python. First, we need to install the library if it's not already installed. Then, we can use the `read_excel` function to read the data from the URL. Here's the code to achieve this:

```
[3]: import pandas as pd

# Read data from the URL
url = "https://github.com/andykir/DataMiningLectures/raw/main/TitanicCleaned.xlsx"
data = pd.read_excel(url)
```

Step 2. Show Titanic Data

Notice that the data is saved into the variable *data* (the variable name may differ in your code). To see the values stored in *data*, type in the Assistant window:

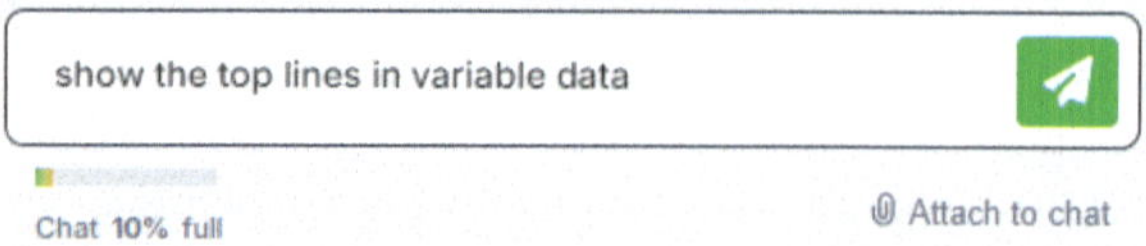

Note: the new cell with the code will be inserted **below the selected cell** in your notebook.

Step 3. Train the Decision Tree Model

Enter your request into the Assistant.

USER:

Your goal is to train a decision tree model. Use variable "data". The target is the "Survived" column. The attributes are 'Sex_Female0Male1', 'Passenger Class', 'Passenger Fare', and 'No of Siblings or Spouses on Board'.

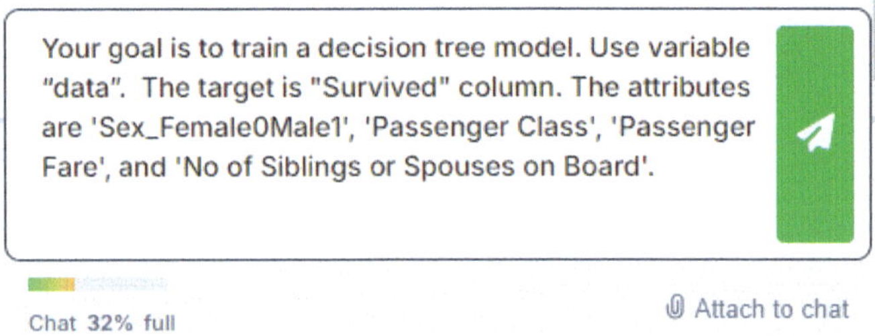

Run the code and observe outcomes. Notice that the code provided by the Assistant is more advanced than ours: it provides the model's accuracy. The model correctly predicts 77% of survivals.

Step 4. Visualize the Decision Tree model

Enter your request into the Assistant window and run the generated code:

USER:

Plot the decision tree model.

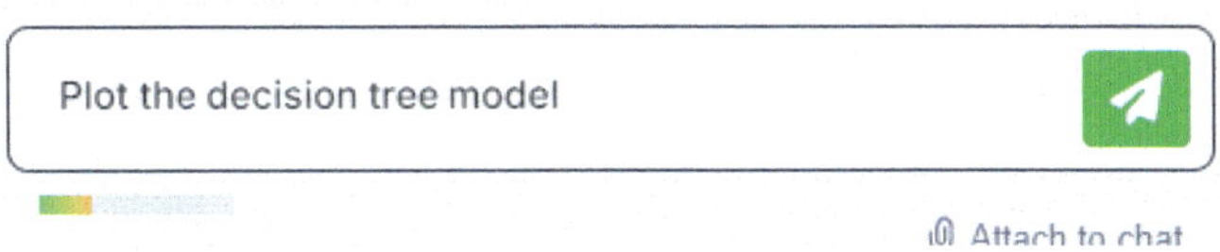

1

Assistant

To plot the decision tree model, we can use the `plot_tree` function from the `sklearn.tree` module. This function allows us to visualize the traine decision tree model. Here's the code to plot the decision tree model:

```
[110]: import matplotlib.pyplot as plt
from sklearn.tree import plot_tree

plt.figure(figsize=(20,10))
plot_tree(model, feature_names=X.columns, class_names=['Not Survived', 'Survived'], filled=True)
plt.show()
```

```
---------------------------------------------------------------------------
InvalidParameterError                     Traceback (most recent call last)
Cell In[110], line 5
      2 from sklearn.tree import plot_tree
      4 plt.figure(figsize=(20,10))
----> 5 plot_tree(model, feature_names=X.columns, class_names=['Not Survived', 'Survived'], filled=True)
      6 plt.show()
```

Attention! There may be an error!

```
---> 95 raise InvalidParameterError(
     96     f"The {param_name!r} parameter of {caller_name} must be"
     97     f" {constraints_str}. Got {param_val!r} instead."
     98 )

InvalidParameterError: The 'feature_names' parameter of plot_tree must be an instance of 'list' or None. Got I
ndex(['Sex_Female0Male1', 'Passenger Class', 'Passenger Fare',
       'No of Siblings or Spouses on Board'],
      dtype='object') instead.
<Figure size 2000x1000 with 0 Axes>
```

....

At the end of the output, you can see that the source of the error is that the Assistant incorrectly specified one of the parameters used in the plot-tree function.

Click "Debug the active code cell" (make sure that the correct notebook cell is selected!).

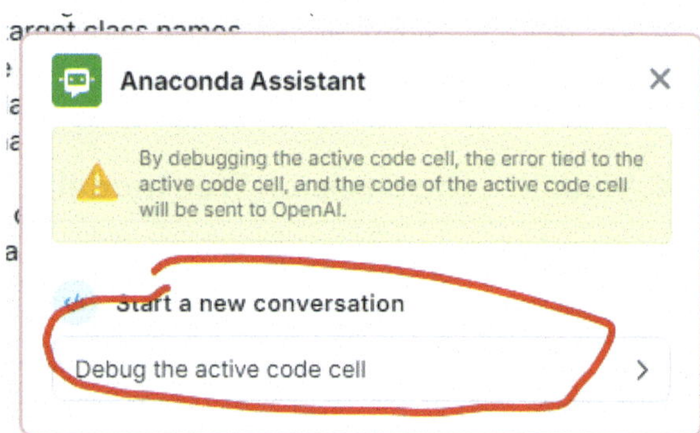

Then, read the explanation for the error in the Assistant chat. Answer "yes". You will see a corrected code in the chat.

Then, export and run the corrected code in the notebook.

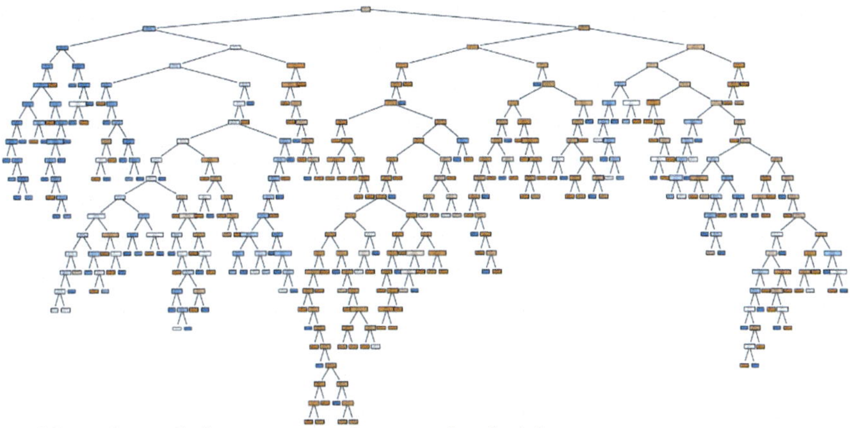

Now, the code is correct, and we see the decision tree.

Step 5. On Your Own

Reduce the Decision Tree model to a depth of 3 and plot it.

Note. What if the Assistant stops working?

Occasionally, the Assistant stops working.

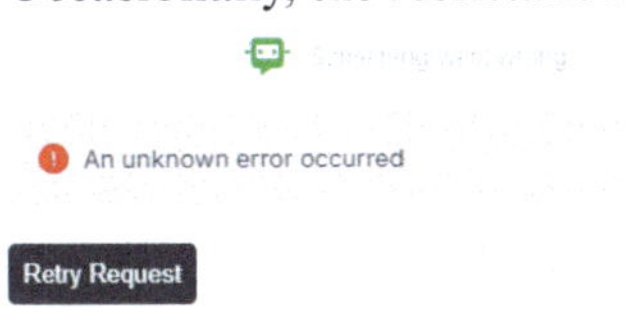

First, you can wait for a few minutes and repeat the request. The Assistant will be back nine times out of ten. If that does not work, you need to refresh your working environment. Here's how.

1

Step 1. Reload the Web Page

This is an example for Google Chrome:

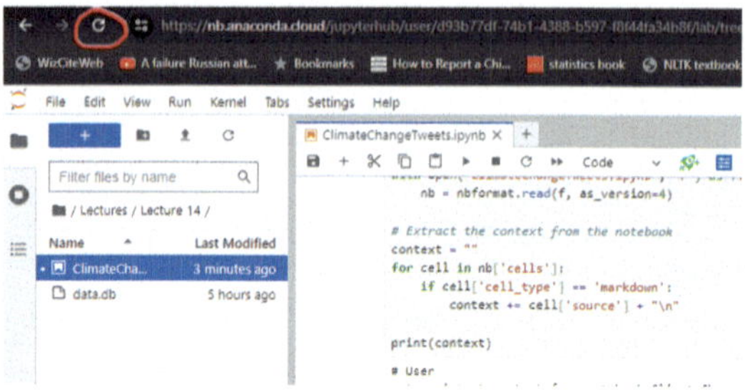

Step 2. Provide context for the Assistant

When you ask the Assistant for help with code development, it uses information from your previous requests to understand the context of your requests. As you refresh the working environment, this context is lost. That means the Assistant forgets your variables' names, the libraries you have already loaded, and so on. This is why, in our requests, we are "reminding" the Assistant of the essential information.

USER:

Get Assistant context from the notebook "Untitled".
Note the name of the notebook. You need to replace this name with your own.

Assistant:

To get the context from the Jupyter Notebook called "Untitled", you can use the *nbformat* package to read the notebook and extract the content. The *nbformat* package allows you to read and manipulate Jupyter Notebooks programmatically. Here's an example of how you can use *nbformat* to get the context from the notebook "Untitled":

```
import nbformat

# Open the notebook file
with open(Untitled.ipynb', 'r') as f:
    nb = nbformat.read(f, as_version=4)

# Extract the context from the notebook
context = ""
for cell in nb['cells']:
    if cell['cell_type'] == 'markdown':
        context += cell['source'] + "\n"

print(context)
```

In this example, the *nbformat* package is used to read the notebook file and extract the content of the markdown cells, which typically contain the context and explanations in a Jupyter Notebook. If you have any specific requirements for extracting context from the notebook, please let me know so I can provide a more tailored solution.

After you execute this cell, the Assistant will get information about your code and data.

The final option is ChatGPT, Gemini, or another generative language model. We will discuss how to do that effectively later in the course.

Improve Python Skills

Chapter 1 of the free book Learn Python with Jupyter by Serena Bonaretti. ► https://www.learnpythonwithjupyter.com/

Note that Bonaretti's book uses Jupyter, while we use Anaconda distribution. From your perspective, there will be little difference.

Additional Reading

Wirth, R., & Hipp, J. (2000, April). CRISP-DM: Towards a standard process model for data mining. In *Proceedings of the 4th International Conference on the Practical Applications of Knowledge Discovery and Data Mining* (Vol. 1, pp. 29–39).

2

Data Pre-processing

A. P. Kirilenko, *Practical Data Mining with AI for Social Scientists*,
Springer Texts in Social Sciences, https://doi.org/10.1007/978-3-031-89689-7_2

Network connectivity

In the previous chapter, we discussed six steps of the Cross Industry Standard Process for Data Mining (CRISP-DM), which together form a systematic framework for data analysis projects. CRISP starts with the business understanding, which determines the needs and objectives of the data mining project. The next steps of CRISP are data understanding and data preparation. This is the topic of the second chapter of the book.

2.1 Data Complexity

Terminology

Hotel review data. Each row represents one review (one *example*). The columns are review *attributes*.

	Attributes		
Examples	ID	Date	Text
	1	1.1.2025	Room is too small
	2	1.1.2025	It's ok
	3	1.2.2025	Far from everything

Contemporary data analytics deals with much more complex data than it used to be at the end of the twentieth century. This complexity comes from high dimensionality, heterogeneity, and hidden relationships among data points.

Data is frequently organized in tables, with each row representing one data point. In different fields of science, these rows may be referred to as records, observations, cases, instances, and so on. In data mining, we use the term "*example*" to refer to one data point (one row of the table). One example may represent one hotel review in a customer satisfaction study or one post in Instagram influencer research.

Data dimensionality

The *dimensionality* of data refers to the number of pieces of information recorded for each example. In simpler terms, the dimensionality of data refers to the number of variables in your dataset.

The dimensionality of data refers to the number of pieces of information recorded for each example (each data point). In statistics, these pieces of information are called variables. In data mining, the terms *features* or *attributes* are commonly used. When data is represented by a table, each attribute is represented by one table column. High-dimensional data, often encountered in social science research, has a high number of attributes. The implications of high dimensionality for data analysis were first realized by a renowned American mathematician, Richard E. Bellman, who called it a "curse of dimensionality."[1] As data dimensionality increases, the volume of the data space expands so rapidly that the data becomes sparse (Fig. 2.1). This sparsity is problematic for traditional analysis methods.

1 Bellman, R. E., & Rand Corporation. (1957). *Dynamic programming* (p. ix). Princeton University Press. ISBN 978-0-691-07,951-6.

Example	every thing	far	from	is	it's	ok	room	small	too
1	0	0	0	1	0	0	1	1	1
2	0	0	0	0	1	1	0	0	0
3	1	1	1	0	0	0	0	0	0

Fig. 2.1 Hotel review data prepared for text analysis. Three examples represent three hotel reviews: (1) Room is too small; (2) It's ok; and (3) Far from everything. The reviews were converted to a document-term matrix where a "1" or a "0" represents a word's presence or absence in a particular review. Notice that the table contains mostly zeros, illustrating the sparsity problem

The term "data heterogeneity" refers to the diversity of data types. For example, a dataset representing an Instagram influencer's posts would include numerical data such as the number of followers, categorical data such as the influencer's gender, date/time data, textual data, images, videos, hyperlinks, and others. The heterogeneity of data requires methods of analysis beyond traditional statistics.

Finally, in social sciences, data points are rarely independent. For example, in social media data, the connections between individuals carry significant information about the network structure. This information is relevant for research on social media communities, news propagation, influencers, data flow, and so on. Specialized methods such as network analysis can help in this type of analysis.

Understanding the complexity of data is the initial step in data pre-processing and scrubbing. It helps in deciding on the methods of data cleaning, reduction, and scaling and in selecting the appropriate data mining techniques. In the following sections, we will recap data types and measurement levels, discuss the sources of data errors and inconsistency, and introduce methods of data quality analysis, data cleaning, reduction, and scaling.

Data types and level of measurement

Data type refers to the way data is represented with numbers, symbols, words, etc. Examples are integer, real (float or floating-point), Boolean, and character. The *level of measurement* refers to the nature of the data. Examples are categorical, ordinal, and ratio data. The same level of measurement can be represented with different data types.

2.2 Data Type and Level of Measurement

Let us briefly recap data types and levels of measurement. Data types refer to the numerical representation of data in a computer. The most frequent data types include Boolean, Float, Integer, Character, and String.

2

Data sparcity

Notice that the tourist/attraction matrix would be predominantly filled with False (zero) values: most tourists would visit only a few of multiple attractions. This *data sparsity* is a common feature. Data sparsity is problematic for traditional analysis methods.

Common levels of measurement

Commonly used measurement levels include nominal, ordinal, interval, and ratio data. Nominal data are also called categorical.

Boolean data can take only two values: True and False. The True value is frequently represented by the number 1, and the False value is represented by the number 0. For example, to represent individual tourist attraction visitation patterns, a researcher may use a matrix of Boolean values where the rows represent individual tourists and the columns represent city attractions. If a particular tourist visited a specific attraction, the corresponding value in the tourist/attraction matrix would have a value of True (1); otherwise, it will be equal to False (0). Internally, a boolean number requires just one bit of computer memory; that is, this data type is highly efficient with huge data sets.

Float (Real) and Integer data represent "common" fractional and whole values. When selecting between float and integer data types, several important considerations must be made. First, calculations with integer numbers are exact, unlike those with float numbers, which commonly return small errors. The downside is that the range of values that integer data types can represent is significantly smaller than the range of float values. For example, a data type that takes 4 bytes (32 bits) of computer memory for a single number can represent 2^{32} different values. Since the integer data type is exact, a 32-bit integer number has a minimum value of $-2^{32}/2=-2,147,483,648$ and a maximum value of $+2^{32}/2-1=2,147,483,647$. Meanwhile, a 32-bit float number does not necessarily have an exact representation in computer memory and hence may have a much wider range in order of $\pm 10^{38}$. In practice, that means that the integer data type is suitable for situations requiring exact numbers: indices, IDs, inventory management, and similar.

Finally, Character and String data types represent a single character or multiple characters. One character takes one to four bytes to store, making string data typically less efficient in terms of computer memory than numerical data types.

As opposed to data type, which relates to how the data is represented numerically or with symbols, the level of measurement refers to the nature of information represented by data. Knowing the level of measurement is important because it defines the operations that can/cannot be used in data processing. In that way, the level of measurement guides the selection of data analysis methods.

The most frequently used levels of measurement include nominal, ordinal, interval, and ratio data. This typology was proposed by psychologist Stanley Stevens in his seminal 1946 article "On the Theory of Scales of Measurement." The *nominal data* differentiate between the values only based on their names and hence does not have any meaningful order. Examples of nominal data include personal name, university ID number, color, and sex. A survey question, "What mode of transportation did you use to arrive in Orlando (airplane, car, train, bus, other/not applicable)?" would return *nominal* data. The responses to this question would break Orlando visitors into five different categories according to their mode of transportation, which is why the *nominal data is also called categorical.*

Ordinal data have a clear order or ranking compared to nominal data, but the intervals between ranks are not necessarily equal. In surveys, the Likert scale responses are represented by ordinal data. For example, a survey question "Are you satisfied with your stay in our hotel (very dissatisfied, somewhat dissatisfied, neutral, somewhat satisfied, very satisfied)" returns ordinal data due to a clear progression in possible responses from the lowest to the highest satisfaction levels. However, it is impossible to compare the difference in satisfaction between very dissatisfied and somewhat dissatisfied customers to, e.g., a difference between somewhat dissatisfied and neutral customers, as the intervals between satisfaction ranks are not equal. That is, one can arrange ordinal data in a clear order but cannot meaningfully define how much one value differs from another.

Compared to ordinal data, the *interval and ratio data* do allow us to measure the difference between the values. The defining property that differentiates between interval and ratio levels of measurement is that the ratio data have a true zero point, which reflects a real absence of the measured quantity. For example, the zero SAT does not constitute an absence of any knowledge; in fact, the minimal SAT is 300 points. As opposed to that, ratio data do have a true zero point. An age of zero years means a person just born; a zero income implies an absence of income, and a zero distance means no separation. An easy test differentiating interval and ratio data is trying to find a ratio of two values. For example, a $200 hotel room is twice as expensive as a $100 hotel room, indicating that hotel room prices have a ratio level of measurement. As opposed to that, a statement that a 40-degree Celsius day

2

is twice as hot as a 20-degree day is meaningless, indicating the interval level of measurement for the temperature.

Frequently encountered nominal data

Binominal: one of two values

Binomial (Boolean): False or True (frequently represented as 0 and 1)
Polynominal data: one of many values

Data type and the level of measurement limit the analysis that can be done with data. For example, many descriptive statistics only apply to interval and/or ratio data; these measures include the mean and median values, standard deviation, and many others. On the contrary, the dependent variable for logistic regression should be categorical. The ordinal variables are somehow "in between" the interval and nominal levels of measurement; they can be used in the analysis that requires nominal variables, but some researchers use them in analyses requiring interval variables as well. The latter is especially true for Likert scale variables taking seven or more possible values.

A common error in data analysis is selecting an inappropriate method for the data type or the level of measurement. For example, one may code car colors with integer indices and then attempt to compute the mean to find the "average" car color. In fact, I observe this particular type of error quite frequently. Few data analysis packages, such as SPSS, control for variable types and levels of measurement, warning researchers trying to use inappropriate analysis.

2.3 Sources of Data Errors and Inconsistency

Data errors and data inconsistency

Data errors refer to values or records that are missing, incomplete, or incorrect.

Data inconsistencies are values that contradict other values, data records, data sources, standards, or general knowledge.

Data quality directly affects the analysis's results. However, data frequently has errors and inconsistencies. Data errors refer to values or records that are missing, incomplete, or incorrect. Data inconsistencies are values that contradict other values, data records, data sources, standards, or general knowledge. Values of "five" or "5 years" for the attribute "Age," which assumes an integer data type, is an example of data error. A record with values "5" and "Married" for the attributes "Age" and "Marriage status" is an example of inconsistent data.

Data errors and inconsistencies can occur during data collection, pre-processing, or storage. Major factors affecting data quality include:

- Measurement errors due to incorrect data collection protocol or faulty instruments. For example, a poorly designed question in a survey on hotel stay satisfaction may lead to ambiguous responses, thus introducing measurement error. In the same survey, the complexity of emotional expression in human language can lead to incorrect recognition of the sentiment expressed in reviews.

- Inconsistencies in data collection methods. For example, if a survey is administered in two different ways, online and in person, the mode of administration will influence responses.
- Inconsistencies in pre-processing methods. For example, in automated sentiment analysis, the outcome depends on the algorithm used. What makes the problem worse is commercial sentiment analysis tools rarely disclose which algorithm they use.
- Data entry errors due to mistyping, incorrect data formatting, and incorrect data type or level of measurement. A frequent error is mistyping the date, using text in place of the numbers, and incorrect placement of the decimal point. For example, a respondent's age might be entered as 25 instead of 52 or as 5.2. A person's height of 6.1 ft may be entered as 6 ft 1 inch, 6′ 1″, or 1.82 [meters].
- Outdated or incompatible records, such as using US statistical areas delineation from the 2010 US Census in conjunction with demographics from the 2020 US Census.
- Failing to account for variability in local and international standards. For example, a ZIP code can be 5- or 9-digit in the US, 6-digit in India, and alphanumeric in the UK. The temperature is measured in degrees Fahrenheit in the US and a handful of other countries, while the rest of the world is using degrees Celsius. Some countries use periods, while others use commas as decimal separators.
- Failing to account for inconsistencies in data transformation during data storage. For example, failing to account for character encoding differences between different software packages could transform the Ciudad Juárez city's name to Ciudad Ju?z.

These sources of data errors and inconsistencies impact the quality of the data and, consequently, the validity of the results obtained from data mining. Therefore, it is crucial to identify and address these issues in the data pre-processing and scrubbing stage.

Text encoding

Recall that text is stored in a computer as a sequence of numeric values. Character encoding is the process of converting text symbols into a machine-readable numeric format. There are multiple standards of text encoding. This calls for frequent errors, especially in scraping web data.

2.4 Data Cleaning

Data cleaning aims to detect errors and repair or remove erroneous values and/or entire records. The data cleaning process includes multiple procedures and tests, which include the following main steps:

- Removing duplicate records;
- Handling missing data;
- Cleaning text data;
- Transformation and standardization of numerical data;
- Detection and cleaning outliers;
- Detecting and cleaning inconsistent data;
- Feature engineering;
- Detection and/or addressing group imbalances.

2.4.1 Duplicates

Duplicates are a frequent data quality concern resulting from errors in data collection, entry, and joining multiple datasets. While identical records are easily detected, partially identical records require additional attention. For example, two records may differ only in their ID field. Of special concern are fuzzy duplicates, that is, records with nearly identical numbers or texts. For example, in text analysis of newspaper publications, near-duplicates appear from news correction, morning/evening issues, or local outlets reprinting news originally appearing in "prestige press."

2.4.2 Missing Data

Missing data is a frequent issue in data mining, especially when data is collected through surveys or interviews due to non-responses. The absence of data can also happen due to measurement problems, data entry mistakes, data loss, or even the data cleaning process, e.g., when a researcher deletes outlier points. Missing data can lead to biased results if not handled appropriately. Several methods exist for dealing with missing data, including:

- *Imputation*: In mean imputation, missing values are replaced with the average of all values for the attribute. One can also use the median ("middle" value) or mode (the most frequent value) imputation. There are more complex imputation methods, such as replacing missing values with the mean of K nearest neighbors (K-NN imputation).

- *Listwise deletion*: removing the records containing missing data from the analyses. This is the most radical method and should be applied only when justified (for example, when only a few records contain missing data);
- *Pairwise deletion*: removing the records containing missing data from *some* analysis. Imagine researching the correlations between an Instagram Influencer's number of followers, engagement (number of likes/followers), and annual compensation for product marketing. There are 100 data records in total, but the Compensation column has only 90 records. In pairwise deletion, the correlation between the number of followers and engagement is computed based on 100 records, while the correlation between the number of followers and compensation is calculated based on 90 records. Compare that to the listwise deletion, where all correlations are computed based on 90 records. Note that pairwise deletion may lead to analysis bias: imagine that missing data mostly relates to mini-influencers who are less likely to report their earnings (Table 2.1).

Finally, the attributes with many missing values may be entirely removed from the analysis.

Solving missing data problem

Imputation: compute a surrogate value

Listwise deletion: do not use an instance with a missing value in analyses

Pairwise deletion: do not use an instance with a missing value in analyses involving the missing value.

2.4.3 Cleaning Text Data

Aside from the obvious problem of typos, slang, and emoticon ("likes") presence, text cleaning includes an involved procedure of preparing human-written texts for automated text analysis, for example, extracting the main topics of online discussion. Generally, the methods of text cleaning may include removing the most frequently used words such as articles ("Stop words"), breaking the text into elementary units of analysis ("Tokenization"), removing special characters non-contributing to text meaning ☺, and lemmatization (reducing the words to their dictionary form) or stemming (reducing the words to their roots). We will discuss text data cleaning later in this course when dealing with text analysis.

English stop words

I	yours
me	yourself
my	yourselves
myself	he
we	him
our	his
ours	himself
ourselves	she
you	her
your	hers
…	…

Table 2.1 Comparison of listwise deletion, pairwise deletion, and mean imputation

Influencer	F: Followers	E: Engagement	I: Income $
Original data			
Anna	1,000,000	0.01	50,000
John	60,000	0.1	1,000
Silvester	1000	5	-
Park	130,000	0.05	30,000
Zixuan	500,000	0.03	40,000
Influencer	**F: Followers**	**E: Engagement**	**I: Income $**
Mean imputation. Pearson's correlation: r(F, E) = −0.46, r(F, I) = 0.75			
Anna	1,000,000	0.01	50,000
John	60,000	0.1	1,000
Silvester	1000	5	30,250
Park	130,000	0.05	30,000
Zixuan	500,000	0.03	40,000
Influencer	**F: Followers**	**E: Engagement**	**I: Income $**
Listwise deletion. Pearson's correlation: r(F, E) = -0.85, r(F, I) = 0.83			
Anna	1,000,000	0.01	50,000
John	60,000	0.1	1,000
Silvester	–	–	–
Park	130,000	0.05	30,000
Zixuan	500,000	0.03	40,000
Influencer	**F: Followers**	**E: Engagement**	**I: Income $**
Pairwise imputation. Pearson's correlation: r(F, E) = −0.46, r(F, I) = 0.83			
Anna	1,000,000	0.01	50,000
John	60,000	0.1	1,000
Silvester	1000	–	–
Park	130,000	0.05	30,000
Zixuan	500,000	0.03	40,000

2.4.4 Data Transformation

Data transformation aims to make data better suitable for further analysis. The categorical and ordinal data are frequently converted into numerical values (*encoded*). Interval and ratio data are frequently scaled into a specific range. For example, the sentiment analysis data ranging from 1 for extremely negative to 5 for very positive may be scaled into the [−1. 1] interval.

Probably the most frequent types of data transformation are *normalization* and *standardization*. Many methods of data mining require all data to be transformed to a uniform scale, which is usually done through one of these two approaches.

Normalization and standardization

Normalization transforms data to fit the range of zero to one. Standardization centers the data at zero value and scales it to have a standard deviation of one.

Normalization converts numerical data into the interval [0, 1] using the following formula:

$$x' = \frac{x - \mu}{\max(x) - \min(x)}$$

where x is the original value, x' is the transformed value, μ is the mean value, and max(x) – min(x) is the range of original data.

Meanwhile, standardization transforms the data so that it is centered at zero value and its standard deviation $\sigma = 1$:

$$x' = \frac{x - \mu}{\sigma}$$

Data scaling, normalization, or standardization ensures uniformity of data for future analysis. Consider the following data on Instagram influencers (Table 2.2).

The linear regression model fitting Income to Followers and Engagement has produced the following results:

$$\text{Income} = 0.042 \times \textit{Followers} - 9516 \times \textit{Engagement} + 12,231$$

Table 2.2 Instagram influencers' number of followers, engagement rate (number of likes per follower), and income (USD)

Influencer	F: Followers	E: Engagement	I: Income $
Anna	1,000,000	0.01	50,000
John	60,000	0.1	1,000
Silvester	1000	5	60,000
Park	130,000	0.05	30,000
Zixuan	500,000	0.03	40,000

2

While this model is moderately good in explaining how influencers' income depends on the number of followers and engagement rate (adjusted R^2=0.59), it is hard to understand the relative influence of independent variables on influencer compensation. Consider the same regression model with standardized data:

The linear regression model with standardized data looks as follows:

$$\begin{aligned} \textit{Standardized Income} = & 0.769 \times \textit{Standardized Followers} \\ & + 0.931 \times \textit{Standardized Engagement} \end{aligned}$$

Increasing the engagement rate by one standard deviation would increase the income by 0.931 standard deviations, and increasing the number of followers by one standard deviation would increase the income by 0.769 standard deviations.

There are many other data transformation methods, including:

Log transformation, Square Root transformation, and Exponential transformation to reduce the skewness of data distribution:

$$x' = \log(x + 1)$$

$$x' = \sqrt{x}$$

$$x' = e^x$$

Box-Cox Transformation to make the data more normally distributed:

$$x' = \begin{cases} \frac{x^\lambda - 1}{\lambda} & \text{if } \lambda \neq 0 \\ \log(x) & \text{if } \lambda = 0 \end{cases}$$

Binning: converting interval/ratio variables into discrete bins or intervals (that is, nominal or ordinal variables) and many others.

2.4.5 Detection and Cleaning Outliers

Outliers are data points that are significantly different from the others. The presence of outliers may be due to data errors or fraud. Outliers can also indicate a rare event ("black swan") or a record coming from a different population. For example, the Instagram influencers'

Table 2.3 Instagram influencers' data with standardized values

Influencer	F: Followers	E: Engagement	I: Income $
Anna	1.77	−0.52	0.69
John	−0.74	−0.47	−1.74
Silvester	−0.90	2.00	1.18
Park	−0.56	−0.50	−0.31
Zixuan	0.43	−0.51	0.19

standardized Engagement rates (Table 2.3) show that Silvester's engagement is an outlier two standard deviations above the mean. A possible reason could be that Silvester is a niche nano-influencer who focuses on a narrow area of interest (for example, extreme ironing). Despite their smaller reach, these influencers have a high engagement rate and can significantly impact their followers' purchasing decisions and loyal community. Brands often collaborate with niche nano-influencers to leverage their authenticity and high level of trust.

Niche influencing: Extreme Ironing

The previous example illustrates a common method of detecting and cleaning the outliers: the values above or below a certain number of standard deviations from the mean (usually 2 or 3 standard deviations) are either removed or replaced. A similar technique uses the interquartile range instead of the standard deviation. Quartiles are three values, Q1, Q2, and Q3, that split sorted data into four parts, each with an equal number of observations, and the interquartile range IQR is the distance between the top and the bottom quartiles: IQR = Q3 − Q1. Typically, the data that falls outside the interval

$$[Q1 - 1.5 \times \text{IQR}, Q3 + 1.5 \times \text{IQR}]$$

are considered outliers.

Windsorizing is another method of outlier removal and replacement. In the windsorizing procedure, the extreme values (for example, the bottom 5% and the top 95% of the values) are replaced with the value at the bottom 5th or top 95th percentile, respectively.

There are more advanced outlier detection and cleaning methods, which we will touch on further in the course.

2.4.6 Inconsistent Data

Inconsistencies can hint at data falsifications

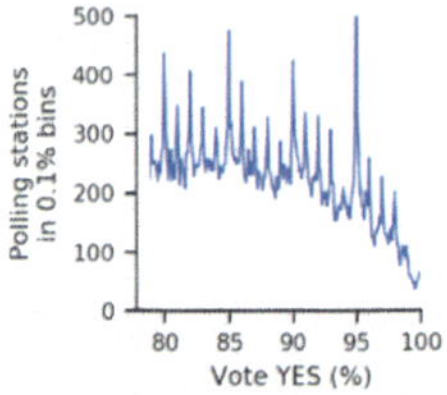

Analysis of the 2020 voting in Russia. Data anomalies at each integer percentage of "Yes" votes and more prominent anomalies at 80%, 85%, 90%, and 95% votes indicate problematic data. Figure modified from Kobak, D., Shpilkin, S., & Pshenitchnikov, M. (2020). Suspect peaks in Russia's "referendum" results. Significance Magazine, 17(5), 8–9.

We have already discussed inconsistent data in ▶ Sect. 2.3. Inconsistent data contains discrepancies and irregularities incompatible with other records—for example, mismatching formats, differences in units of measurement, unreasonable or impossible values, and incompatible attribute values in the same records. Inconsistent data is observed as data anomalies, but some inconsistencies are very hard to detect; investigating data frequency distribution and data visualization is highly helpful in this respect.

2.4.7 Feature Engineering

Raw data attributes are frequently highly correlated, making them redundant in data analysis. On the other hand, critical attributes may be missing and must be generated from other features. *Feature engineering* refers to transforming raw data into a new, frequently reduced representation that is easier to manipulate and interpret and more suitable for analysis. Feature engineering is especially important when large datasets are challenging to analyze without appropriate reduction.

Data reduction aims to decrease the number of attributes (or records) while preserving the relevance of the original dataset. The primary goal of data reduction is to create a condensed representation of the data that maintains the essential characteristics of the original dataset. One set of methods, generally called *dimensionality reduction*, transforms the original high-dimensional data (that is, the datasets containing multiple attributes) into a lower-dimensional space. These methods include Principal Component Analysis (PCA), Singular Value Decomposition (SVD), Nonnegative matrix factorization (NMF), and many others, each with its own advantages and limitations and hence more suitable with different data mining methods.

Normalization, standardization, binning, and other data transformations, discussed in ▶ Sect. 2.4.4, are also methods of feature engineering. Their goal is to make features more suitable for analysis rather than to reduce the volume of data. Finally, feature selection methods help to select the most influential factors affecting the target variable. This is particularly useful when dealing with datasets with many variables, as it can help identify the most significant variables and eliminate those that contribute little to the overall analysis. In the ▸ Chapter 1 lab exercise, we developed the decision tree model to explain factors essential for the survival of Titanic passengers. The Information Gain method used in the decision tree model is an example of a feature selection algorithm and will be discussed in ▸ Chapter 4.

Overall, feature selection helps reduce the data's volume and complexity, improve the efficiency of data analyses, and return more meaningful insights.

2.4.8 Group Imbalances

Nominal and ordinal attributes in imbalanced datasets break data into groups of radically different sizes. For instance, the "star rating" attribute in the hotel review data typically breaks the review dataset into two very large groups of 4* and 5* "good" reviews, a much smaller group of 3* "moderate" reviews, and a very small group of 1* and 2* "bad" reviews. This imbalance makes the direct application of machine learning methods challenging. For example, a Naïve Bayes model, which we will learn further in the course, trained to predict hotel ratings from review texts, will target optimal prediction of the large 4* and 5* groups, paying little attention to small 1* and 2* reviews. Essentially, data imbalance tends to render small groups irrelevant for analysis.

Group imbalance

Group imbalance refers to a radically different membership in data point groups. Many data mining methods can lead to biased conclusions in presence of group imbalance.

One way of dealing with imbalanced groups is data undersampling. This method removes records from the dominant groups (in our example, from the 4* and 5* reviews), making all groups similar in size. The downside of this method is that information from removed records is lost for future analysis.

The opposite of the undersampling method is data oversampling. This method artificially generates additional records in the underrepresented groups. The easiest oversampling method is record duplication; more involved methods allow synthetic data generation.

Under- and oversampling

The goal is to make the number of records for each group similar. Undersampling would reduce the number of records in the larger group. Oversampling would artificially generate additional records for the smaller group.

Finally, some data mining algorithms have been specifically developed to mitigate the problem of group imbalance. For example, the Naïve Bias with attribute weighting allows the assignment of different weights to records from specific groups. The Random Forest method, discussed in ▸ Chapter 4, accommodates small groups by generating multiple decision tree models. Still, group imbalance is easy to miss, yet it can critically impact the quality of data mining outcomes while giving an illusion of well-made analysis.

2.5 Conclusion

Raw data tends to be messy, full of typos, errors, incompatible records, and missing values. Because of that, data cleaning is a critical preliminary step in the data mining pipeline. Yet many junior researchers omit this step, rushing to start analyzing the data they just collected. That puts their results on a shaky foundation and devalues their contribution to learning data insights.

2.6 Review Questions

1. What are the main sources of errors and inconsistencies in raw data, and how do they impact analysis?
2. Describe the difference between the data types and the levels of measurement. Give examples of nominal, ordinal, interval, and ratio data.
3. Explain the main parts of data cleaning together with specific problems frequent in raw data they attempt to resolve.
4. What do we mean by "high-dimensional data," and what are its challenges for data analysis?
5. What are the main approaches for dealing with high-dimensional data?
6. Discuss the main methods to deal with missing data.
7. Explain the utility of data standardization and normalization. What are other ways of data transformation during pre-processing?
8. Explain the main methods of detecting and cleaning data outliers.
9. What is "feature engineering"? Why feature engineering is important?
10. Give a real-world example of imbalanced groups other than those in the Chapter. Discuss the methods for handling group imbalances in datasets and the potential impacts of not addressing this issue.

2.7 Lab

Objectives

1. Practice detection and cleaning of missing and erroneous data in Python
2. Practice data transformation
3. Analyze real-world survey data.

Data

This lab utilizes data from a basketball game visitors' survey. Survey data is simplified. The variables are as follows:

Variable	Level of measurement	Type	Explanation
ID	Nominal	Integer	Unique ID
Gender	Nominal	Integer	Gender (1: Male, 2: Female)
Age	Ratio	Float	Age (years)
Attend_this	Ratio	Integer	The number of this team's games you attend this season
Attend_next	Ratio	Integer	The number of this team's games you intend to attend next season
Spend_this_season	Ratio	Float	The total dollar amount you spent this season
Spend_next_season	Ratio	Float	The total dollar amount you plan to spend next season
Donate	Nominal	Integer	Do you donate to this team? (1: Yes; 2: No)
How_much	Ratio	Float	If yes, how much did you donate last year?

Part 1. Plan for Data Cleaning

The cleaning should include the following steps

1. Import raw data from GitHub (file Gators_Basketball_Survey_Subset1.xlsx)
2. Check columns against the data description. Are there any missing, mistyped, or extra variables?
3. Check variable types. Any errors?
4. Check the values of variables.
   a. Check rows for duplicate participants (Duplicate ID number)
   b. Remove the values outside the possible range
5. Check variable distribution.
   a. Transform highly skewed variables
   b. Remove the outliers
   c. Normalize variables
6. Save cleaned data.

Part 2. Clean Data

Step 1. Import Raw Data

We will write the code using the AI assistant.

Start a new empty notebook. If you don't see the file browser on the left of the screen, open it. Right-click in the file browser window and select "New notebook".

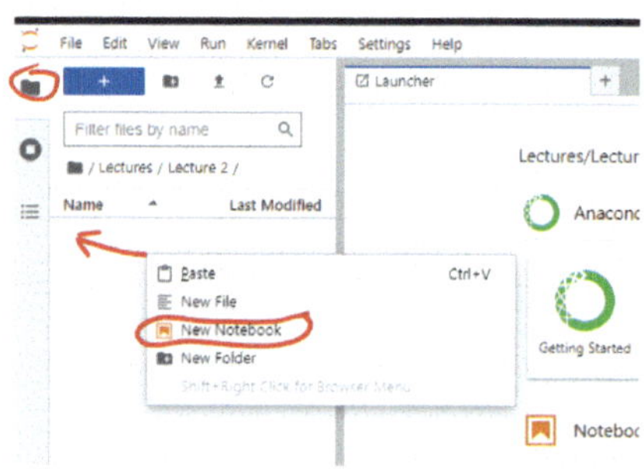

Enter the following text to the Anaconda Assistant window on the low right and start the chat:

USER:

Read data from ▶ https://github.com/andykir/DataMiningLectures/raw/main/Gators_Basketball_Survey_Subset1.xlsx

Assistant:

To read data from the provided URL, we can use the `pandas` package in Python. We will use the `read_excel` function from the `pandas` package to read the data from the Excel file. Here's the code to achieve this:

```
import pandas as pd

url = "https://github.com/andykir/DataMiningLectures/raw/ma
data = pd.read_excel(url)
```

This code will read the data from the provided URL and store it in a pandas DataFrame called `data`.

Click "Run the code in the Notebook window." The code will appear in a cell of the notebook window and be executed.

Step 2. Check Columns Against the Data Description. Are There Any Missing, Mistyped, or Extra Variables?

USER:

Print information about data columns.

Assistant:

To print information about the columns in the data, we can use the *info*() method of the pandas DataFrame. This method provides a concise summary of the columns, including the column names, data types, and non-null counts. Here's the code to achieve this:

```
print(data.info())
```

When you run this code, it will display information about the data columns, including their names, data types, and non-null counts.

Result:

```
<class 'pandas.core.frame.DataFrame'>
RangeIndex: 283 entries, 0 to 282
Data columns (total 9 columns):
 #   Column                     Non-Null Count  Dtype
---  ------                     --------------  -----
 0   ID                         283 non-null    int64
 1   Gender                     283 non-null    int64
 2   Age                        282 non-null    float64
 3   AttendedGamesThisSeason    283 non-null    int64
 4   WillAttendGamesNextSeason  282 non-null    float64
 5   Spend_this_season          283 non-null    int64
 6   Plans_Spend_next_season    102 non-null    float64
 7   Donate                     283 non-null    int64
 8   How_much                   151 non-null    float64
dtypes: float64(4), int64(5)
memory usage: 20.0 KB
None
```

Notice missing values ("Non-Null Count" lesser than 283). Some variables have names that differ from the data description. Finally, notice that some column types differ from the intended ones. Identify these columns and correct data types.

USER:

In the DataFrame data, change the names of the following columns: [AttendedGamesThisSeason, WillAttendGamesNextSeason, Plans_Spend_next_season] to [Attend_this, Attend_next, Spend_next_season].

Assistant:

```
data.rename(columns={'AttendedGamesThisSeason': 'Attend_this',
'WillAttendGamesNextSeason': 'Attend_next', 'Plans_Spend_next_season':
'Spend_next_season'}, inplace=True)
```

Explanation:

In this code, we use the *rename*() method to specify the new column names for the columns *AttendedGamesThisSeason*, *WillAttendGamesNextSeason*, and *Plans_Spend_next_season*. The *inplace=True* parameter ensures the changes are made directly to the original DataFrame.

NOTE: Even though the Assistant uses *inplace=True*, it is not considered good practice: it is potentially confusing and does not always behave as it should. Instead, you can use:

```
data = data.rename(columns={'AttendedGamesThisSeason': 'Attend_this',
'WillAttendGamesNextSeason': 'Attend_next', 'Plans_Spend_next_season':
'Spend_next_season'})
```

After running this code, the specified column names in the DataFrame will change.

To check on changes, click the empty space below the last cell in the Notebook window. A new empty cell will appear. Enter the code *print(data.info())* and then click "Run the cell and advance." Alternatively, you can use the Assistant.

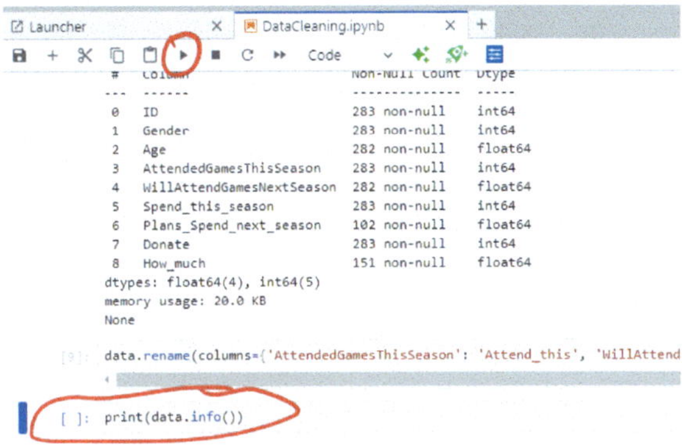

Result:

```
<class 'pandas.core.frame.DataFrame'>
RangeIndex: 283 entries, 0 to 282
Data columns (total 9 columns):
 #   Column              Non-Null Count  Dtype
---  ------              --------------  -----
 0   ID                  283 non-null    int64
 1   Gender              283 non-null    int64
 2   Age                 282 non-null    float64
 3   Attend_this         283 non-null    int64
 4   Attend_next         282 non-null    float64
 5   Spend_this_season   283 non-null    int64
 6   Spend_next_season   102 non-null    float64
 7   Donate              283 non-null    int64
 8   How_much            151 non-null    float64
dtypes: float64(4), int64(5)
memory usage: 20.0 KB
None
```

Step 3. Check Variable Types. Any Errors?

USER:

In the DataFrame data, change the types of the following columns: [*Attend_next, Spend_this_season*] to [int64, float64]. Print data information.

Assistant:

```
data['Attend_next'] = data['Attend_next'].astype('int64')
data['Spend_this_season'] =
data['Spend_this_season'].astype('float64')

print(data.info())
```

Explanation:

In this code, we first use the *astype*() method to change the data types of the columns *Attend_next* and *Spend_this_season* to 'int64' and 'float64', respectively. Then, we use the *info*() method to print the information about the columns in the data, including the updated data types. When you run this code, it will display the information about the columns in the data, including the updated data types for the specified columns.

Result:

There is an error. The end of the output gives an idea that there are missing values for which the type could not be converted:

```
IntCastingNaNError: Cannot convert non-finite values (NA or inf) to integer
```

We can deal with this problem in multiple ways.

- Change the data type of all columns to float. This data type has a special NA (or Nan) symbol for missing values.
- Leave data as is (not advised: for example, the types of attributes *Attend_this* and *Attend_next* would differ)
- Replace missing values with a valid value (-9999 is frequently used)
- Remove rows with missing values, and so on (see the lecture).

We will use the first option.

USER:

In the DataFrame "data", change the types of all columns to float64. Print "data" information.

Assistant:

```
data = data.astype('float64')
print(data.info())
```

Explanation:

In this code, we use the *astype*() method to change the data types of all columns in the DataFrame to 'float64'. Then, we use the *info*() method to print the information about the columns in the data, including the updated data types for all columns. When you run this code, it will display the information about the columns in the data, showing that all columns have been converted to the 'float64' data type.

Result:

```
<class 'pandas.core.frame.DataFrame'>
RangeIndex: 283 entries, 0 to 282
Data columns (total 9 columns):
 #   Column              Non-Null Count  Dtype
---  ------              --------------  -----
 0   ID                  283 non-null    float64
 1   Gender              283 non-null    float64
 2   Age                 282 non-null    float64
 3   Attend_this         283 non-null    float64
 4   Attend_next         282 non-null    float64
 5   Spend_this_season   283 non-null    float64
 6   Spend_next_season   102 non-null    float64
 7   Donate              283 non-null    float64
 8   How_much            151 non-null    float64
dtypes: float64(9)
memory usage: 20.0 KB
None
```

Step 4a. Check Rows for Duplicate Participants (Duplicate ID Number)

USER:

Print the number of duplicated values in data.ID.

Note: *data.ID* specifies that we want to use the column *ID* of the DataFrame *data*. An identical specification is *data['ID']*.

Assistant:

To print the number of duplicated values in the 'ID' column of the DataFrame, we can use the following code:

```
num_duplicates = data['ID'].duplicated().sum()
print("Number of duplicated values in data.ID:", num_duplicates)
```

Explanation:

In this code, we use the '*duplicated*()' method to identify duplicate values in the *ID* column, and then use the *sum*() method to count the number of duplicated values. When you run this code, it will display the number of duplicated values in the *ID* column of the DataFrame *data*.

Result:

```
Number of duplicated values in data.ID: 0
```

Step 4b. Remove the Values Outside the Possible Range

USER:

In the DataFrame 'data', remove all rows with cells outside the following ranges but keep the rows with missing data. The ranges are as follows:
Age: [0,100]
Attend_this: [0,12]
Attend_next: [0,12]
Spend_this_season: [0,5000]
Spend_next_season: [0,5000]
How_much: [0, 5000]

Save the result to DataFrame data. Print DataFrame data information.

Assistant:

```
# Remove rows with column values outside the specified ranges, keeping
NaN values
data = data[(data['Age'].between(0, 100, inclusive='both') |
data['Age'].isnull()) &
         (data['Attend_this'].between(0, 12, inclusive='both') |
data['Attend_this'].isnull()) &
         (data['Attend_next'].between(0, 12, inclusive='both') |
data['Attend_next'].isnull()) &
         (data['Spend_this_season'].between(0, 5000, inclusive='both')
| data['Spend_this_season'].isnull()) &
         (data['Spend_next_season'].between(0, 5000, inclusive='both')
| data['Spend_next_season'].isnull()) &
         (data['How_much'].between(0, 5000, inclusive='both') |
data['How_much'].isnull())]

# Print DataFrame information
print(data.info())
```

Explanation:

In this code, we use the *between*() method to filter the DataFrame based on the specified ranges for each column. The *inclusive*="*both*" parameter ensures that the range includes the specified values. After filtering, we print the information about the columns in the data to show the updated DataFrame information. When you run this code, it will display the information about the columns in the data, showing the updated DataFrame after removing rows with column values outside the specified ranges.

Result:

```
<class 'pandas.core.frame.DataFrame'>
Index: 256 entries, 1 to 282
Data columns (total 9 columns):
 #   Column             Non-Null Count  Dtype
---  ------             --------------  -----
 0   ID                 256 non-null    float64
 1   Gender             256 non-null    float64
 2   Age                256 non-null    float64
 3   Attend_this        256 non-null    float64
 4   Attend_next        255 non-null    float64
 5   Spend_this_season  256 non-null    float64
 6   Spend_next_season  84 non-null     float64
 7   Donate             256 non-null    float64
 8   How_much           131 non-null    float64
dtypes: float64(9)
memory usage: 20.0 KB
None
```

Step 5a. Check the Variables' Distribution. Transform Highly Skewed Variables

We start with a visualization of variable distributions.

USER:

Draw histograms for the distribution of the following columns of DataFrame 'data': Age, Attend_this, Attend_next, Spend_this_season, Spend_next_season, How_much.

Assistant:

```
import matplotlib.pyplot as plt

# Assuming 'data' is the name of the DataFrame
# Replace 'data' with the actual name of your DataFrame

# Define the columns to draw histograms for
columns_to_plot = ['Age', 'Attend_this', 'Attend_next',
'Spend_this_season', 'Spend_next_season', 'How_much']

# Set up the subplots
fig, axes = plt.subplots(nrows=2, ncols=3, figsize=(15, 10))

# Flatten the axes for easy iteration
axes = axes.flatten()

# Plot histograms for each column
for i, col in enumerate(columns_to_plot):
    data[col].plot(kind='hist', ax=axes[i], title=col,
color='skyblue', edgecolor='black')
    axes[i].set_xlabel(col)
    axes[i].set_ylabel('Frequency')

# Adjust layout
plt.tight_layout()
plt.show()
```

Explanation:

In this code, we use matplotlib to create subplots for each column and then plot histograms for the specified columns in the DataFrame 'data'. The histograms will display the distribution of each column's values.

Result:

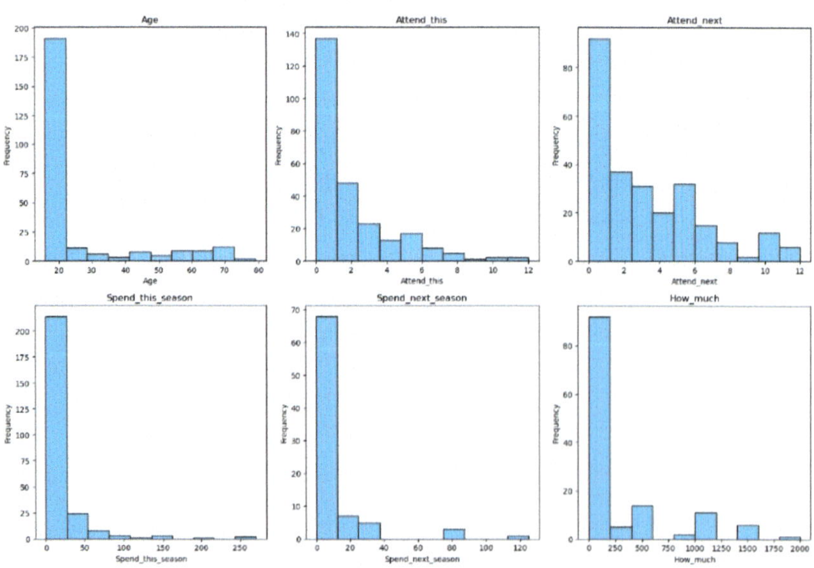

Notice that data distributions are left-skewed. To deal with a potential problem, we will apply the logarithmic transformation.

USER:

Apply logarithmic transformation to the following columns of DataFrame 'data': Age + 1, Attend_this + 1, Attend_next + 1, Spend_this_season + 1, Spend_next_season + 1, How_much + 1. Save results as new columns in the DataFrame data: AgeLog, Attend_thisLog, Attend_nextLog, Spend_this_seasonLog, Spend_next_seasonLog, How_muchLog.

Assistant:

```
# Assuming 'data' is the name of the DataFrame
# Replace 'data' with the actual name of your DataFrame

# Define the columns to apply a logarithmic transformation
columns_to_transform = ['Age', 'Attend_this', 'Attend_next',
'Spend_this_season', 'Spend_next_season', 'How_much']

# Apply logarithmic transformation after adding 1 to each value and
save results as new columns
for col in columns_to_transform:
    data[col + 'Log'] = np.log(data[col] + 1)

# Print the first few rows of the DataFrame to verify the new columns
print(data.head())
```

Result:

```
   ID  Gender   Age  Attend_this  Attend_next  Spend_this_season  \
1  2.0     1.0  20.0          2.0          2.0                0.0
2  3.0     1.0  19.0          1.0          1.0                0.0
3  4.0     1.0  19.0          1.0          2.0                0.0
4  5.0     1.0  19.0          1.0          5.0                0.0
5  6.0     1.0  18.0          0.0          0.0                0.0

  Spend_next_season  Donate  How_much    AgeLog  Attend_thisLog  \
1               NaN     1.0    1000.0  3.044522        1.098612
2               NaN     1.0     800.0  2.995732        0.693147
3               NaN     2.0       0.0  2.995732        0.693147
4               NaN     2.0       0.0  2.995732        0.693147
5               0.0     2.0       0.0  2.944439        0.000000

  Attend_nextLog  Spend_this_seasonLog  Spend_next_seasonLog  How_muchLog
1       1.098612                   0.0                   NaN     6.908755
2       0.693147                   0.0                   NaN     6.685861
3       1.098612                   0.0                   NaN     0.000000
4       1.791759                   0.0                   NaN     0.000000
5       0.000000                   0.0                   0.0     0.000000
```

Observe the distributions of the transformed variables.

USER:

Draw histograms for the distribution of the following columns of DataFrame 'data': AgeLog, Attend_thisLog, Attend_nextLog, Spend_this_seasonLog, Spend_next_seasonLog, How_muchLog.

Assistant:

```
import matplotlib.pyplot as plt

# Assuming 'data' is the name of the DataFrame
# Replace 'data' with the actual name of your DataFrame

# Define the columns to draw histograms for
columns_to_plot = ['AgeLog', 'Attend_thisLog', 'Attend_nextLog',
'Spend_this_seasonLog', 'Spend_next_seasonLog', 'How_muchLog']

# Set up the subplots
fig, axes = plt.subplots(nrows=2, ncols=3, figsize=(15, 10))

# Flatten the axes for easy iteration
axes = axes.flatten()

# Plot histograms for each column
for i, col in enumerate(columns_to_plot):
    data[col].plot(kind='hist', ax=axes[i], title=col,
color='lightgreen', edgecolor='black')
    axes[i].set_xlabel(col)
    axes[i].set_ylabel('Frequency')

# Adjust layout
plt.tight_layout()
plt.show()
```

Result:

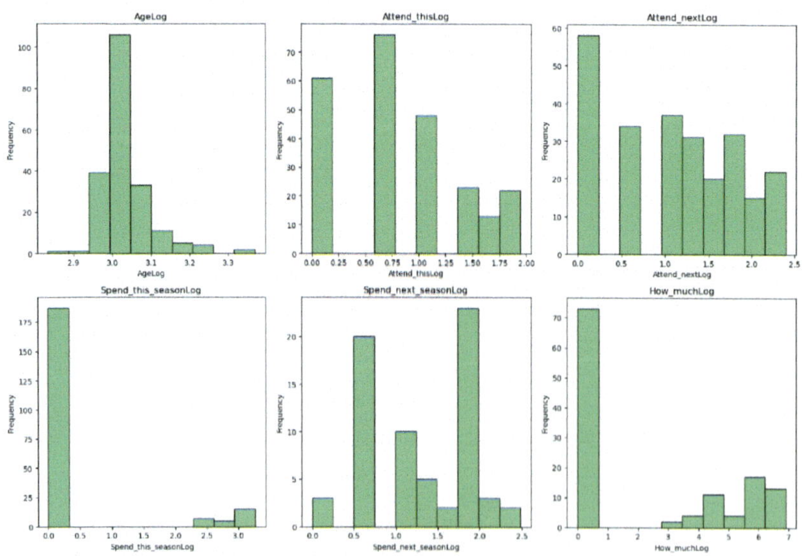

The transformed data does not contain outliers. However, the *Spend_this_seasonLog* attribute shows an interesting pattern: a considerable number of people reported no spending. This could be due to how the survey question was formulated or how survey participants were selected. The only candidate for outlier removal is the *AgeLog* variable.

Step 5b. Remove the Outliers

We will use the interquartile range method to remove the outliers. All values outside the range will be removed, but we will ignore the missing values (NaN).

USER:

In DataFrame 'data', find the outliers in the following column: AgeLog. Replace outliers with NaN values. Save the result in DataFrame data. Draw a histogram for the distribution of the column AgeLog.

Assistant:

```
import matplotlib.pyplot as plt

# Assuming 'data' is the name of the DataFrame
# Replace 'data' with the actual name of your DataFrame

# Create a copy of the original DataFrame
data_copy = data.copy()

# Calculate the 25th and 75th percentiles for AgeLog
Q1 = data_copy['AgeLog'].quantile(0.25)
Q3 = data_copy['AgeLog'].quantile(0.75)

# Calculate the interquartile range (IQR) for AgeLog
IQR = Q3 - Q1

# Define the lower and upper bounds for outliers
lower_bound = Q1 - 1.5 * IQR
upper_bound = Q3 + 1.5 * IQR

# Replace outliers in AgeLog with NaN values
data_copy['AgeLog'] = data_copy['AgeLog'].mask((data_copy['AgeLog'] <
lower_bound) | (data_copy['AgeLog'] > upper_bound))

# Save the result as DataFrame data
data = data_copy

# Draw a histogram for the distribution of AgeLog
plt.figure(figsize=(8, 6))
data['AgeLog'].plot(kind='hist', color='lightblue', edgecolor='black')
plt.title('Distribution of AgeLog')
plt.xlabel('AgeLog')
plt.ylabel('Frequency')
plt.show()
```

Explanation:

In this code, we first create a copy of the original DataFrame to avoid modifying the original data. We then calculate the 25th and 75th percentiles for the *AgeLog* column and use these to determine the lower and upper bounds for outliers. We remove the outliers and save the result as DataFrame data. Finally, we draw a histogram for the *AgeLog* column distribution.

Result:

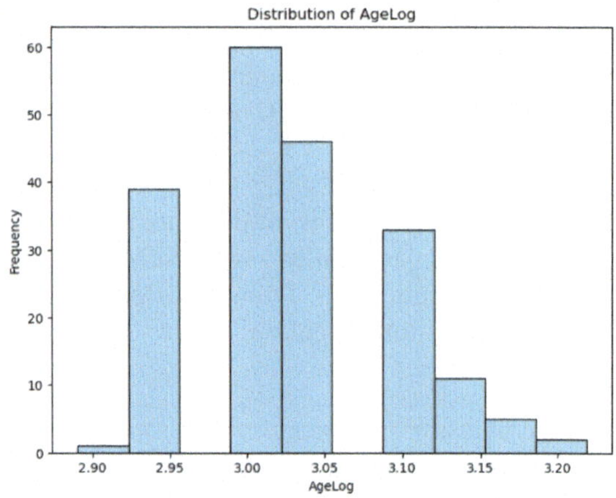

Notice that one outlier was removed.

Part 3. On Your Own

1. Why did we transform the variables first and remove the outliers second? Change the code to remove the outliers first and then convert the variables. Elaborate on the outcomes.
2. The column *Donate* shows if a respondent donated to the team, while the column *How_much* shows the donation amount. Modify the code to check for consistency between these two columns. That is, for those who don't donate, the donation amount should be 0.
3. Normalize the attributes responsible for the amount of money spent, planned to spend, and donated. Remove the values from the cells with zero donations.
4. Export the cleaned data into an Excel file.

Build your Python skills.

Chapters 3–5 of the free book Learn Python with Jupyter by Serena Bonaretti. ▶ https://www.learnpythonwithjupyter.com/

Additional Reading

Ilyas, I. F., & Chu, X. (2019). *Data cleaning*. Morgan & Claypool.

Introduction to Data Mining Methods. Association Rules

A. P. Kirilenko, *Practical Data Mining with AI for Social Scientists*,
Springer Texts in Social Sciences, https://doi.org/10.1007/978-3-031-89689-7_3

3.1 Statistics and Data Mining

Data analysis encompasses many fields, all contributing to the broader goal of making sense of data. These fields largely overlap in practice yet frequently have their distinct terminology. In this section, we attempt to cover some of these fields, pointing at their interconnectivity.

Statistics is the discipline fundamental to all data-related methodologies. Traditionally, statistics deals with:

- data description, such as summarization and visualization, and
- inferences: making informed conclusions about populations based on their samples.

These fields include hundreds of statistical methods of data analysis developed in dozens of scientific disciplines. Examples of fundamental statistical methods include correlation, regression analysis, sampling, time series, and hypothesis testing. At the same time, new developments in data science and computer processing capabilities have extended the data analysis toolkit (for example, with *machine learning* methods) so much that it calls for a new name: data mining.

Machine learning

Machine learning is a group of methods targeting automated learning from data without explicit instruction on the learning process. A classical example of these methods is an artificial neural network (ANN), a computational model inspired by the network of neurons in animals. Machine learning is also a subfield of the Artificial Intelligence (AI); frequently, these terms are used interchangeably.

Data mining is the discipline of discovering patterns and extracting knowledge from large amounts of data. While the goal of data mining is similar to that of statistics, the nature of data brings different methods. The most frequently used data mining methods include:

- Classification: categorization of data points into pre-defined groups based on data attributes;
- Clustering: finding clusters of data points based on data attributes. Note that, distinctly from classification, clustering is applied when there is no prior knowledge about the groups exists;
- Association rule mining: finding patterns of co-occurring data points. Association rule mining can be applied to search for groups of products likely to be purchased together. The learned associations can then be used to suggest additional products to a customer based on his/her previous purchases or inquiries;
- Decision trees: building tree-like graphical models representing the decision process. We used a decision tree method to analyze Titanic data in ▸ Chapter 1
- Anomaly detection (outlier analysis): identification of unusual data points;

- Text mining: extraction of knowledge from unstructured textual data;
- Image recognition: identification of objects on photographs and videos.

Since data mining deals with large volumes of data, data organization and storage are essential. Because of that, a database system is an indispensable component of data mining. Overall, data mining draws ideas from machine learning, AI, statistics, and database systems.

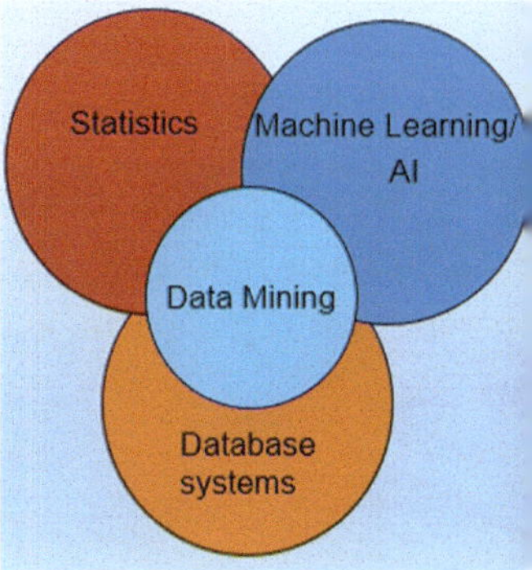

3.2 Informal Definition of Association Rules

Association rules are a machine learning method that seeks to find connections between data attributes, that is, a pattern in which the attributes tend to appear together. The discovered connections are called *rules*. The rules essential for the goals of data analysis are called *interesting rules*. For instance, for someone interested in superstitions, an observation that crossing the path with a black cat tends to coincide with an accident later in that day, the rule {black cat} → {accident} is interesting.

Shopping basket analysis

Shopping basket analysis seeks to identify the associations between the products bought together in a single transaction

Consider one application of association rules: shopping basket analysis. Shopping basket analysis (also called market basket analysis) is a data mining technique used to understand customers' purchase behavior. Specifically, shopping basket analysis seeks to identify the associations between the products bought together in a single transaction. Shopping basket analysis helps businesses optimize their inventory, marketing strategies, and store layout.

Consider ten hypothetical purchases from a grocery store:

Transaction	Items purchased
1	{milk, bread, butter}
2	{bread, butter, jam}
3	{milk, bread}
4	{milk, jam}
5	{bread, butter}
6	{milk, butter}
7	{jam, butter}
8	{milk, bread, butter}
9	{bread, coffee}
10	{milk, bread, jam}

A casual analysis of this table reveals a frequent pattern of milk being purchased together with bread. Specifically, out of six transactions containing milk, four also contain bread. This pattern is expressed as a rule:

$\{\text{Milk}\} \rightarrow \{\text{Bread}\}$,

That is, a purchase of milk leads to a purchase of bread. Our *confidence* in this rule can be computed as the number of transactions where the rule works divided by the sum of the number of transactions where the rule works and does not work. That is, the number of transactions that contain milk and bread together (4) divided by the total number of transactions that contain milk but not bread (6):

$$\text{Confidence}(\{\text{Milk}\} \rightarrow \{\text{Bread}\}) = 4/6 = 0.67.$$

Another noticeable pattern is that a purchase of milk and bread tends to coincide with a purchase of butter in the same transaction. This pattern is expressed as a rule:

$$\{\text{Milk, Bread}\} \rightarrow \{\text{Butter}\}.$$

The confidence for this rule is:

$$\text{Confidence}(\{\text{Milk, Bread}\} \rightarrow \{\text{Butter}\}) = 2/4 = 0.5$$

How frequently the rule appears in our dataset reflects how well supported by data the rule is. The rule $\{\text{Milk, Bread}\} \rightarrow \{\text{Butter}\}$ is supported by two observations only, while the rule $\{\text{Milk}\} \rightarrow \{\text{Bread}\}$ is supported by four observations and hence is better supported by data. This is expressed as a ratio of the number of transactions containing Milk, Bread, and Butter divided by the total number of transactions:

$$\text{Support}(\{\text{Milk, Bread}\} \rightarrow \{\text{Butter}\}) = 2/10 = 0.2$$

Comparatively,

$$\text{Support}(\{\text{Milk}\} \rightarrow \{\text{Bread}\}) = 4/10 = 0.4$$

These examples are extremely simplistic. In real life, business analysts deal with millions of transactions covering thousands of products. The next section will discuss additional indicators that help define interesting rules. After that, we will discuss approaches that allow us to find the rules that maximize these indicators.

3.3 Metrics of Rule Interestingness

3.3.1 Association Rules Formalized

Consider customers' transaction database:

Transaction	item 1	item 2	item 3	item 4	item 5	item 6	...
T_1	1	0	1	1	0	0	...
T_2	0	1	0	1	0	0	...
T_3	1	1	1	0	1	0	...
T_4	0	1	0	1	0	1	...
...	...	...	...	...	...	...	...

In this table, the rows represent individual transactions, the columns represent an entire inventory of products, and a cell value of 1 means that a particular transaction contained the respective item. For example, transaction T1 contained a set of items 1, 3, and 4. Any combination of items is called an *itemset.*

$$X \rightarrow Y,$$

An association rule is then expressed as an implication. where X and Y are *disjoint itemsets*. Disjoint itemsets are itemsets that don't have any shared items. Left-hand itemset X is called an *antecedent,* and the right-hand itemset Y is called a *consequent*.

Definitions

Itemset: any set of items.

Support: how frequently an itemset is observed in a transaction database.

Confidence: how frequently an association rule is correct among all transactions containing the rule antecedent.

The support and confidence should be above a certain threshold for a rule to be interesting.

3.3.2 Association Rules' Support and Confidence

There is a very high number of possible rules: the number of association rules that can be potentially generated from an itemset of 1000 products is well exceeding the number of atoms in the Universe. However, the majority of these rules are useless.

The strength of an association rule is measured by two metrics: *support* and *confidence*. The **support** of a rule is the proportion of transactions in the dataset that contain all items in both the antecedent (X) and the consequent (Y) of the rule. In other words, support is a probability P that both the X and Y itemsets appear in one transaction:

$$\text{supp}(X \rightarrow Y) = P(\{X, Y\}) = \text{count}(\{X, Y\})/N,$$

where N is the total number of transactions. Support favors the items that are frequently purchased, uncovering the patterns that are worth taking advantage of.

Confidence of a rule is the conditional probability of having the consequent in a transaction given that the antecedent is present:

$$\text{conf}(X \rightarrow Y) = P(X \mid Y\}) = P(\{X, Y\}) / P(X)$$

We can use the following formulas to calculate the confidence:

$$\text{conf}(X \rightarrow Y) = P(X \mid Y\}) = \text{count}(\{X, Y\}) / \text{count}(X)\text{, or}$$

$$\text{conf}(X \rightarrow Y) = \text{supp}(X \rightarrow Y) / \text{supp}(X).$$

Notice that the calculation of confidence returns an invalid result when the frequency of itemset X is 0. This happens when purchases of specific products are missing in the analyzed transaction dataset. The *Laplace correction* resolves this problem with a minor modification of confidence calculation:

$$\text{conf}(X \rightarrow Y) = (\text{count}(\{X, Y\})+1) / (\text{count}(X) + 2)\text{,}$$

thus avoiding division by zero and providing a more robust measure.

Confidence provides the reliability measure of the rule. Now, we can formulate the initial criteria for a rule to be interesting: Both the support and the confidence of an interesting rule should be above a certain threshold. The exact value of these thresholds, minSupp and minConf, is set by the analyst:

$$\text{supp}(X \rightarrow Y) > \text{minSupp; conf}(X \rightarrow Y) > \text{minConf,}$$

where minSupp and minConf values are greater than 0 and lesser than 1.

Most frequently used measures of interestingness

Support: how frequently an itemset is observed in a transaction database.

Confidence: how frequent an association rule is correct among all transactions containing the rule antecedent.

Both the support and confidence should be above a certain threshold for a rule to be interesting.

Lift: Rule confidence divided by the expected confidence, assuming there is no association between the purchased items.

3.3.3 Lift and Other Important Metrics of Interestingness

High support and confidence are necessary but not enough for an association rule to be interesting. Returning to the store purchase example, the rule bread → milk has very high support since those items are frequently available, yet there is no causality in this purchase. Bread and milk are purchased together coincidentally because they are staple foods (at least, in Western cuisine), not because people like to eat bread with milk. Imagine randomly shuffling every column in the transaction database. This shuffling eliminates all causality in individual transactions. Yet, the rule bread → milk will still have a high confidence just because they are present in most purchases.

Meanwhile, businesses are interested in uncovering causal, unexpected associations among the items in their inventory. One approach in developing a metric prioritizing such an association is to compare the *observed* probability of two itemsets appearing together in the transaction database with the *expected probability, assuming no association between the itemsets.* The random column shuffling in the transaction database we discussed in the previous paragraph illustrates how the items' association can be eliminated. This metric is called *Lift*:

$$Lift(X \rightarrow Y) = \mathrm{P}(\{\mathrm{X},\ \mathrm{Y}\})/(\mathrm{P}(\mathrm{X})\mathrm{P}(\mathrm{Y}))$$

A mathematically equivalent formulation of Lift metric is as follows:

$$Lift(X \rightarrow Y) = supp(X \rightarrow Y)/(supp(X)supp(Y))$$

An *interesting* rule should have a Lift value greater than 1; a value of Lift ≤ 1 indicates that having the itemset X in the shopping basket does not increase the chances of purchasing the itemset Y. In other words, such a rule has no skill.

Returning to ► Sect. 3.1 example,

supp({Milk} → {Bread}) = 4/10 = 0.4

supp({Milk}) = 6/10 = 0.6

supp({Bread}) = 6/10 = 0.6

Lift({Milk} → {Bread}) = supp({Milk} → {Bread})/(supp({Milk})*supp({Bread}) = 1.11

The rule {Milk} → {Bread} is slightly greater than 1, showing a weak association between these products. A lift value of 1.11 implies that customers who have purchased bread are 11% more likely to buy milk.

On the other hand, the rule {Bread} → {Butter} has a Lift value of 0.74, revealing no interesting association between these two products. In fact, a Lift value lesser than 1 signifies a negative association between the products, implying that the presence of one antecedent itemset reduces the probability of the consequent itemset (recall that data in this example is fictionary).

Leverage (Piatetsky-Shapiro index) is very similar to Lift, but instead of the ratio, it uses a difference:

$$Leverage(X \rightarrow Y) = Support(X \rightarrow Y) - Support(X)Suport(Y).$$

3

Additional measures of interestingness

Leverage: rule's strength to make a *correct* prediction compared with a random guess.

Conviction: rule's strength *not* to make an *incorrect* prediction compared with a random guess.

Zhang et al. measure: a combination of association (buying A makes me buying B) and disassociation (buying A makes me not buying B) rule strength.

A value of leverage ≤ 0 indicates that having the itemset X in the shopping basket decreases the chances of purchasing the itemset Y.

Both the Lift and Leverage metrics compare how more often the itemsets X and Y occur together compared to what we could expect if they were statistically independent. In other words, Lift and Leverage measure the rule's strength to make a *correct* prediction compared with a random guess. A complementary metric, the **conviction** of an association rule, is a measure of the rule's strength *not* to make an *incorrect* prediction compared with a random guess.

Conviction is defined as the ratio of the expected frequency that itemset X is purchased *without* the itemset Y (assuming that X and Y are independent) to the observed frequency of incorrect predictions:

$$Conv(X \rightarrow Y) = \mathrm{P(X)P(\overline{Y})/P(X, \overline{Y})}$$

A mathematically equivalent formulation of conviction metric is as follows:

$$Conv(X \rightarrow Y) = (1 - supp(X)/(1 - conf(X \rightarrow Y))$$

Robust rules should have the conviction value greater than 1; a rule that always holds has Conv $= \infty$.

Zhang's is another metric that includes both product presence and absence in a single transaction. Similar to Leverage, positive values of Zhang's index indicate a positive association between itemsets in one transaction. Values close to zero indicate no association and negative values suggest avoidance.

The discussed measures are complementary. Each metric has its own strengths and weaknesses. For example, conviction is undefined for perfect rules, leading to a division by zero in the formula. Similarly, the gain can be misleading when the support of the antecedent or consequent is very low, resulting in an inflated lift value. In addition, there are many other metrics of rule interestingness. In practice, the question of retaining or discarding a tentative rule is decided based on a trade-off among multiple criteria. For example, one may require rule support and confidence to exceed 0.1 and lift and conviction to exceed 2. The specific values of these thresholds are decided individually for each transactional database.

3.4 Approaches to Mine Association Rules

Any real-world transactional database contains a vast number of association rules reflecting potential associations between products (items) that tend to occur together in the same transaction. Very few of these rules are helpful (interesting). In the previous section, we reviewed some famous metrics that help to differentiate interesting rules from useless ones. An interesting rule should have high support, confidence, lift, and conviction. This section briefly reviews the methods of searching such rules in an almost infinite number of worthless ones.

The most rudimentary approach to mining association rules is the brute force method. This technique involves generating all possible rules and then calculating their support, confidence, and other measures. However, this method is computationally expensive and inefficient, making it impractical for any database containing more than 10–15 products.

The **Apriori approach** significantly reduces the number of candidate itemsets in interesting rules. A rule with high support should include frequently appearing itemsets. The Apriori approach is based on an observation that *a subset of a frequent itemset must also be a frequent itemset*. For example, for an itemset {Bread, Milk, Butter} to be frequent, the following itemsets must be frequent as well: {Bread, Milk}, {Milk, Butter}, {Bread, Butter}, {Bread}, {Milk}, {Butter}.

Reversing this observation, the Apriori approach posits that *a superset of a frequent itemset will also be infrequent and, therefore,* can be eliminated from further consideration. The approach starts with observing all itemsets containing only one item (1-itemsets); infrequent itemsets are removed from all further consideration. Then, a similar principle is applied for 2-itemsets, and so on. For example, 1-itemset {coffee} appears only once in ► Sect. 3.1 transaction database. That means all itemsets containing coffee are infrequent as well and can be excluded from future consideration. The Apriori approach allows for a radical reduction of the number of potential itemsets.

Despite its strength, the Apriori approach is computationally expensive and still leaves a large number of potential interesting rules. Another approach is the **FP-Growth** (Frequent Pattern Growth) algorithm, which converts the database table into a tree-like structure (FP-tree). This method eliminates the need for candidate generation and allows more efficient data parsing. In addition, there are also partition-based methods, such as the **Savasere, Omiecinski, and Navathe (SON) algorithm**, which divides the database into smaller, more manageable partitions and mines each partition independently, and many others. Covering each of these approaches is beyond this book, but it is important to keep in mind that there are multiple algorithmic solutions fitting different kinds of data. For instance, the Apriori algorithm is effective for smaller datasets, but its performance degrades as the size of the dataset increases. Conversely, the FP-Growth algorithm performs well with large datasets but is computer memory-hungry.

3.5 Data Considerations

Data considerations

Amount of data

Measurement level

Nature of data

Distribution of values

Is time important?

We covered multiple metrics and algorithms for association rules mining. The selection of a specific analysis method depends on its purpose and the characteristics of the transaction database. The items are typically categorical, but association rules can also be applied to continuous data. The continuous data can be discretized (binned) to convert it into a categorical level of measurement.

The first consideration is the **size and measurement level** of the data. Large data volumes potentially provide more interesting and robust rules reflecting data patterns and associations. However, large data volumes come with computational challenges. Conversely, smaller datasets may lead to potentially biased or incomplete results.

The second consideration is the **nature of the data**. Association rules are typically applied to transactional data, where each individual transaction is represented by a set of products or services. Yet they are also used in applications as broad as assisting in medical diagnostics, biochemistry, processing of population census data, and

Customer Relationship Management (CRM). In tourism, association rules are extremely useful in building recommender systems. An association rules analysis of attraction points visited by past tourists within one visit allows for individualized recommendations on travel itineraries based on the traveler's profile.

The third consideration is the **distribution** of the data. Rare items may be interesting from a business perspective, e.g., because of their high price. Still, they would be excluded from further analysis if they don't appear frequently enough to meet the minimum support threshold. On the other hand, the presence of highly frequent items may include these items in a large number of rules, potentially obscuring more rare, unusual, and, hence, interesting associations. To address these issues, an analyst may want to remove the most frequent items from the dataset and also reduce the support threshold.

The fourth consideration is the **temporal dimension** of the data. In our grocery store transaction example, the order of individual items in a single transaction is irrelevant. However, in some applications, such as recommender systems, the transactions occur over time. For example, tourists do not visit attraction points on their wish list randomly. The order of visitation heavily depends on the attractions' proximity. A modification of association rule mining, sequential pattern mining, takes into account the sequential patterns of data.

3.6 Conclusion

The association rules method illustrates the differences between "traditional" statistics methods and data mining. Note that association rules mining emphasizes discovering hidden patterns and relationships within datasets without formulating prior hypotheses. While traditional statistics often rely on predefined models and hypotheses testing to validate theories about data, data mining focuses on exploring data to find unexpected associations and correlations. This shift from *hypothesis-driven to data-driven discovery* highlights the strengths of data mining in uncovering insights that existing theories and conventional statistical methods might miss.

3

3.7 Review Questions

1. What is the primary goal of data mining? How does it differ from traditional statistics?
2. List and briefly describe commonly used data mining methods.
3. What is the role of a database system in data mining?
4. Define association rules and explain their typical applications.
5. Describe the purpose of shopping basket analysis and its application in retail.
6. How does support differ from confidence among the association rules metrics?
7. Why might high support and confidence alone be insufficient to identify interesting association rules?
8. Explain the purpose of the lift metric and its main difference from rule support.
9. Discuss the difference between the lift and conviction metric in evaluating association rules.
10. Formulate the main idea of the Apriori algorithm.
11. What factors should be considered when selecting a method for association rule mining?

3.8 Lab

Objectives

1. Practice data descriptive analysis
2. Practice association rules mining
3. Practice writing analysis report.

Data

This lab aims to find sets of products typically purchased together from a bakery. A typical application is a system that recommends additional products to customers based on their existing product selection. The dataset provided by Akashdeep Kuila (▶ https://www.kaggle.com/datasets/akashdeepkuila/bakery) is from the Bread Basket, a bakery located in Edinburgh. It describes details of over

9,000 online transactions of customers, who ordered over 20,507 products in total. The data is stored as ▸ https://github.com/andykir/DataMiningLectures/raw/main/Bakery.xlsx.

Variables

Variable	Explanation	Notes
TransactionNo	Unique identifier for every single transaction	ID
Items	Items	Item name (text). Example: Bread
DateTime	Date and time stamp of the transactions	Example: 10/30/2016 9:58:11 AM
Daypart	Part of the day when a transaction is made	Morning, afternoon, evening, or night
DayType	Day of the week	Weekend or weekday

Part 1. Find Patterns in Bakery Purchases

Step 1. Read Data

We will write the code using the AI assistant. Initiate a new empty notebook.

Enter the following text to the Anaconda Assistant window on the low right and start the chat.

USER:

Read data from ▸ https://github.com/andykir/DataMiningLectures/raw/main/Bakery.xlsx

Assistant:

```
import pandas as pd

# Read the data from the URL
url =
"https://github.com/andykir/DataMiningLectures/raw/main/Bakery.xlsx"
bakery_data = pd.read_excel(url)

# Display the first few rows of the DataFrame to verify the data has
been read successfully
print(bakery_data.head())
```

Explanation:

In this code, we use the *read_excel* function from pandas to read the data from the provided URL and store it in the *bakery_data* DataFrame. We then print the first few rows of the DataFrame *bakery_data* to verify that the data has been read successfully.

Copy and run the code in the notebook you just created.

Result:

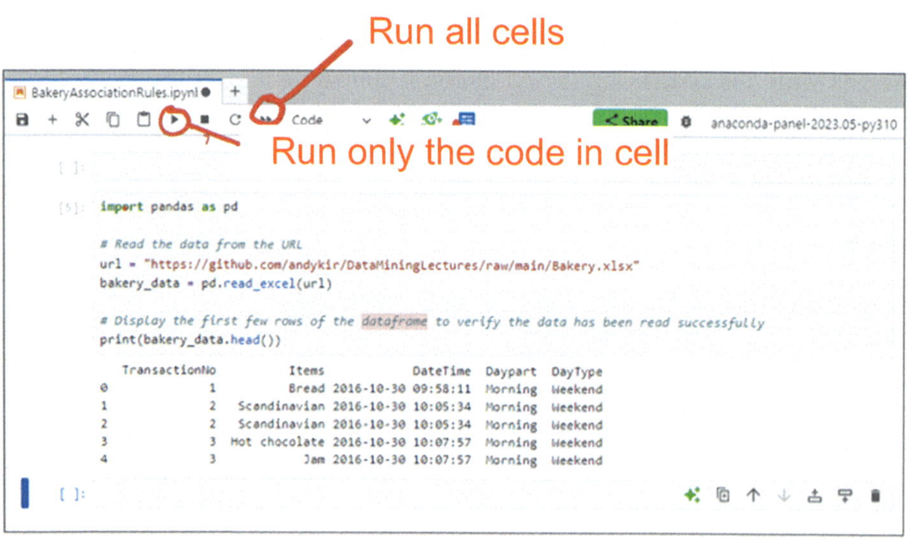

Recall that when the assistant stops, you may reload the web page. After that, run all cells (see the picture above). That will give the assistant the context for your requests.

Step 2. Descriptive Analysis

Explore your data. Ask the AI assistant:

USER:

How many unique IDs are there in the TransactionNo variable?

Assistant:

```
# Count the number of unique IDs in the "TransactionNo" variable
unique_ids = bakery_data['TransactionNo'].nunique()
print("Number of unique IDs in TransactionNo variable:", unique_ids)
```

Explanation:

In this code, we use the *nunique*() function to count the number of unique IDs in the *TransactionNo* attribute and then print the result.

Result:

```
[9]: # Count the number of unique IDs in the "TransactionNo" variable
     unique_ids = bakery_data['TransactionNo'].nunique()
     print("Number of unique IDs in TransactionNo variable:", unique_ids)

     Number of unique IDs in TransactionNo variable: 9465
```

USER:

What are the top three most frequent values in 'Items'?

Assistant:

```
# Find the top three most frequent values in the "Items" column
top_three_items = bakery_data['Items'].value_counts().nlargest(3)
print("Top three most frequent values in Items:")
print(top_three_items)
```

Explanation:

In this code, we use the *value_counts*() function to count the occurrences of each unique value in the *Items* column and then use the *nlargest*(3) function to find the top three most frequent values. Finally, we print the result.

3

Result:

```
Top three most frequent values in Items:

Items
Coffee     5471
Bread      3325
Tea        1435
```

Let us describe the distribution of the number of items in one transaction.

USER:

For each TransactionNo, count the number of duplications. Save the result in the variable "Duplications".

Assistant:

```
# Count the number of duplications for each TransactionNo
duplications =
bakery_data['TransactionNo'].value_counts().reset_index()
duplications.columns = ['TransactionNo', 'Duplications']

# Display the first few rows of the duplications DataFrame
print(duplications.head())
```

Explanation:

In this code, we use the *value_counts*() function to count the number of duplications for each *TransactionNo*. We then reset the index and rename the columns to *TransactionNo* and *Duplications*. Finally, we display the first few rows of the *duplications* DataFrame to verify the results.

Result:

```
   TransactionNo  Duplications
0           6716            11
1           6279            11
2           6474            11
3           6412            11
4           6045            10
```

In the following dialog, notice how we specify the column *Duplications* in the DataFrame *duplications*.

USER:

Compute the minimum, maximum, mean, median, and mode values for duplications.Duplications.

Note: Instead of *duplications.Duplications*, we can use the *duplications*[“*Duplications*”] notation. Notice that double and single quotation marks both can be used: *duplications*[‘*Duplications*’].

Assistant:

```
# Compute the minimum, maximum, mean, median, and mode values for the
"Duplications" variable
min_duplications = duplications['Duplications'].min()
max_duplications = duplications['Duplications'].max()
mean_duplications = duplications['Duplications'].mean()
median_duplications = duplications['Duplications'].median()
mode_duplications = duplications['Duplications'].mode()[0]

# Print the results
print("Minimum value of duplications:", min_duplications)
print("Maximum value of duplications:", max_duplications)
print("Mean value of duplications:", mean_duplications)
print("Median value of duplications:", median_duplications)
print("Mode value of duplications:", mode_duplications)
```

Explanation:

In this code, we use the *min*(), *max*(), *mean*(), *median*(), and *mode*() functions to compute the minimum, maximum, mean, median, and mode values for the *Duplications* attribute and then print the results.

Result:

```
Minimum value of duplications: 1
Maximum value of duplications: 11
Mean value of duplications:
2.1666138404648705
Median value of duplications: 2.0
Mode value of duplications: 1
```

USER:

Build a histogram for ‘Duplications’ distribution.

Assistant:

```
import matplotlib.pyplot as plt

# Build a histogram for the "Duplications" variable
plt.hist(duplications['Duplications'], bins=10, color='skyblue',
edgecolor='black')

# Add labels and title
plt.xlabel('Number of Duplications')
plt.ylabel('Frequency')
plt.title('Histogram of Duplications')

# Display the histogram
plt.show()
```

Explanation:

This code uses the *hist* function from *matplotlib.pyplot* to create a histogram for the *Duplications* attribute. We specified the number of bins, color, and edgecolor for the histogram, and then added labels and a title before displaying the histogram.

Result:

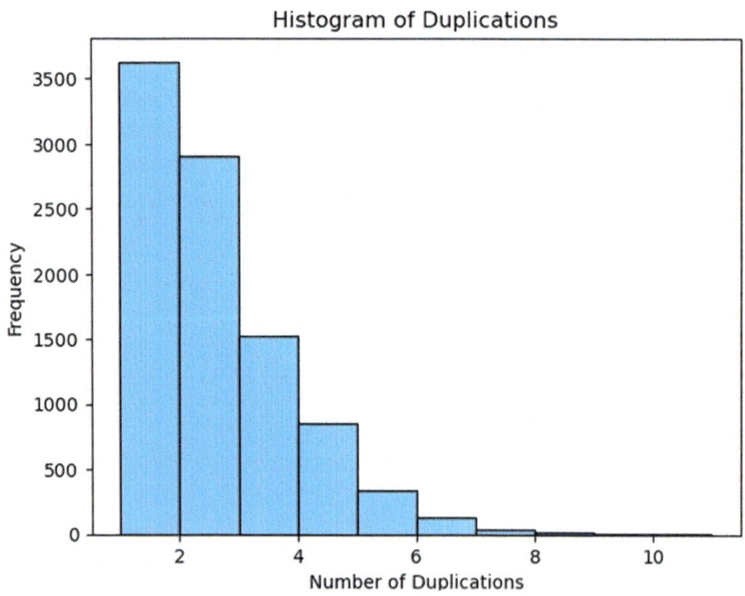

Conclusion of the Descriptive Analysis

There are 9465 unique transactions in the database. Half of the customers would purchase only one or two products, but distribution is highly skewed. The maximum number of transactions is 11, and the mean number of products in one transaction is 2.17. The most purchased products are coffee, tea, and bread, hinting that there are two groups of customers: those eating in and those buying takeout. In the first group, there are those with tea and coffee preferences. It makes sense to explore whether these groups prefer additional products differently.

Step 3. Association Rules

USER:

In the DataFrame bakery_data, column TransactionNo represents an ID of unique transactions, and column Items represents transaction items. Build association rules for Items. Use the Apriory association rule algorithm. Use minimum support minSupp=0.01 and minimum confidence minConf=0.01.

Assistant:

```
import pandas as pd
from mlxtend.frequent_patterns import apriori, association_rules

# Convert the Items column to a one-hot encoded DataFrame
one_hot_encoded = bakery_data.groupby(['TransactionNo',
'Items'])['Items'].count().unstack().reset_index().fillna(0).set_index
('TransactionNo')
one_hot_encoded = one_hot_encoded.applymap(lambda x: 1 if x > 0 else
0)

# Build frequent itemsets using the Apriori algorithm with specified
minimum support
frequent_itemsets = apriori(one_hot_encoded, min_support=0.01,
use_colnames=True)

# Build association rules with specified minimum confidence, lift, and
conviction
association_rules_df = association_rules(frequent_itemsets,
metric="confidence", min_threshold=0.01)

# Display the association rules
print(association_rules_df)
```

Explanation:

To build association rules for the Items column using the Apriori association rule algorithm with specific parameters, we can use the *mlxtend* library in Python. First, we need to install the library. Then, we can use the *apriori* and *association_rules* functions from *mlxtend.frequent_patterns* to generate the association rules with the specified parameters. In the code, we first convert the *Items* column to a one-hot encoded DataFrame; then we use the Apriori algorithm to build frequent itemsets with a minimum support of 0.01. Next, we use the *association_rules()* function to generate association rules with a minimum confidence as specified. Finally, we display the association rules that meet the specified criteria.

Result:

```
---------------------------------------------------------------------------
ModuleNotFoundError                       Traceback (most recent call last)
Cell In[14], line 1
----> 1 from mlxtend.frequent_patterns import apriori, association_rules
      3 # Convert the Items column to a one-hot encoded DataFrame
      4 one_hot_encoded = bakery_data.groupby(['TransactionNo', 'Items'])['It
ems'].count().unstack().reset_index().fillna(0).set_index('TransactionNo')

ModuleNotFoundError: No module named 'mlxtend'
```

We have already encountered this error: a required module is missing. It has to be installed.

USER:

Download the module mlxtend.

Assistant:

To download the *mlxtend* module, you can use the following command in your Python environment:

```
!pip install mlxtend
```

Result:

```
...
Installing collected packages: mlxtend.
Successfully installed mlxtend-0.23.1
```

At this point, you can re-run the previous code.

Result:

```
         antecedents        consequents  antecedent support  consequent support  \
0            (Bread)        (Alfajores)            0.327205            0.036344
1        (Alfajores)            (Bread)            0.036344            0.327205
2           (Coffee)        (Alfajores)            0.478394            0.036344
3        (Alfajores)           (Coffee)            0.036344            0.478394
4            (Bread)          (Brownie)            0.327205            0.040042
..               ...                ...                 ...                 ...
69       (Tea, Cake)           (Coffee)            0.023772            0.478394
70    (Coffee, Cake)              (Tea)            0.054728            0.142631
71             (Tea)     (Coffee, Cake)            0.142631            0.054728
72          (Coffee)        (Tea, Cake)            0.478394            0.023772
73            (Cake)      (Tea, Coffee)            0.103856            0.049868

     support  confidence      lift  leverage  conviction  zhangs_metric
0   0.010354    0.031644  0.870657 -0.001538    0.995145      -0.180870
1   0.010354    0.284884  0.870657 -0.001538    0.940818      -0.133570
2   0.019651    0.041078  1.130235  0.002264    1.004936       0.220910
3   0.019651    0.540698  1.130235  0.002264    1.135648       0.119574
4   0.010777    0.032935  0.822508 -0.002326    0.992651      -0.242849
..       ...         ...       ...       ...         ...            ...
69  0.010037    0.422222  0.882582 -0.001335    0.902779      -0.119934
70  0.010037    0.183398  1.285822  0.002231    1.049923       0.235157
71  0.010037    0.070370  1.285822  0.002231    1.016827       0.259266
72  0.010037    0.020981  0.882582 -0.001335    0.997149      -0.203223
73  0.010037    0.096643  1.937977  0.004858    1.051779       0.540090
```

Notice there are 74 rules, but most have low lift, conviction, leverage, and/or Zhang index. That means these rules are no better than a random guess and their high support and confidence are based solely on products' popularity. For example, rules #0 and 1 suggest that bread is frequently purchased together with alfajores (Latine America specialty confection), but this association is, in fact, worse than expected due to chance (Lift < 1). When the popularity of both products is considered, customers who buy bread avoid purchasing alfajores and vice versa.

Compare that with the rules #2 and 3. Lift > 1, leverage > 0, conviction > 1, and Zhang > 0 values suggest that alfajores are unusually popular with coffee purchases. That hints that the alfajores product is a good recommendation for a new coffee customer.

Step 3 (On Our Own)

1. We analyzed only the first four generated rules. Save DataFrame *association_rules_df* as an Excel file and download it to your computer.

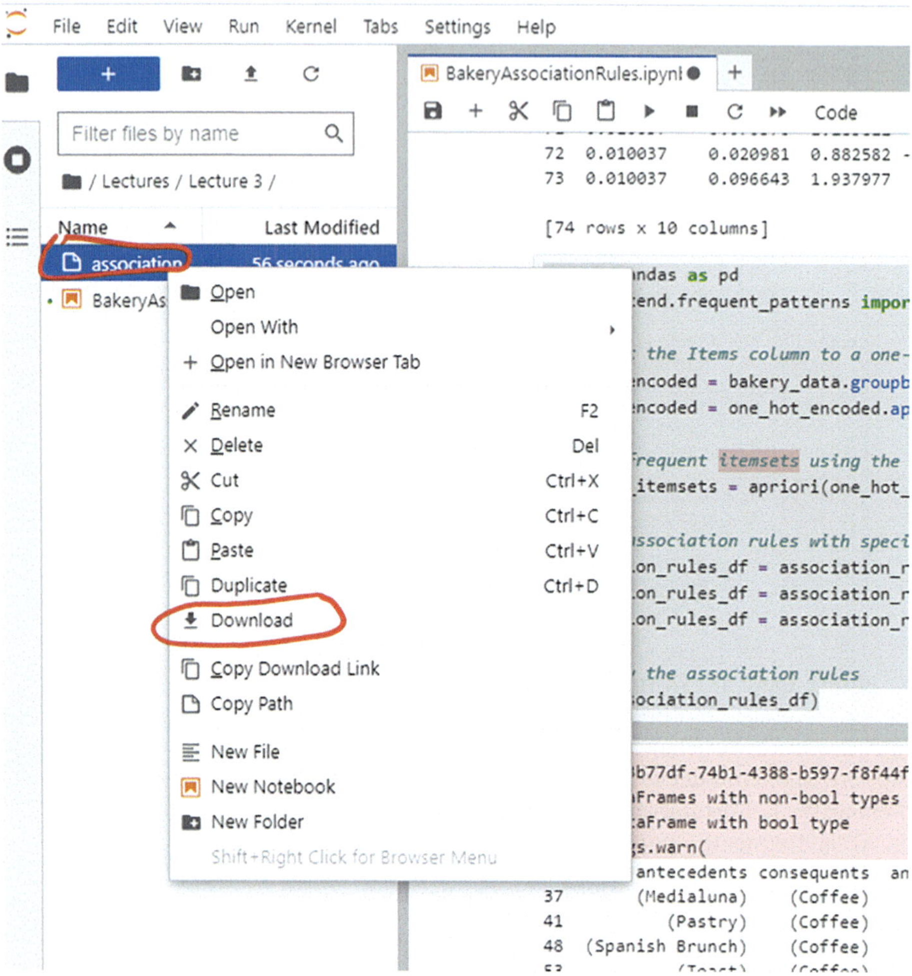

2. Open the file in Excel and analyze the rules with Lift > 1.1 (that is, with the association between the antecedent and consequent itemsets at least 10% higher than due to chance alone). What will be your recommendation for those customers who order tea and coffee? Will you recommend the alfajores to those customers ordering tea? Write a report based on your analysis.
3. Modify your code so that only the rules with conviction > 1.05 and lift > 1.1 are left, and other rules are filtered out.

Build Your Python Skills

Chapters 6 and 7 of the free book Learn Python with Jupyter by Serena Bonaretti. ► https://www.learnpythonwithjupyter.com/.

Additional Reading

Hahsler, M. (2015). A probabilistic comparison of commonly used interest measures for association rules. ► https://mhahsler.github.io/arules/docs/measures

Jah, J., Kamber, M., & Pei, J. (2012). Data mining concepts and techniques. simon fraser university (Chapter 6). ► https://myweb.sabanciuniv.edu/rdehkharghani/files/2016/02/The-Morgan-Kaufmann-Series-in-Data-Management-Systems-Jiawei-Han-Micheline-Kamber-Jian-Pei-Data-Mining.-Concepts-and-Techniques-3rd-Edition-Morgan-Kaufmann-2011.pdf

Decision Trees

A. P. Kirilenko, *Practical Data Mining with AI for Social Scientists*,
Springer Texts in Social Sciences, https://doi.org/10.1007/978-3-031-89689-7_4

4.1 Introduction

Decision trees

A decision tree is a graphical representation of the decisions taken in predictive or explanatory data analysis.

The tree grows upside down; the root is on the top.

A fork of the tree is a test on an attribute.

A branch is the outcome of the test.

Tree leaves are the final decisions.

The paths from the root to the leaf shows how decisions are made.

4

Decision trees are a popular tool in data analytics, primarily due to their ability to clearly illustrate how the final decision was made. They are a type of predictive or explanatory modeling approach that is used for both classification and regression tasks. The decision tree model predicts the value of a target variable (dependent variable) by learning simple decision rules inferred from the data features (independent variables, also called attributes).

A decision tree approach builds a flowchart-like "tree" structure. Each internal node of the tree represents a test on an attribute (e.g., whether a tourist is male or female), each branch represents the outcome of the test, and each leaf node represents a class label (a final decision taken after computing all attributes). The paths from the root to the leaf represent the decision-making mechanism at work.

Decision trees stand out for their simplicity and explanatory power. These models mimic human decision-making processes by branching out from simple questions to more and more complex queries, leading to a final decision. Theoretically, decision trees are a non-parametric non-linear model used for both classification and regression tasks. They are called "trees" because they resemble an upside-down tree, with the root at the top and the leaves at the bottom. This structure makes the model decision-making process intuitive and easily interpretable by humans. For instance, in tourism data analysis, decision trees can be used to predict traveler's behavior, optimize pricing strategies, improve market segmentation, and much more.

We will start explaining decision trees applied to categorical data only, but the approach can handle both categorical and numerical data, making it versatile for various data types. Furthermore, decision trees require relatively little data preparation, making them a convenient option for quick and effective data analysis.

4.2 Components of Decision Trees

Decision trees are represented by a tree-like structure of nodes connected by *branches*. Further, there are three kinds of nodes: *root node, decision (internal) nodes, and leaf (terminal) nodes* (■ Fig. 4.1).

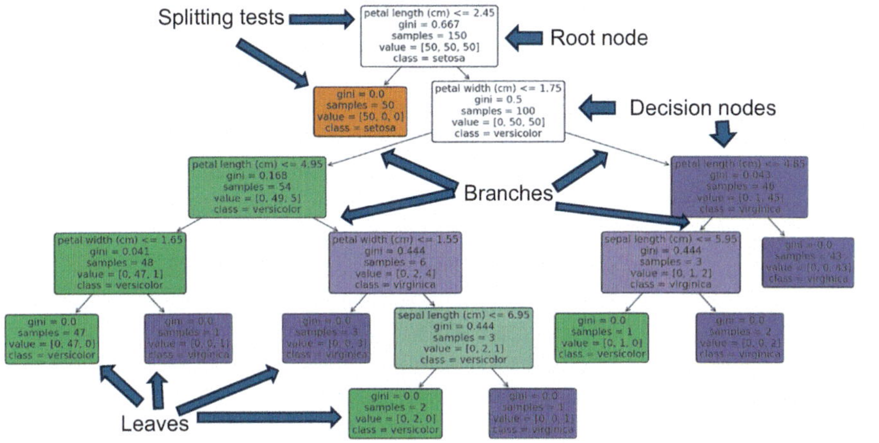

Fig. 4.1 Classification of iris flowers into species based on the dimensions of their petal and sepals

Root Node: This element represents the beginning of the decision tree. The root node represents the entire dataset at the beginning of the analysis. It is interpreted as the initial decision point from which data subdivision starts. The root node selects an independent variable (attribute) that has the most significant impact on the dependent variable (target). Then, the dataset is *split* into two or more subsets of data based on the value of the selected most influential attribute. This splitting is a result of a test on the attribute's value.

Decision tree components

Root: starting point

Decision nodes: internal nodes of a tree

Leaves: final nodes of a tree

Branches: connections between the nodes

Splitting: a decision made at the decision node leading to building the branches.

Splitting in Decision Trees: Splitting is the process of dividing a node into multiple sub-nodes according to the value of one of the attributes. The exact way of dividing the dataset is determined by criteria such as Gini impurity, information gain, and others. The splitting breaks down the dataset into smaller, more focused subsets for further analysis in downstream nodes of the tree.

Decision Nodes: Branching from the root, the data is further divided into smaller and smaller subsets in decision nodes. Similar to the root node, the decision nodes select the most influential attribute and apply a test on its value. The goal of the test is to make the "downstream nodes" more and more "pure" in terms of the target variable values. While the root node has a mix of records with different target values, the downstream nodes are progressively more homogenous. In ► Chapter 1 example with Titanic data, the root node contained a mix of survived and perished passengers, while the decision nodes were dominated by either survived or dead.

Branches in Decision Trees: The branches connect the nodes, representing the outcomes of tests at each node. Each branch depicts a potential decision pathway resulting from a test at a node, directing data analysis.

Leaf Nodes: The final node of a tree is called a leaf. Leaves are termination points of the decision-making process represented by the decision tree. Leaf nodes represent the final classification or prediction based on the attribute values.

The following sections go deeper into the principles underlying building the decision trees and parameter selection for the decision tree algorithms.

4.3 Splitting Decisions

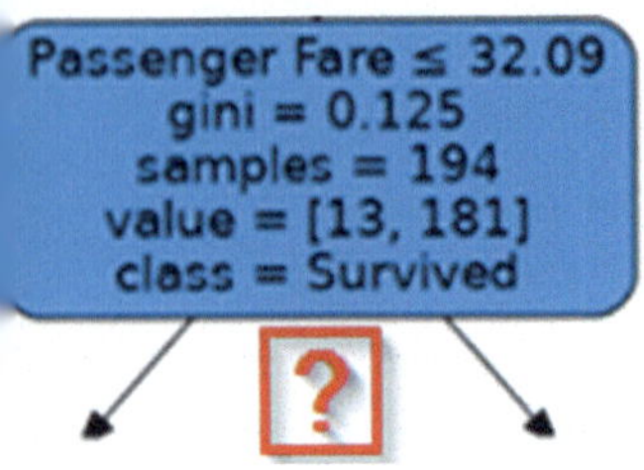

The heart of decision tree model functionality is the mechanism of making splitting decisions. Splitting decisions partition the dataset into subsets based on certain conditions. Thereby, the splitting decisions directly influence the path that leads to the final decision by building decision nodes, branches, and leaves in the tree. There are multiple ways of selecting splitting criteria, which in turn impact the tree's depth, complexity, and performance accuracy. Several statistical measures serve as the basis for making splitting decisions, each with its own approach to evaluating the best split at any given node. The most frequently used measures include Gini impurity, Entropy and information gain, Variance reduction, and Chi-Square.

Gini Impurity: A measure of how often a randomly chosen element from the set would be incorrectly labeled if it was randomly labeled according to the distribution of labels in the subset. The Gini impurity is minimized when nodes are most homogeneous.

Entropy and Information Gain: Entropy is a metric of randomness or uncertainty. Information gain, derived from entropy, measures the reduction in entropy before and after the split. Maximizing information gain equates to achieving the most significant reduction in randomness with each split.

Variance Reduction: This criterion is mainly used in decision regression trees, which are discussed later in this chapter. The goal of algorithms based on variance reduction is to reduce the variance of the target variable within each node. A split that results in subsets with lower variance is preferred, as it signifies more homogenous groups.

Chi-square: Chi-square measure evaluates the statistical significance of the differences between sub-nodes and parent node distributions. This criterion is used to assess whether splits effectively capture the distribution of different outcomes in the categorical target variable. Historically, this is the oldest measure used in decision tree models.

There are other methods such as Hellinger Distance, Minimum Description Length, Reduced Error Pruning, Gain Ratio, etc.

Latte is an illustration of a low entropy system. Coffee is separated from milk.

4.3.1 Gini Impurity

Gini impurity measures the probability of incorrectly classifying a randomly chosen element in the dataset if it were randomly labeled according to the distribution of classes in the subset. The Gini impurity reaches its maximum value when the classes are perfectly mixed in a node, indicating maximum uncertainty or disorder. Conversely, the minimum value of Gini impurity is zero, occurring when all elements in a node belong to a single class, signifying no uncertainty or pure homogeneity.

Flat white is an illustration of a low entropy system. Coffee is perfectly mixed with milk.

Mathematically, for a categorical dependent variable that can take m different values, Gini is expressed as the following:

$$Gini = 1 - \sum_{i=1}^{m} p_i^2,$$

where p_i is the proportion of class i in the node.

Let us consider an example. In ▶ Chapter 2, we developed a decision tree model explaining factors relevant to surviving the Titanic catastrophe. ◘ Figure 4.2 shows the top nodes of the tree. Notice that each node contains information on the Gini index for this node.

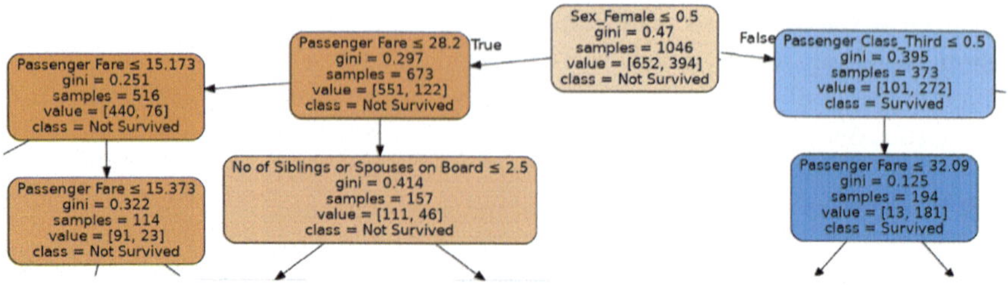

Fig. 4.2 Building a decision tree using the Gini measure of node impurity

The root node has the maximum Gini index: gini = 0.47. It reflects a high degree of impurity in the node: 652 deaths (62%) and 394 survivals (38%).

$$Gini_{beforesplit} = 1 - \left(\left(\frac{652}{1046} \right)^2 + \left(\frac{394}{1046} \right)^2 \right) = 0.47$$

The two child nodes of the tree have much smaller levels of impurity. The left child (male passengers) has 652 deaths (82%) and 394 survivals (18%) with a corresponding Gini = 0.297. The right child (female passengers) has 101 deaths (27%) and 272 survivals (73%) with a corresponding Gini of 0.395. The overall Gini index after the split is calculated as a weighted average of the Gini indices of child nodes:

$$Gini_{aftersplit} = \tfrac{652}{1046}0.297 + \tfrac{394}{1046}0.395 \; = \; 0.33$$

After splitting the dataset on the Sex variable, the Gini index has reduced from 0.47 to 0.33, reflecting more homogeneous data in each child node of the tree. Of course, we could try splitting our data on other variables: passenger fare, cabin class, and so on. The decision to split the tree based on the value of the Sex variable (and not on, for example, the passenger fare) was made because the impurity reduction based on the passenger sex is the highest.

4.3.2 Entropy and Information Gain

Entropy is a measure of the randomness in data. Similar to Gini, it quantifies the impurity in the classification variable. A high entropy value indicates a mixture of different classes in a node, while a low entropy signifies that a node is primarily homogeneous, containing instances of the same class. Even though the entropy is similar to Gini, mathematically, it is expressed differently:

$H = -\sum_{i=1}^{m} p_i log_2 \llbracket(p\rrbracket_i)$

Note that the entropy value is positive, and the data with a smaller impurity has a smaller entropy. Reduction in entropy resulting from splitting the data based on a specific attribute (for example, Sex) is measured by information gain. After computing the entropy for the parent node and each of the child nodes, the information gain is calculated as the difference between the parent's entropy and the weighted average of entropy of all child nodes:

$$IG = H(D) - \sum_{j=1}^{k} \frac{|D_j|}{|D|} H(D_j),$$

where $|D|$ is the total number of instances in the decision node D; D_j are the child nodes, and H(D) is the entropy of a node. Essentially, it calculates how much more ordered (or less random) the dataset becomes after a split. The attribute that results in the highest information gain is chosen for the split at each node in the tree.

4.3.3 Splitting Decisions for Continuous Attribute Variables

In ▶ Sect. 4.3, we described two methods for selecting the attribute variable on which to make a split. Did you notice that both methods assumed that the variables are categorical? Information gain and Gini methods are designed to be applied when the attribute variables are categorical, but there are ways to use both methods with continuous attribute variables as well. For example, we can convert the continuous variable into discrete bins and use the frequencies of each class to compute the entropy. This procedure is called *discretization*.

For example, the Titanic dataset (refer to ▶ Chapter 1) contains a variable Passenger Fare. This is a continuous variable as it can technically take any value between zero and the maximum fare for a sea cabin at the time.[1] By

Categorical and continuous data

Categorical variables can take a limited number of possible values. The sex of Titanic passenger since it takes only two values. Similarly, the cabin class is a categorical variable since it can take only three values: First, Second, and Third. In comparison, Passenger Fare is a continuous variable and can take any value.

1 The fare is not truly continuous (say, a $53.123 cabin fare is impossible), but it is frequently useful to treat discrete ratio, interval, or even ordinal measures as continuous.

partitioning the range of passenger fares into two intervals, below and above a certain threshold, we can convert the fare into a discrete variable. For example, those fares below $20 would be placed in a bin called "cheap," and the fares $20+would be placed in a bin called "expensive." This "threshold" fare of $20 is called the *cut point*. In effect, we have replaced the continuous variable "passenger fare" with a new categorical variable with just two values, "cheap" and "expensive."

You may have noticed that to apply this discretization method, we need to set the threshold. One easy-to-understand way to do that is to minimize the weighted average of the entropy of the resulting child nodes IG. The cut point for which the IG value is minimal compared to all other possible cut points is then selected as the best cut point.

4.4 Pruning the Trees

Overfitting

Increasing model complexity may lead to model overfitting. Overfitted models are not generalizable, that is, they cannot be applied to a new dataset.

By design, decision trees are prone to *overfitting*. An overfitted model is highly accurate on the data it was trained on but low accurate on the test data. In other words, overfitted models are not generalizable and cannot be applied to unseen data.

Overfitting occurs when the model's excessive complexity is not justified by the data, so the model captures noise or random patterns in the training data. Overfitted decision trees (Fig. 4.3) tend to be overly complex and hard to understand, which kills their most significant advantage. As an illustration, recall the first decision model we built in ▶ Chapter 1 lab. This model perfectly predicts which Titanic passengers survived, yet it is highly unhelpful in determining the factors relevant to their survival.

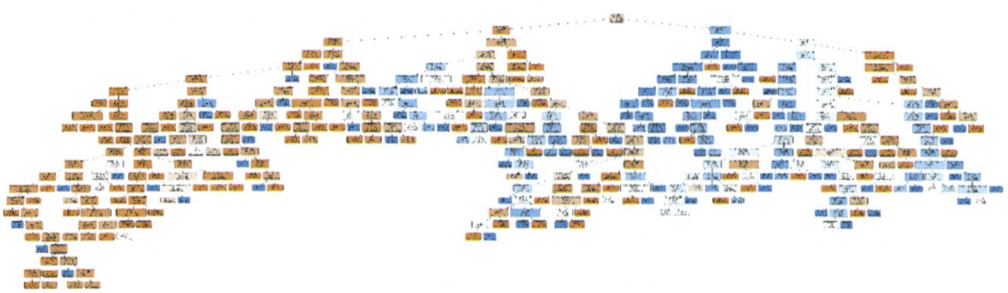

Fig. 4.3 An overfitted decision tree

Pruning, or cutting back the tree, addresses the overfitting issue by simplifying the model, making it more robust. One way of tree pruning is to stop it from growing beyond a certain point. This method is called *pre-pruning* or *early stopping*. Pre-pruning can be achieved by setting thresholds on the minimum number of samples per leaf, the maximum depth of the tree, the minimum gain in impurity required for a split, or using other criteria limiting tree growth. In the Titanic example, we could do the following:

- Stipulate that each leaf of the tree should have at least five passengers;
- Stipulate the death of the tree of three at most;
- Stipulate that a node is split only if the information gain is 0.05 at least.

Pruning

Simplifying decision trees is called pruning. Pre-pruning is applied during the process of tree growing. Post-pruning allows the tree to fully grow followed by cutting tree branches.

Another way of decision tree pruning involves first allowing the tree to grow to its full complexity and then simplifying it. This method is called *post-pruning* or *cost complexity pruning*. One common approach is to evaluate the effect of removing tree branches based on a cost-complexity criterion, which balances the tree's accuracy against its complexity. Branches that contribute little to predictive accuracy are removed from the tree.

Pruning reduces the model's complexity, making it less likely to fit noise in the training data and more likely to generalize well to new data. Pruned trees are also better interpretable, making it easier for humans to understand the decision-making process and the model's predictions.

4.5 Decision Trees with Continuous Target Variable

So far, we have applied decision trees to classification or prediction problems when the target variable is categorical. For example, we used decision trees to predict the survival of a Titanic passenger based on his or her characteristics (attributes). Survival is a categorical variable with two possible values: Yes (that is, survived) or No (not survived). Decision tree methodology, however, can also be used with continuous targets.

Consider the following example. Dataset Tourism to Adversary Country represents a survey of Russian people asked about their willingness to take a tourism trip to the

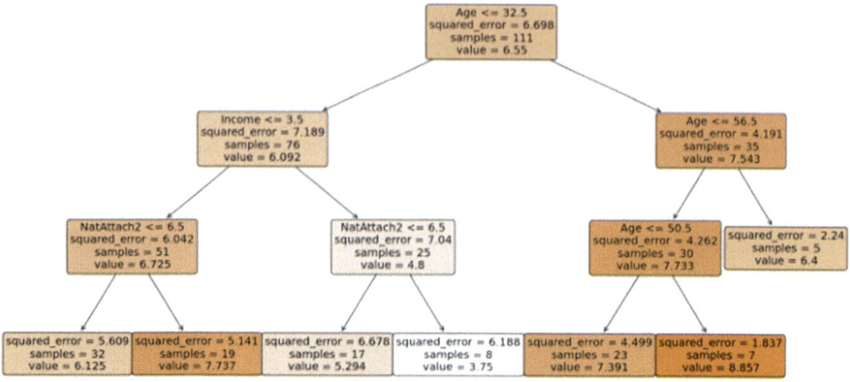

Fig. 4.4 Pruned decision tree. The original tree is shown in Fig. 4.3

USA.[2] The survey included multiple other questions, including demographics, national attachment, a feeling of animosity towards the United States, and many others. You will work with the abridged survey data in the lab.

In one of the survey questions, the participants were asked to evaluate how world instability (terrorism, military and ethnic conflicts, epidemics, etc.) influenced their desire to vacation abroad. The corresponding Unstable World construct was measured using a 10-point Likert scale, with a value of 1 representing no effect and 10 representing extremely high effect. This is an ordinal variable with ten levels; in social science research, such variables are frequently treated as continuous interval/ratio variables.

The following decision tree (Fig. 4.4) illustrates the relationship between the participant's age, income, and feeling of attachment to Russia and the effect of global instability on their willingness to spend their vacation while traveling abroad. Lighter colors represent low values of the Unstable World variable; these participants are not likely to worry about global instability when selecting between domestic and international vacations. Conversely, darker colors represent high values of the Unstable World variable; these participants have very high concerns about global risks in their vacation planning.

2 Stepchenkova, S., Su, L., & Shichkova, E. (2019). Marketing to tourists from unfriendly countries: Should we even try? *Journal of Travel Research*, *58*(2), 266–282.

The smallest levels of concern about global instability (mean value = 3.75) are associated with younger age (32 y.o. or younger), relatively low income (Income < 3.5 corresponds to less than $5,500 annually), and moderately low level of national attachment. On the contrary, the highest level of concern (mean value = 8.857) is associated with the older generation (age group between 51 and 56). Overall, the most important variable affecting respondents' concerns about world instability when deciding to travel abroad is age.

However, there are methods specifically designed to deal with continuous variables. One of these methods is Variance reduction. Just like the Information gain and Gini, Variance reduction is used to evaluate and select the best split during the construction of the tree, aiming to produce subtrees that are more homogenous in terms of the target variable's values.

Variable variance is a statistical measure that quantifies the spread or dispersion of a set of data points around their mean value. Mathematically, the variance σ^2 is calculated as follows:

$$\sigma^2 = \frac{1}{N}\sum_{i=1}^{N}(x_i - \mu)^2$$

Here, x_i—all values that the target variable x has in a decision node, for example, the answers to the world instability question. The mean value for the instability question answers is μ. Finally, N is the total number of answers.

The goal of the variance reduction method is to split the data at a decision node in a way that minimizes the variance of the target variable in each of the child nodes. Specifically, the variance reduction seeks to maximize the difference between the variance at the parent node and the weighted average of the variances in the child nodes:

$$VarianceReduction = \sigma^2(D) - \sum_{j=1}^{k}\frac{|D_j|}{|D|}\sigma^2(D_j)$$

where $\sigma^2(D)$—variance of the target variable at the parent node; $\sigma^2(D_j)$ variance of the target variable at the child node j; $|D_j|$—the number of records at the child node j.

4.6 Method comparison

With many methods in existence, selecting the one to use is not trivial. One consideration is the type of data to be analyzed. The variance reduction method is suitable for continuous target variables rather than categorical ones. On the opposite, Information gain, Gini impurity, as well as Gain ratio, Chi-Squared, and Hellinger distance are suitable for categorical target variables. Another consideration is computational costs. For example, Information gain is more complex computationally than most other methods and may take longer to complete.

Further, there is a consideration of data quality. For example, some methods, such as Chi-squared, are sensitive to irrelevant variables. Information gain may prioritize the importance of categorical attributes with many categories. Gini impurity may produce biased trees when a dominant categorical attribute is present.

Finally, the method should reflect the focus of the analysis. The Gini measure is frequently preferred in predictive models, while the Information gain is used in exploratory analysis. Gini returns larger partitions, while Information gain tends to create more balanced splits. One important feature of Gini is that it can produce *binary trees* only. Binary trees have only two branches coming from the decision nodes.

Note that the process of selecting the best split involves evaluating every possible division across all features based on the chosen criterion, which makes *brute force algorithms* extremely expensive in terms of the number of computations to perform. This is why there are multiple algorithms that attempt to radically reduce the number of computations required to produce decision trees. Each algorithm can use one or more (but not all) measures to make splitting decisions. For example, algorithms like ID3 and C4.5 use entropy and information gain for this purpose, whereas CART (Classification and Regression Trees) employs Gini impurity or variance reduction depending on the nature of the target variable. Overall, the selection of either method is determined by the data, analysis goals, and available software.

4.7 From Decision Trees to a Random Forest

In the previous chapters, we covered the shortcomings of the decision tree models, the most significant of which is probably the tendency to overfit, that is, to learn highly irregular data patterns. Overfitted decision trees tend to grow extremely large; they can also be recognized for their tendency to return very accurate predictions when applied to training data yet being very imprecise when applied to new datasets.

Random Forests tackle the decision trees' shortcomings by using samples of data and/or samples of attributes to grow multiple trees. The fundamental principle behind Random Forests is that a group of "weak learners" can come together to form a "strong learner." Each tree in the forest makes an independent prediction, and the final output is determined by averaging the predictions (for regression tasks) or by a majority vote (for classification tasks). In effect, the Random Forest uses the strengths of multiple decision trees to increase the accuracy and robustness of classification or prediction. Note, however, that this improvement comes at a cost: Random Forest models are very costly computationally. They are also not entirely free of the overfitting problem. Overall, selecting specific parameters of the Random Forest method should be automatic but based on the analyzed data and the project goals.

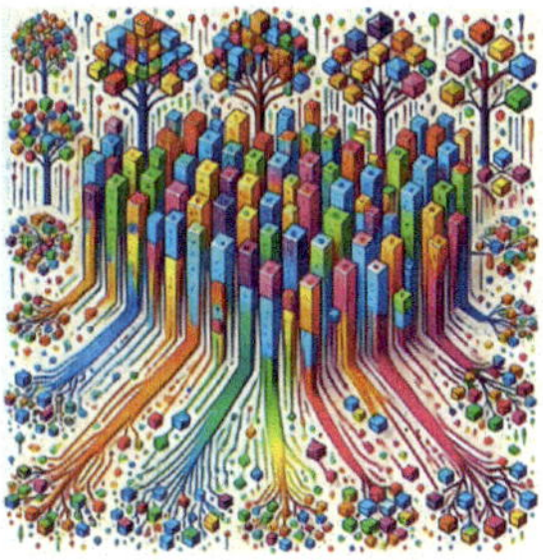

4.8 Review Questions

1. Define Decision Trees and explain their advantages in data analytics.
2. Describe the Components of a Decision Tree. Elaborate on the roles and characteristics of root nodes, decision nodes, leaf nodes, branches, and splitting in the structure of a decision tree.
3. Explain the concept of splitting in decision trees. How is it determined, and what criteria (e.g., Gini impurity, information gain) are commonly used to decide how to split the data at each node?
4. Illustrate the decision-making process of a decision tree using a real-world example. You may refer to any suitable example from the tourism and hospitality sector.

5. Discuss the methods of evaluating splitting decisions, such as Gini impurity, entropy and information gain, variance reduction, and Chi-square. Include examples to illustrate when one measure might be preferred over the other.
6. Explain the process of handling continuous and categorical variables in decision trees. How does the splitting approach differ between these types of variables?
7. Describe the concept of tree pruning, including the differences between pre-pruning and post-pruning. How does pruning help prevent overfitting, and how does it impact the model's complexity and interpretability?
8. Discuss the application of decision trees in predicting continuous target variables. To illustrate this application, provide an example, such as the use of decision trees to understand the influence of world instability on travel planning.
9. Evaluate the methods of splitting decisions for continuous attribute variables. Discuss how continuous variables are handled differently from categorical variables in the context of decision trees.
10. Reflect on the advantages and limitations of decision trees in data analytics. Consider their simplicity, interpretability, and susceptibility to overfitting in your response.
11. Explain the transition from decision trees to random forest. What are the shortcomings of decision trees that random forests aim to overcome, and how does the random forest algorithm improve upon the decision tree model?

4.9 Lab

Objectives

1. Confirm data cleaning skills
2. Apply decision trees analysis to explain tourists' intention to travel to an adversary country
3. Practice writing analysis report

Data

The data was collected in a 2017 survey of prospective Russian travelers to the USA, a project led by University of Florida professor Svetlana Stepchenkova as part of her research on motivations to travel to adversary countries.[3] The participants were asked about their worldviews, political opinions, travel habits, and perceptions of the US as a travel destination. The dataset presents an abridged version of the survey.

The data was collected by surveying travelers at the main travel station of Nizhnii Novgorod, a city in the European part of Russia, in 2017. The dataset has 139 entries.

Variables

Variable	Explanation	Code explanation
ID	ID	Identifier of respondent
Gender	Gender	1—Male; 2—Female
Age	Age	Years
Educ	Education	1—incomplete high school; 2—High school; 3—Incomplete Bachelor's; 4—Bachelor's; 5—Postgraduate; 6—Other
Income	Income	1—<10 K; 2—22-20 K; 3—21-30 K; 4—31-40 K; 5—41-50 K; 6—51 + K (Russian rubles/month)
TravelExp	Trips abroad last 3 yr	N times
UnstableWorld	Effect of Instability in the World on my desire to travel abroad	1—extremely low; 10—extremely high
NatAttach	A feeling of connection and pride towards Russia	1—extremely low; ... 7—extremely high
Attitude	Interest in the US as a country	1—extremely negative; ... 10—extremely positive
Animosity	Animosity towards the US	0—No animosity; ... 9—the most hated country
IntentionToVisit	Intention To Visit the US	0—No; 1—Yes

3 Stepchenkova, S., Su, L., & Shichkova, E. (2019). Intention to travel internationally and domestically in unstable world. *International Journal of Tourism Cities, 5*(2), 232–246.

Part 1. Build the Initial Model

Step 1. Read Data

We will write the code using the AI assistant. Initiate a new empty notebook.

Enter the following text to the Anaconda Assistant window and start the chat:

USER:

Read data from ▶ https://github.com/andykir/DataMiningLectures/raw/main/AnimosityTowardsUSsurveyAdriged.xlsx

Assistant:

```
import pandas as pd

url =
"https://github.com/andykir/DataMiningLectures/raw/main/AnimosityTowar
dsUSsurveyAdriged.xls"
data = pd.read_excel(url)

# Now the data is stored in the 'data' variable and can be used for
further analysis.
```

Explanation:

In this code, we use the *read_excel* function from the pandas library to read the data from the provided URL. The data will be stored in a DataFrame *data*, which can be used for further analysis.

Step 2. Data Description and Cleaning

USER:

Show basic statistics of the DataFrame 'data'.

Assistant:

```
import pandas as pd

basic_statistics = data.describe()
print(basic_statistics)
```

Explanation:

To show the basic statistics of a DataFrame in Python, we use the *describe*() method from the pandas library. This method provides a summary of the central tendency, dispersion, and shape of the dataset's distribution, excluding the missing values (NaN).[4]

Note that in the previous labs, we used the methods *DataFrame.info*() and *DataFrame.head*() to get the initial data understanding. In this lab, we introduce a new method, *DataFrame.describe*(). The first method focuses on data types, the second prints the first few rows of data, and the third outputs basic descriptive statistics. A good idea is to edit the code in the notebook window to include all three methods:

```
print(data.info())
print(data.head())
print(data.describe())
```

Result:

print(data.info())

```
RangeIndex: 139 entries, 0 to 138
Data columns (total 11 columns):
 #   Column            Non-Null Count  Dtype
---  ------            --------------  -----
 0   ID                139 non-null    int64
 1   Gender            139 non-null    int64
 2   Age               139 non-null    int64
 3   Educ              139 non-null    int64
 4   Income            139 non-null    int64
 5   TravelExp         139 non-null    int64
 6   UnstableWorld     139 non-null    int64
 7   NatAttach         139 non-null    float64
 8   Attitude          139 non-null    int64
 9   Animosity         139 non-null    float64
 10  IntentionToVisit  139 non-null    int64
dtypes: float64(2), int64(9)
memory usage: 12.1 KB
```

4 NaN stands for Not a Number. It is a special value, used to represent missing or invalid values in a dataset.

Note that some variables have integer data types while others have float data types. The latter variables were computed by averaging answers to multiple questions of a construct. All variables are numerical; this conforms with the requirements of the decision tree algorithm we will use. There is no missing data, which makes missing data processing unnecessary.

print(data.head())

```
   ID  Gender  Age  Educ  Income  TravelExp  UnstableWorld  NatAttach  \
0  1       2   24     4       3          2             10   7.000000
1  2       2   24     4       3          2              9   6.000000
2  3       2   24     4       3          3             10   5.333333
3  4       2   26     4       2          0              9   6.666667
4  5       1   29     4       2          0              8   7.000000

   Attitude  Animosity  IntentionToVisit
0         4        7.0                 0
1         7        4.0                 1
2         2        6.0                 0
3         6        1.0                 1
4         4        6.0                 0
```

print(data.describe())

```
               ID      Gender         Age        Educ      Income   TravelExp
count  139.000000  139.000000  139.000000  139.000000  139.000000  139.000000
mean    70.000000    1.633094   30.964029    3.784173    3.510791    1.798561
std     40.269923    0.483704   11.422968    0.787241    1.661006    1.900343
min      1.000000    1.000000   18.000000    2.000000    1.000000    0.000000
25%     35.500000    1.000000   22.000000    4.000000    2.000000    0.000000
50%     70.000000    2.000000   28.000000    4.000000    3.000000    1.000000
75%    104.500000    2.000000   36.500000    4.000000    5.000000    3.000000
max    139.000000    2.000000   67.000000    6.000000    6.000000    9.000000

       UnstableWorld   NatAttach    Attitude   Animosity  IntentionToVisit
count     139.000000  139.000000  139.000000  139.000000        139.000000
mean        6.474820    5.683453    6.035971    3.402878          0.697842
std         2.596907    1.191978    2.093503    1.591278          0.460854
min         1.000000    1.000000    1.000000    1.000000          0.000000
25%         5.000000    5.000000    5.000000    2.000000          0.000000
50%         7.000000    6.000000    6.000000    3.000000          1.000000
75%         9.000000    6.666667    8.000000    4.500000          1.000000
max        10.000000    7.000000   10.000000    7.000000          1.000000
```

The variable's minimum and maximum values are within reasonable ranges: the gender takes values of 1 and 2; the ages are between 18 and 67, and so on. There are no "Other" responses among the values of the variables *Income*, *TravelExp*, *UnstableWorld*, *NatAttach*, *Attitude*, and *Animosity*, which allows us to treat these variables as interval measures. However, the Likert scale variables *TravelExp*, *UnstableWorld*, *NatAttach*, *Attitude*, and *Animosity* have different scales (1—7, 1—10, or 0—9); it makes sense to transform these variables to fit a uniform range from 1 to 10 for a more straightforward interpretation. To better understand the variables, we also need to analyze their distributions.

USER:

Draw a histogram for columns Gender, Age, Educ, Income, TravelExp, UnstableWorld, NatAttach, Attitude, Animosity, and IntentionToVisit in DataFrame "data".

Assistant:

```
import matplotlib.pyplot as plt

# Select the columns for which you want to draw histograms
columns = ['Gender', 'Age', 'Educ', 'Income', 'TravelExp',
'UnstableWorld', 'NatAttach', 'Attitude', 'Animosity',
'IntentionToVisit']

# Create subplots for each column
fig, axes = plt.subplots(nrows=5, ncols=2, figsize=(15, 20))

# Flatten the axes array for easy iteration
axes = axes.flatten()

# Plot histograms for each column
for i, col in enumerate(columns):
    data[col].plot(kind='hist', ax=axes[i], title=col)
    axes[i].set_ylabel('Frequency')

# Adjust layout and display the plots
plt.tight_layout()
plt.show()
```

Result:

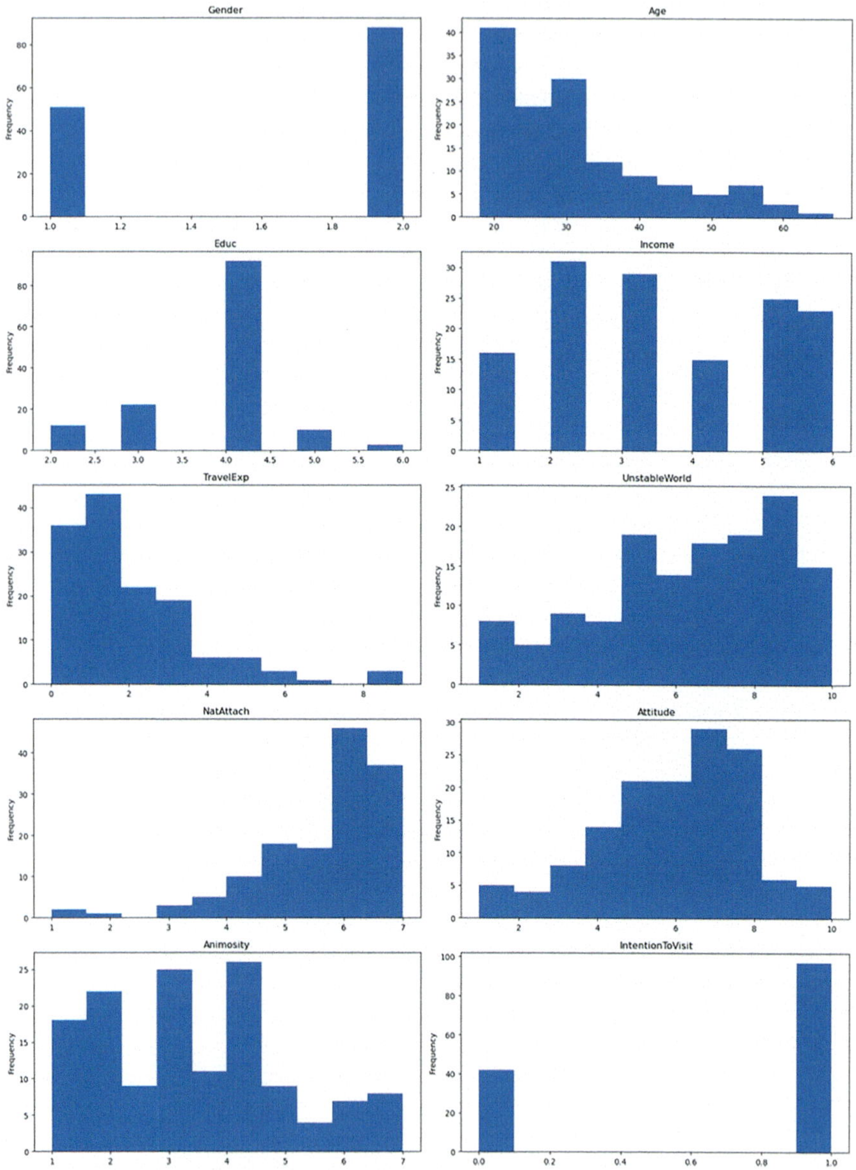
Gender
Age
Educ
Income
TravelExp
UnstableWorld
NatAttach
Attitude
Animosity
IntentionToVisit
Frequency

The distribution of variables *Age*, *TravelExp*, and *NatAttach* is skewed, which may be problematic with some analytical methods, but not for decision trees. Notice that both males and females are well represented in the dataset, which is good for the generalizability of findings. Similarly, the target variable *IntentionToVisit* has a balanced representation of both classes. The respondents tend to be relatively young (median age 28 y.o.) and have a university degree. Half of the respondents have little experience traveling abroad (0–1 trip/year); their concern about global instability is an important factor affecting their decision to travel. Finally, notice a very high level of national attachment among the respondents.

USER:

Rescale columns Income, TravelExp, UnstableWorld, NatAttach, Attitude, and Animosity in the DataFrame data to the range 1–10.

Assistant:

```
from sklearn.preprocessing import MinMaxScaler

# Select the columns to be rescaled
columns_to_rescale = ['Income', 'TravelExp', 'UnstableWorld',
'NatAttach', 'Attitude', 'Animosity']

# Initialize the MinMaxScaler
scaler = MinMaxScaler(feature_range=(1, 10))

# Fit and transform the selected columns using the scaler
data[columns_to_rescale] =
scaler.fit_transform(data[columns_to_rescale])

# Display the first few rows of the DataFrame to verify the rescaling
print(data.head())
```

Explanation:

To rescale the specified columns in the DataFrame *data* to the range 1–10, we use the Min–Max scaling technique. Specifically, we use the *MinMaxScaler* from the *sklearn.preprocessing* module. After rescaling, the first few rows of the DataFrame are printed to verify the changes.

4

The code suggested by Assistant then uses the method *data.head*() to verify that the Likert scale variables have been rescaled properly:

```
# Display the first few rows of the DataFrame to verify the rescaling
print(data.head())
```

However, the *data.describe*() method is more advantageous in this context. Rescaling should change the maximum values of the transformed variable. These statistics are provided by *data.describe*(), but not by the *data.head*() method.

In the notebook page, replace the code *print(data.head( ))* with *print(data.describe( ))*.

```
# Display the statistics of the DataFrame to verify the rescaling
print(data.describe())
```

Result:

	ID	Gender	Age	Educ	Income	TravelExp
count	139.000000	139.000000	139.000000	139.000000	139.000000	139.000000
mean	70.000000	1.633094	30.964029	3.784173	5.519424	2.798561
std	40.269923	0.483704	11.422968	0.787241	2.989811	1.900343
min	1.000000	1.000000	18.000000	2.000000	1.000000	1.000000
25%	35.500000	1.000000	22.000000	4.000000	2.800000	1.000000
50%	70.000000	2.000000	28.000000	4.000000	4.600000	2.000000
75%	104.500000	2.000000	36.500000	4.000000	8.200000	4.000000
max	139.000000	2.000000	67.000000	6.000000	10.000000	10.000000

	UnstableWorld	NatAttach	Attitude	Animosity	IntentionToVisit
count	139.000000	139.000000	139.000000	139.000000	139.000000
mean	6.474820	8.025180	6.035971	4.604317	0.697842
std	2.596907	1.787967	2.093503	2.386917	0.460854
min	1.000000	1.000000	1.000000	1.000000	0.000000
25%	5.000000	7.000000	5.000000	2.500000	0.000000
50%	7.000000	8.500000	6.000000	4.000000	1.000000
75%	9.000000	9.500000	8.000000	6.250000	1.000000
max	10.000000	10.000000	10.000000	10.000000	1.000000

Step 3. Build an Initial Decision Tree Model

USER:

In the DataFrame 'data', column IntentionToVisit is the target, and columns Gender, Age, Educ, Income, Travel-

Exp, UnstableWorld, NatAttach, Attitude, and Animosity are attributes. Build a decision tree.

Recall that in data mining, the dependent variable is called the Target, and the independent variables are called Attributes.

Assistant:

```
from sklearn.tree import DecisionTreeClassifier
from sklearn.model_selection import train_test_split
from sklearn.metrics import accuracy_score, classification_report,
confusion_matrix

# Define the features (labels) and the target attribute
features = ['Gender', 'Age', 'Educ', 'Income', 'TravelExp',
'UnstableWorld', 'NatAttach', 'Attitude', 'Animosity']
target = 'IntentionToVisit'

# Split the data into features and target attribute
X = data[features]
y = data[target]

# Split the data into training and testing sets
X_train, X_test, y_train, y_test = train_test_split(X, y,
test_size=0.2, random_state=42)

# Initialize the decision tree classifier
clf = DecisionTreeClassifier(random_state=42)

# Fit the classifier to the training data
clf.fit(X_train, y_train)

# Make predictions on the testing data
y_pred = clf.predict(X_test)

# Evaluate the model
accuracy = accuracy_score(y_test, y_pred)
print("Accuracy:", accuracy)
print("\nClassification Report:")
print(classification_report(y_test, y_pred))
print("\nConfusion Matrix:")
print(confusion_matrix(y_test, y_pred))
```

Explanation:

In this code, we use the *DecisionTreeClassifier* from *scikit-learn* library to build a decision tree model. We split the data into training and testing sets, fit the classifier to the training data, make predictions on the testing data, and evaluate the model using accuracy, classification report, and confusion matrix.

Result:

```
Accuracy: 0.75

Classification Report:
              precision    recall  f1-score   support
           0       0.33      0.40      0.36         5
           1       0.86      0.83      0.84        23
    accuracy                           0.75        28
   macro avg       0.60      0.61      0.60        28
weighted avg       0.77      0.75      0.76        28

Confusion Matrix:
[[ 2  3]
 [ 4 19]]
```

Let us go through the code more thoroughly. First, notice that the dataset was randomly split into two parts: train (80% of data) and test (20% of data). The parameter *random_state* sets the initial value for the random number generator:

```
# Split the data into training and testing sets
X_train, X_test, y_train, y_test = train_test_split(X, y,
test_size=0.2, random_state=42)
```

The *train* part of the dataset is used to train the decision tree algorithm. The trained model is then applied to the *test* part to compute the accuracy and other metrics of model performance. Always evaluate model performance on data not used in model training! We will cover model evaluation in detail in ► Chapter 8.

Accuracy = 0.75 means that the model correctly predicts the value of *IntentionToVisit* in 75% of all cases. Precision, recall, and F1 measure the proportion of correct predictions of the *IntentionToVisit* value. Refer to ► Chapter 8 for a detailed description of these metrics.

Two lines in the classification report describe model success in predicting the decision not to travel (*IntentionToVisit* = 0) and to travel (*IntentionToVisit* = 1). Notice that the model is more successful in predicting the decision to travel than the decision not to travel. Nevertheless, only five participants in the test dataset don't intend to travel to the US: support = 5 for the *IntentionToVisit* = 0 prediction. That makes this evaluation less reliable.

Finally, the confusion matrix compares predicted and true values of the *IntentionToVisit* variable. The diagonal values of the matrix represent the correct predictions: two participants were correctly predicted as not intending to travel, and 19 participants were correctly predicted as intending to travel. The off-diagonal values represent model failures. Three participants were predicted as not intending to travel while they did intend to travel. Four participants were predicted as intending to travel while they did not.

Step 4. Visualize the Decision Tree

USER

Visualize the decision tree "clf".

Assistant

```
from sklearn import tree
import matplotlib.pyplot as plt

# Visualize the decision tree
plt.figure(figsize=(20,10))
tree.plot_tree(clf, feature_names=features, class_names=clf.classes_,
filled=True)
plt.show()
```

Explanation:

To visualize the decision tree classifier *clf* that we built using the *scikit-learn* library, we use the *plot_tree* function from the *tree* module of the *scikit-learn* library.

An attempt to run this cell returns an error:

```
InvalidParameterError: The 'class_names' parameter of plot_tree must be an instance of 'list' or None. Got array([0, 1]) instead.
```

The parameter *class_names*, which contains the names for the values of the target variable, is an array (with the values 0 and 1), while the *list* type is expected. We can make a simple in-place code correction: replace *class_names = clf.classes_* with *class_names* = ["No", "Yes"].

Corrected code:

```
from sklearn import tree
import matplotlib.pyplot as plt

# Visualize the decision tree
plt.figure(figsize=(20,10))
tree.plot_tree(clf, feature_names=features, class_names=["No", "Yes"],
filled=True)
plt.show()
```


Result:

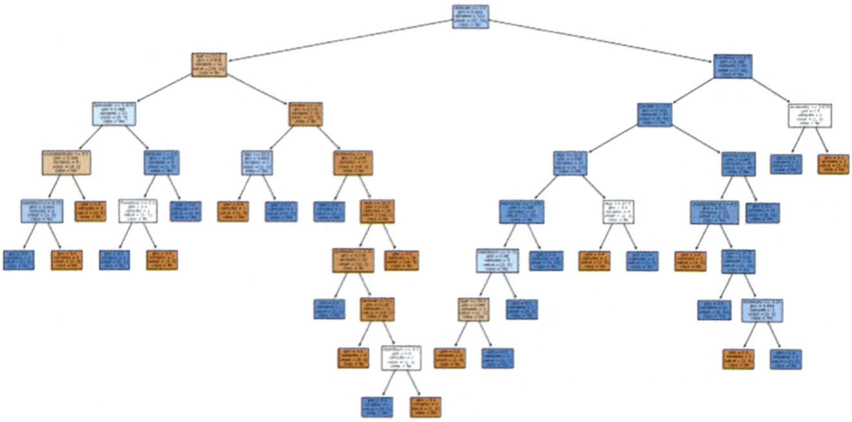

Part 2. Modify the Initial Model

The initial model has multiple problems:

1. The test data is imbalanced: only five records have the value of *IntentionToVisit* attribute equal to zero. It is desirable to have a proportion of 0 and 1 values similar to that in the original dataset (40–60%).
2. It is desirable to have a similar proportion of successful predictions of *IntentionToVisit*=0 and *IntentionToVisit*=1 values, even though it might reduce model accuracy.
3. The model is too large, making it hard to understand. Reducing the number of levels in the decision tree is desirable, even though it might decrease model accuracy.

Instead of starting a new code, we will directly modify the cells of the notebook, using the Assistant to learn about the code that implements these modifications.

Step 1. Balance the Test Dataset

Recall that the initial dataset was split into the train and test parts with the following code:

```
# Split the data into training and testing sets
X_train, X_test, y_train, y_test = train_test_split(X, y,
test_size=0.2, random_state=42)
```

The code used the function *train_test_split*, which was imported from the machine learning library *scikit-learn*:

```
from sklearn.model_selection import train_test_split
```

We can ask the Assistant about additional parameters.

USER:

What are the parameters of *train_test_split* function?

Assistant:

The *train_test_split* function from the *sklearn.model_selection* module splits arrays or matrices into random train and test subsets. The function takes several parameters:
1. *arrays*: This parameter represents the input data to be split. This could be a single array or a list of arrays.
2. *test_size*: This parameter represents the proportion of the dataset to include in the test split. It can be a float (0.0–1.0) to indicate the proportion of the dataset, or an integer to represent the absolute number of test samples.
3. *train_size*: This parameter represents the proportion of the dataset to include in the train split. It is the complement of the *test_size*.
4. *random_state*: This parameter represents the seed the random number generator uses. If an integer is provided, it will set the seed. If None, the random number generator is the *RandomState* instance used by *np.random*.
5. *shuffle*: This parameter represents whether or not to shuffle the data before splitting. It is set to *True* by default.
6. *stratify*: This parameter represents the array containing the class labels or the target variable. If specified, the data is split in a stratified fashion, using this as the class labels.

These parameters allow for flexibility in splitting the data into training and testing sets for machine learning models.

Code modification:

The following part of the Assistant's advice is relevant to balancing the data:

6. *stratify*: This parameter represents ... the target variable. If specified, the data is split in a stratified fashion ...

4

The name of the target variable is *y*. You can find it in the following code lines.

```
target = 'IntentionToVisit'
...
y = data[target]
```

We will use the parameter *stratify* to ensure the balance of the test dataset. Modify the code line as follows:

```
X_train, X_test, y_train, y_test = train_test_split(X, y,
test_size=0.2, random_state=42, stratify= y)
```

Result:

```
Accuracy: 0.75

Classification Report:
              precision    recall  f1-score   support
           0       0.56      0.62      0.59         8
           1       0.84      0.80      0.82        20

    accuracy                           0.75        28
   macro avg       0.70      0.71      0.70        28
weighted avg       0.76      0.75      0.75        28

Confusion Matrix:
[[ 5  3]
 [ 4 16]]
```

In addition to better support for the *IntentionToVisit*=0 value of the target variable, model performance for the *IntentionToVisit*=0 prediction has also increased. Meanwhile, the overall model accuracy has not decreased.

Step 2. Reduce the Size of the Decision Tree

The following line of the code initializes the decision tree model:

```
clf = DecisionTreeClassifier(random_state=42)
```

This time, we will modify the existing cell code without asking the Assistant for help with the code.[5] Instead, we will use the Assistant to learn about the parameters.

USER:

What are the parameters of "DecisionTreeClassifier"?

Assistant:

The *DecisionTreeClassifier* in *scikit-learn* is a class that represents a decision tree classifier. It has several parameters that can be used to customize the behavior of the decision tree. Some of the key parameters include:

1. criterion: The function to measure the quality of a split. It can be set to "*gini*" value for the Gini impurity or "*entropy*" for the information gain.
2. *splitter*: The strategy used to choose the split at each node. It can be set to "*best*" to choose the best split or "*random*" to choose the best random split.
3. *max_depth*: The maximum depth of the tree. If set, the tree will be pruned to limit the maximum depth.
4. *min_samples_split*: The minimum number of samples required to split an internal node.
5. *min_samples_leaf*: The minimum number of samples required to be at a leaf node.
6. *max_features*: The number of features to consider when looking for the best split.
7. *random_state*: The seed used by the random number generator.

These are just a few parameters in the *DecisionTreeClassifier*. Each parameter allows for customization of the decision tree model to suit the specific requirements of the problem at hand.

---

5 Alternatively, you could also ask the Assistant "Cut the decision tree clf to 4 levels.".

Code modification:

To simplify the tree, we will use the following parameters:

$max\_depth = 4$: limit the size of the tree to four levels;

$min\_samples\_split = 4$: do not split the nodes with fewer than 4 participants;

$min\_samples\_leaf = 2$: the leaves should represent at least 2 participants.

The modified line of the code is as follows:

```
clf = DecisionTreeClassifier(max_depth = 4, min_samples_split = 4,
min_samples_leaf = 2, random_state=42)
```

Run the modified cell.

Result:

```
Accuracy: 0.8214285714285714

Classification Report:
              precision    recall  f1-score   support
           0       0.64      0.88      0.74         8
           1       0.94      0.80      0.86        20

    accuracy                           0.82        28
   macro avg       0.79      0.84      0.80        28
weighted avg       0.85      0.82      0.83        28

Confusion Matrix:
[[ 7  1]
 [ 4 16]]
```

Notice that the Step 2 model performance is very similar to the performance of the Step 1 model.

Part 3. On Your Own

1. Visualize the reduced decision tree model.
2. Experiment with the model, further simplifying the decision tree to increase the requirements for node split and minimum leaf size.
3. Write a report including:
   a. Performance of your model.
   b. Discussion of model results.

The discussion should explain the main factors contributing to positive and negative intentions to visit the adversary country. For example, among the youngest population ($Age < 20$), the major factor is low travel experience.

Build Your Python Skills

Chapters 8 and 9 of the free book Learn Python with Jupyter by Serena Bonaretti. ▶ https://www.learnpythonwithjupyter.com/.

Additional Reading

Smith, C. (2017). Decision trees and random forests: A visual introduction for beginners. Blue Windmill Media.

Clustering Techniques: K-means and DBSCAN

A. P. Kirilenko, *Practical Data Mining with AI for Social Scientists*,
Springer Texts in Social Sciences, https://doi.org/10.1007/978-3-031-89689-7_5

5.1 Basics of Cluster Analysis

Why manual cluster recognition may be misleading

Humans tend to imagine connections between unrelated data points (events, random drawing patterns, numbers, and so on). This tendency, called apophenia, is visible in relying on patterns in gambling, conspiracy theories, seeing faces and animals in passing clouds, and similar pattern recognition in random data.

One of the frequent aims of data analysis is grouping the data points into larger clusters or categories of somehow similar objects. For example, we might want to group survey participants according to their sociodemographics, group customers according to their purchasing habits, travelers according to the types of attractions they like to visit, destinations according to their road distance, and so on. Grouping of data records may be the final goal of the analysis, but more frequently, this is just a step leading to an independent analysis of each of the groups or an in-depth study of one of the groups. For example, in the analysis of the University of Florida Gators basketball game attendants survey (▶ Chapter 2 lab), one may pay a particular interest in a deeper analysis of the group likely to bring donations to the team.

Humans are naturally gifted in recognizing data patterns. This gift, however, is frequently taken to an extreme when the human brain connects factually unrelated data points into a group. On the other hand, humans have problems recognizing patterns in multidimensional data. Finally, humans may have problems explaining their decision to organize certain data points into one group. This is why cluster analysis, which is a field of data mining dealing with assembling objects into groups (clusters), is an important tool for a quantitative social scientist.

Cluster analysis deals with the problem of organizing objects into clusters such that objects within a cluster have high similarity in comparison to one another but are *dissimilar* to objects in other clusters. This informal definition illustrates the significance of the concept of similarity to cluster analysis. The similarity is usually quantified in terms of the distance between the objects: that is, the similarity is the highest when the distance is zero and the dissimilarity is the highest when the distance between the data points is infinite. In spatial clustering, a subfield of cluster analysis dealing with finding clusters of objects on a map, the similarity includes the geographical distance between the objects. For example, in urban planning, cluster analysis helps identify "similar" adjacent areas of the city.

Similarity measures depend on the nature of the data and research question. Socio-demographic data may quantify similarity among the data points based on age, income, and educational attainment. The hospitality industry may cluster the hotels based on their financial performance. Conventions and Visitors Bureau may cluster the attraction point based on the co-visitation data, and so on.

Even the simplest example with clustering of objects based on their geographical location leads to multiple ways of measuring the distance and even more ways of forming the clusters. Consider clustering points of interest in a city for a travel recommendation tool. To measure the similarity between the points, one may use the Euclidean, Manhattan, or the road distance between the map points. The Euclidean ("how the crow flies") distance is simply the length of the straight map connecting two points. The straight line is, however, not how people travel in the majority of real-life situations. Another useful measure of the distance between the points of interest is the Manhattan (taxi cab) distance, which measures the distance on a square grid. This is a reasonable approximation of a travel path in many North American cities with a rectangular network of streets, but not in the European tourism centers, many of which have inherited their radial street arrangement from medieval history. An even better approximation is a road distance, which reflects the shortest distance over the actual network of city streets. Further improvement of this similarity measure takes into consideration a strong human preference for traveling along a straight line, minimizing the number of turns, even though this may lengthen the path.

When locations are polyghons than points, it further complicates decision-making on the similarity measure. Consider clustering city neighborhoods according to their proximity to city parks. One may measure the distance between the closest edges of a particular neighborhood and a park, the distance between the centroids, the median distance between neighborhood houses and a park, and so on.

Finally, the number of clusters is also an arbitrary choice. A general plan for city attraction visitation may prefer a few large clusters for tourist exploration. Meanwhile, many more detailed clusters would be required to map neighborhood clusters according to their park accessibility.

In the subsequent sections, we will discuss the principles of two popular partitioning clustering techniques: K-means and DBSCAN. Then, we will introduce measures

Clustering is somewhat arbitrary

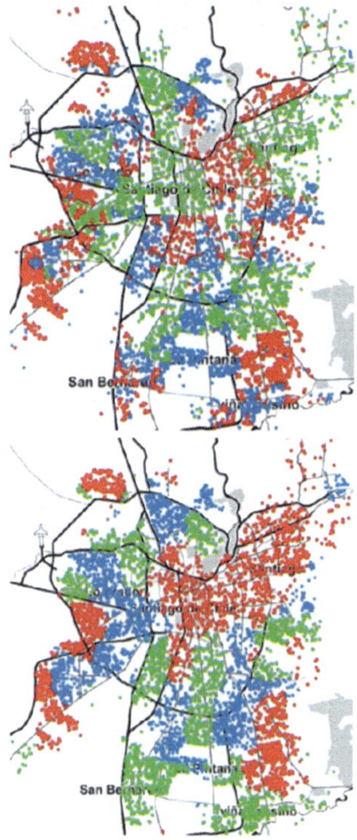

Two alternative ways of clustering Santiago according to the socioeconomic status of residents, voting, aesthetics, etc. From: Vera, C., et al. "Learning to cluster urban areas: two competitive approaches and an empirical validation." EPJ Data Science 11.1 (2022): 62.

of cluster evaluation. Furthermore, we will consider the critical question of how many clusters to select, introducing the Elbow method and Silhouette method as potential solutions. We will finish with cluster visualization methods.

5.2 Partitional Clustering: K-means

The K-means (together with a connected method, K-medoid) is a very popular method of separating the data into non-intersecting clusters. The idea of K-means is to partition data points based on their distance from cluster geometrical centers (formally centroids). The algorithm works in iterations. First, each data point is assigned to the nearest point cluster. This assignment changes cluster configuration. Then, the algorithm calculates the centroids of the updated clusters and the distances from each data point to the new centroids. After that, these steps are repeated.

The K-means algorithm begins with a random initialization of K centroids, which become the seeds of provisional clusters. The parameter K represents the number of clusters and is defined by the analyst. We will cover the methods of optimal K setting later in the chapter. Each data point is then assigned to the nearest centroid based on a distance metric, typically Euclidean distance. Following this, the centroids are recalculated as the mean (for K-means) or median (for K-median) of all data points assigned to the respective cluster.

▫ Figure 5.1 example uses an artificially generated dataset of average ratings for restaurants and hotels for 50 different destinations. The goal is to cluster the data points into three groups according to their ratings. Five plots represent the first five steps of K-means clustering.

Notice the initial random placement of the centroid centers (marked with a red cross) with the corresponding designation of clusters (marked with different colors). The cluster designation of data points is adjusted with every step. In the fourth step, the clusters are finalized: there is no cluster change in step 5.

One important issue with the K-means algorithm is the random placement of the initial centroids. ▫ Figure 5.2 illustrates this "initial centroid problem." An occasional unfortunate position of the seed centroids leads to non-optimal clusters even though this example uses the date data as the previous one. Objectively, the quality of these two solutions differentiates in how close data points are to their respective centroids. The measures of clustering quality will be discussed in the next section; meanwhile, a simple solution to the initial centroid problem is multiple runs of the clustering algorithm, followed by a selection of the cluster configuration with the best quality. There are other approaches to solving the initial centroid problem (including the "greedy k-means++" implemented as a default in the algorithm you will use in this chapter's lab); an important point, however, is that the K-means method may result in suboptimal clusters if not used in conjunction with one of those methods.

5.3 Evaluating the Clusters

5.3.1 Euclidean Distance: Sum of Squared Errors

Differentiation between multiple ways of clustering data points is impossible without clustering performance evaluation. The most frequently used measure is the Sum of Squared Errors (SSE, also called the *inertia*), which computes clustering performance as the total variance within the clusters. The SSE is calculated by summing the squared Euclidean "how the crow flies" distances between each data point and the centroids of the corresponding cluster. A lower SSE indicates that the data points are closer to their respective centroids. This happens when clusters are more compact and well-defined. Conversely, a higher SSE suggests that the data points are spread out, indicating less distinct clusters. For instance, the inertia of the review rating data clustering illustrated by ◘ Fig. 5.1 decreases from 48.6 in the first step to 46.6 in the last step, demonstrating progressively improving clustering performance. In comparison, the suboptimal selection of the initial centroids illustrated by ◘ Fig. 5.2 results in inertia decreasing from 59.9 to 46.7, that is, in a larger error in both the first and the final steps of the clustering algorithm.

The process of centroid and SSE calculation is also straightforward for multidimensional data. Consider five destinations, each characterized by average hotel price, average restaurant rating, and the number of tourist attractions:

Popular way of measuring the distance on the map

The Euclidian distance: the straight-line distance between two points.

The Manhattan distance: the distance for traveling on a rectangular grid

The road distance: the real-world distance while traveling on the road network. Calculating the road distance requires a Geographical Information System (GIS).

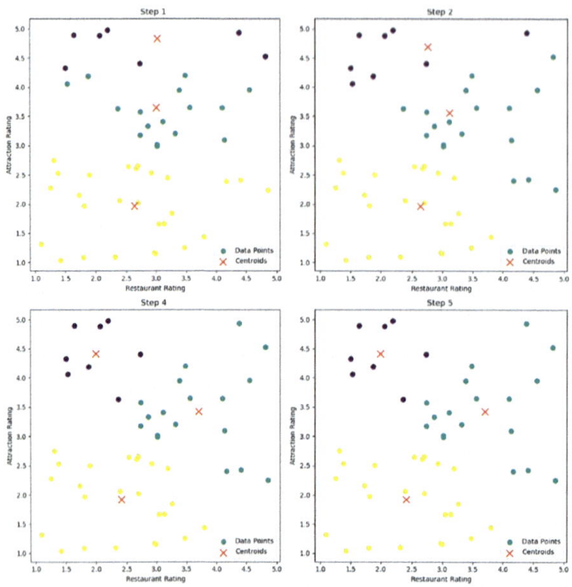

Fig. 5.1 Progressive steps of K-means clustering

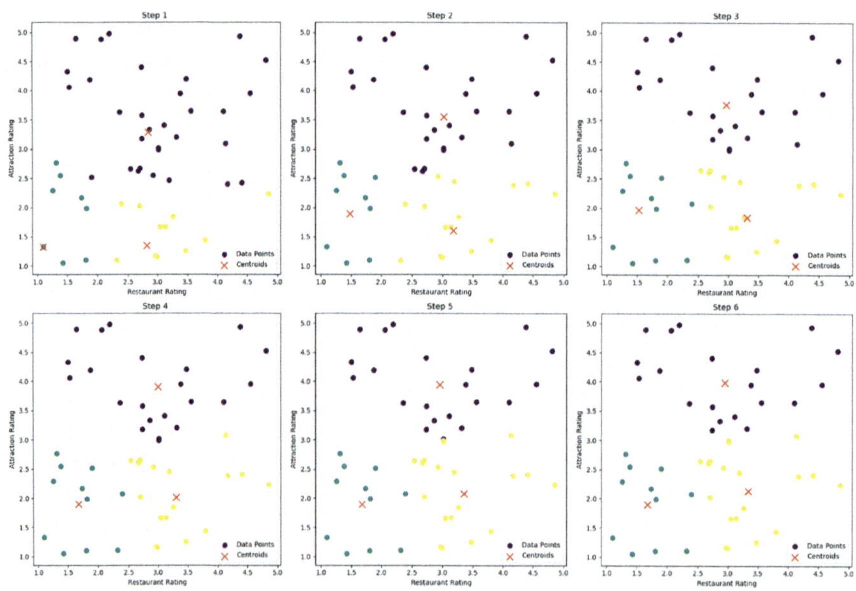

Fig. 5.2 Initial centroid problem: a poor selection of the seed centroids may result in suboptimal clusters

Destination	Hotel price (USD)	Restaurant rating	N attractions
Destination A	120	4.5	15
Destination B	150	4	10
Destination C	100	4.2	12
Destination D	130	4.7	18
Destination E	110	4.1	14

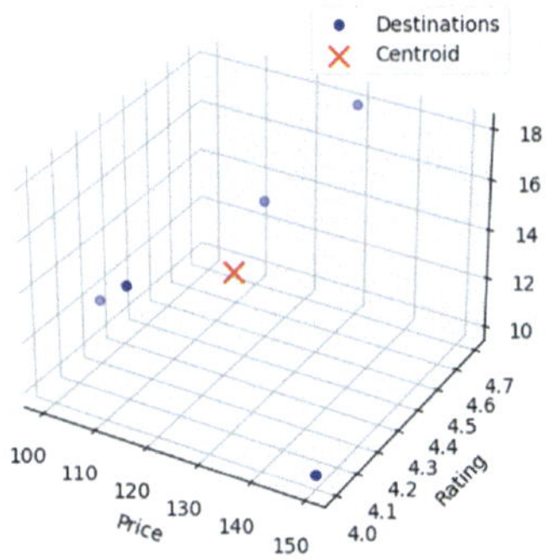

Three attributes, hotel price, restaurant rating, and the number of tourist attractions define the positions of these destinations in a three-dimensional space with coordinates Price, Rating, and Attractions. Each coordinate of the centroid of these points can be found as the mean of the respective data point coordinate:

$$\text{Price} = (120 + 150 + 100 + 130 + 110)/5 = 122$$

$$\text{Rating} = (4.5 + 4.0 + 4.2 + 4.7 + 4.1)/5 = 4.3$$

$$\text{Attractions} = (15 + 10 + 12 + 18 + 14)/5 = 13.8$$

Then, the squared distance from the destination A to the centroid $d_a = (120-122)^2 + (4.5-4.3)^2 + (15-13.8)^2 = 5.48$.

Similarly, $d_B = 798.53$; $d_C = 487.25$; $d_D = 81.8$; and $d_E = 144$.

Finally, the SSE = 5.48 + 798.53 + 487.25 + 81.8 + 144 = 1517.06.

Clustering geographical locations

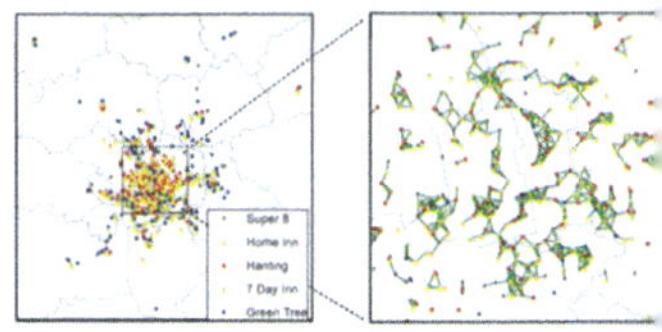

Clustering Beijing hotels. From Su, L., Kirilenko, A., & Stepchenkova, S. (2024). Compete or Avoid? Assessing Brand Competition Strategies with Spatial Colocation Analysis. Tourism and Hospitality, 5(1), 160-166.

5.3.2 Manhattan Distance

The SSE computation in the previous example used the Euclidean distance to measure the distance between the data points. Sometimes, however, the Euclidean distance is not appropriate. Consider a dataset of hotel locations expressed as a list of geographical latitude and longitude pairs. An appropriate similarity measure for city locations is frequently the "taxi cab" Manhattan distance, calculated as the sum of **absolute differences** between the data points and their respective cluster centroids. Compare that to SSE computed as the sum of **squared differences.**

5.3.3 Cosine Distance

Clustering methods can be applied to a wide variety of objects, including texts. Consider a simple example of three textual documents:

- Document 1: "I love traveling to new places."
- Document 2: "Traveling to new places is exciting."
- Document 3: "I enjoy reading books."
- Document 4: "Books are a great way to learn."
- Document 5: "I travel to new places every year."
- Document 6: "Reading new books is a wonderful hobby."

After removing the words inessential for understanding the text meaning ("stop words") and changing the inflected forms of words, these documents can be transformed into the following sets of 15 unique tokens:

- Document 1: ["love", "travel", "new", "place"]
- Document 2: ["travel", "new", "place", "exciting"]
- Document 3: ["enjoy", "read", "book"]
- Document 4: ["book", "great", "way", "learn"]
- Document 5: ["travel", "new", "place", "every", "year"]
- Document 6: ["read", "new", "book", "wonderful", "hobby"]

which in turn can be formalized as the following "one-hot" presence/absence matrix:

ID	Book	Enjoy	Every	Exciting	Great	Hobby	Learn	Love	New	Place	Read	Travel	Way	Wonderful	Year
D1	0	0	0	0	0	0	0	1	1	1	0	1	0	0	0
D2	0	0	0	1	0	0	0	0	1	1	0	1	0	0	0
D3	1	1	0	0	0	0	0	0	0	0	1	0	0	0	0
D4	1	0	0	0	1	0	1	0	0	0	0	0	1	0	0
D5	0	0	1	0	0	0	0	0	1	1	0	1	0	0	1
D6	1	0	0	0	0	1	0	0	1	0	1	0	0	1	0

In this table, "1" represents the presence of a token in the respective document. Hence, each of the documents can be represented as a vector in a 15-dimensional token space. The similarity between the documents can be expressed as a cosine of the angle between these vectors:

ID	D1	D2	D3	D4	D5	D6
D1	1	0.75	0.00	0.00	0.67	0.22
D2	0.75	1	0.00	0.00	0.67	0.22
D3	0.00	0.00	1	0.29	0.00	0.52
D4	0.00	0.00	0.29	1	0.00	0.22
D5	0.67	0.67	0.00	0.00	1	0.20
D6	0.22	0.22	0.52	0.22	0.20	1

The distance between the documents is computed as 1—similarity. Notice a high similarity between documents 1, 2, and 5. In addition, there is a moderate similarity between documents 3, 4, and 6. The respective documents are likely to form two clusters: one discussing travel and the other discussing books.

Interestingly, only six short texts have resulted in a 15-dimensional token space. This is typical of text analysis clustering applications. The cosine metrics are a typical measure of computing the distance between the vectors representing the documents in the multi-dimensional space for future clustering applications.

Cosine similarity

Cosine similarity measures the similarity (distance) between two documents represented by their vectors in token space as as the cosine of the angle between the vectors. Practically, it shows if the vectors are pointing in a similar direction.

5.4 How Many Clusters to Select?

5.4.1 Elbow Method

One of the principal questions to ask in cluster analysis is how many clusters are there. This is especially important in methods like K-means, which require the analyst to set the number of clusters manually. Sometimes, we are able to deduce the number of clusters from theory. An analyst performs cluster analysis multiple times, comparing the emerging separation of data points with his or her expert understanding of the field. Then, the optimal number of clusters is set to match the parameters of the cluster analysis model with outcomes best fitting the theory.

While it is a good idea always to perform manual cluster validation against the theory, there are approaches that aid the process. This section discusses a very popular approach among practitioners: the elbow method. The next section deals with another popular approach, the silhouette method.

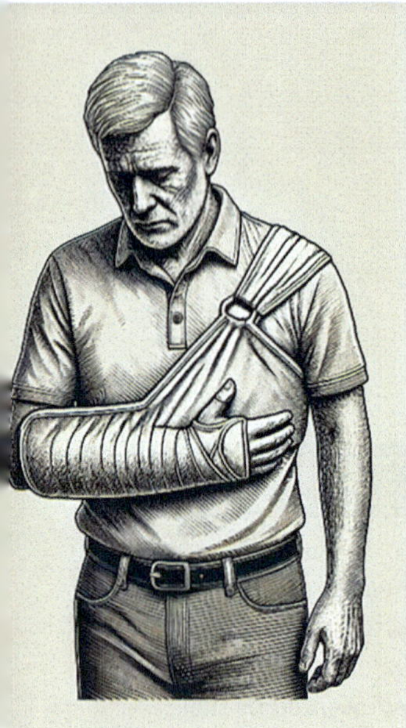

The idea of the elbow method is to run a clustering algorithm multiple times, keeping track of one or more cluster evaluation methods. Frequently, the measure of choice is the within-cluster sum of squares (WCSS) measure, discussed in ▶ Sect. 5.3.1. Recall that WCSS measures the compactness of the clusters. Lower WCSS indicates tighter, more defined clusters. As the number of clusters increases, the WCSS decreases; when the number of clusters equals the number of data points, the WCSS equals zero: every data point becomes its own cluster. However, the rate of decrease is not constant. Initially, the decrease is steep as each introduction of new centroids reduces the nearest centroid distance for the most remotely located data points. This improvement, however, tends to saturate as more and more centroids appear. As a result, the WCSS vs. the number of clusters plot forms an 'elbow' (◘ Fig. 5.3). The number of clusters at this 'elbow' point is considered optimal.

In addition to WCSS, the elbow method can accommodate many other distance measures. One useful measure is the Between Cluster Sum of Squares (BCSS), also known as the inter-cluster variance. While WCSS measures clusters' tightness, the BCSS measures how well the clusters are separated from one another. Since there is no holistic measure accounting for all desirable properties of clustering outcomes, it makes sense to use multiple measures prior to making a decision on the "best" number of clusters.

Despite its popularity, the elbow method has limitations. First, the elbow point is not always as distinct as shown in ◘ Fig. 5.3. This leads to ambiguity in the selection of the optimal number of clusters. Second, the elbow method is largely subjective, relying on visual plot inspection rather than a quantitative metric. Finally, despite its intuitive clarity, the method is not supported the-

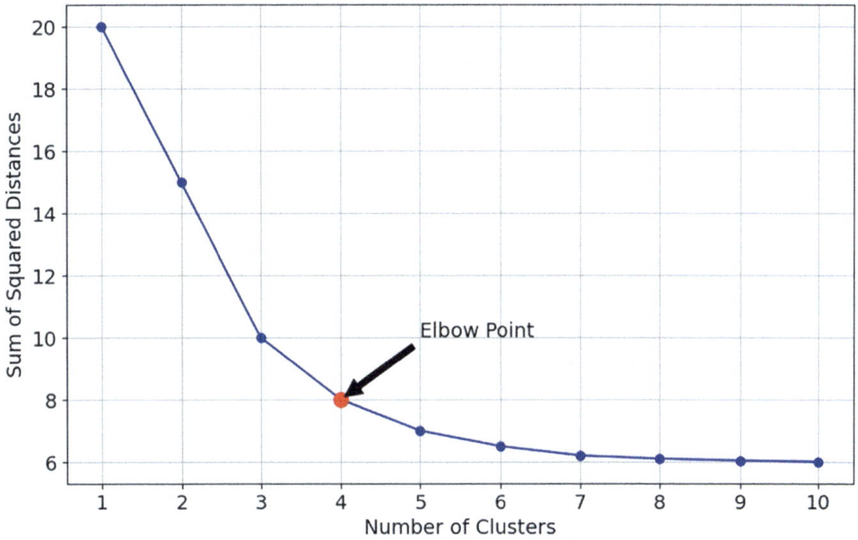

Fig. 5.3 Elbow method for cluster number optimization. Note that the exact location of the elbow point is somewhat subjective

oretically. This is why, while the elbow method can give some indications of the number of clusters, expert judgment has to be applied before making the final decision on K-means parameters.

5.4.2 Silhouette Method

Compared to the elbow method, the silhouette method of determining the optimal number of clusters is more quantitative. The silhouette score measures how close each point in one cluster is to the points in the neighboring clusters. Score values range from −1 to 1. The high positive score values indicate tight clusters, well separated from other clusters, while negative scores indicate that the point is assigned to the wrong cluster. As a rule of thumb, the mean Silhouette scores above 0.7 indicate strong clustering, above 0.5 medium, and above 0.25 weak clustering.

Similar to the elbow method, the silhouette method requires tracing the mean silhouette score over multiple clustering algorithm runs. The run which returns the maximum score then determines the optimal number of clusters. The silhouette method also offers visualization in the form of a silhouette plot, which shows how well

Silhouette scores

- Above 0.7: strong clustering
- 0.5–0.7: medium clustering
- 0.25–0.5: weak clustering
- Below 0.25: no clustering

each data point is located within its cluster. The silhouette plot is useful in identifying outliers or misclassified points.

Consider the review ratings clustering example in ▶ Sect. 5.2. After running the K-means algorithm for the number of clusters ranging from 2 to 10, the following mean silhouette scores were obtained:

N clusters	2	3	4	5	6	7	8	9	10
Score	0.368	0.387	0.372	0.423	0.459	0.474	0.454	0.458	0.449

The silhouette scores are maximized when the number of clusters is seven. We can accept this number of clusters to be optimal for our data. The silhouette score of 0.47 indicates that clustering is weak to marginally average. Figure 5.4 shows a silhouette plot for seven clusters.

Each colored polygon on the plot indicates a separate cluster. The polygon's vertical dimension indicates cluster size. Within each polygon, individual silhouette scores for all data points in that cluster are plotted. For example, in Cluster 0 (black color), the scores are distributed from approximately 0.3 to 0.7, indicating a moderately tight cluster. Compared to Cluster 0, Cluster 2 is loose. Finally, the dashed line shows the mean silhouette score value. The silhouette plot provides a comprehensive overview of clustering algorithm performance.

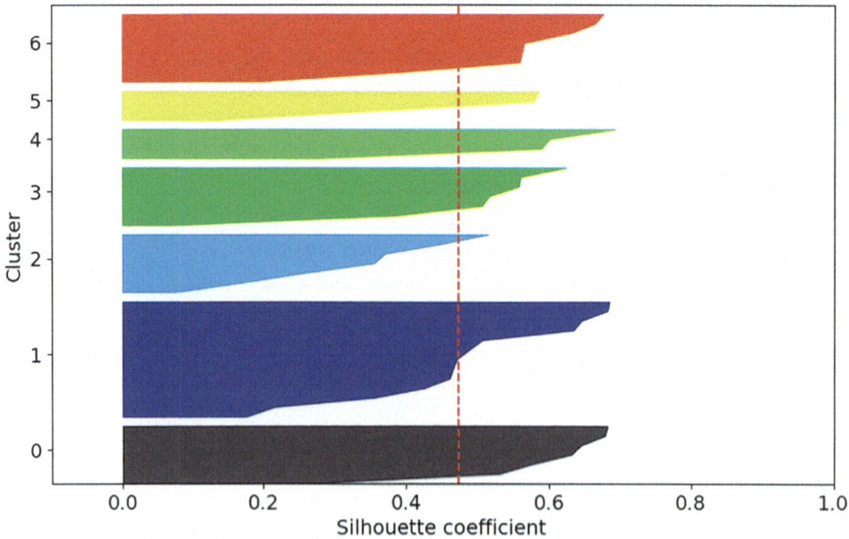

Fig. 5.4 Silhouette plot for the review ratings clustering example in ▶ Sect. 5.2

5.5 Data Considerations and Limitations

5.5.1 Levels of Measurement

The idea of the K-means algorithm is to manipulate cluster assignments based on the distance between the data points. An implication of that is that the distance measure should be meaningful. This is usually the case when all attributes involved in clustering have at **least an interval level of measurement**. This is why a frequent recommendation for K-means is not to be used with nominal data. Then, the most frequently used Euclidean distance would be appropriate for cluster assignments as long as the attributes have a comparable scale range.

Consider clustering survey participants according to their annual income (USD), Age (years), Educational Attainment (coded as 1: no high school; 2: some college; 3: Bachelor or Associate degree; 4: Masters degree; 5: PhD; 6: Other), Sex (1: Male; 2: Female), and Employment status (1: Full time; 2: Part-time; 3: Unemployed or not in the workforce). First, consider the first two data attributes: income and age. While the participants' age range is unlikely to exceed 100 years, the income difference would typically exceed tens of thousands USD. One implication is that WCSS calculation would involve adding "apples and oranges". Yet even disregarding this problem, the WCSS value will be disproportionally affected by income, rendering the age nearly irrelevant. This is because a 10-year difference in age and a $10 difference in annual income would result in an identical difference in the WCSS, while in practice, the effect of a 10-year difference in age on a lifestyle is much higher than the effect of a $10 difference in salary.

This is why, in terms of data preparation, the K-means algorithm requires that **all attributes of the data points have the same scale**. This is a critical prerequisite, failing which is likely to reduce the input of attributes with a smaller scale range into the clustering decisions. In addition, there should be no missing data. That means that, aside from regular data cleaning, data preparation for cluster analysis should include data normalization or standardization. The only exception from this requirement is when all attributes have the same scale, like the ones in the ▶ Sect. 5.2 example.

One caveat of making the decision on data normalization or standardization is that it still assumes that in

K-means considerations

- No categorical features; ordinal data may be problematic
- Same scale for all features
- Clusters are globular in shape, with similar density, and non-intersecting.

the transformed data one unit of change in every attribute has the same importance for cluster assignment. Only expert knowledge of the field can help in making the final decision on data transformation.

The third attribute, Educational Attainment, can be considered ordinal if the data points with this attribute's "Other" value are removed from the clustering. In practice, the ordinal scale variables are frequently treated like the interval and ratio variables. However, treating the Education attainment attribute as an interval variable assumes that the change from a high school to a Bachelor's degree has the same effect on cluster assignment as the change from a Bachelor's to a Master's.

The fourth variable, Sex, is binary. Some publications treat binary variables the same way as ordinal ones. Note, however, that such treatment assumes that the effect of the difference between the two sexes on cluster assignment is the same as the effect of a one-unit change in educational attainment, age, or income (after data transformation). In my view, binary variables should be treated as nominal except when there is a theoretical justification for the otherwise.

Finally, the Employment status is clearly a nominal attribute. One way of dealing with nominal attributes was considered in ▶ Sect. 5.3.3, which applied a "one-hot" transformation to text documents, effectively transforming texts into a set of binary "presence-absence" attributes. This idea is utilized in the K-modes algorithm, a K-means modification specifically designed for handling a mix of interval and nominal data. K-modes transforms categorical attributes into a set of dummy binary attributes and uses a Hamming distance as a dissimilarity measure.

5.5.2 Limitations

The K-means algorithm targets the minimization of the distance within the clusters and/or maximization of the distance to other clusters. In other words, the algorithm searches for clusters that are *convex* and *isotropic*, which are the fancy words for globular, uniform, and symmetrical in all directions shapes. However, in the real world, the clusters can be of various shapes. It is important to keep in mind that K-means performance is best when data naturally forms clusters that are:

- Roughly globular in shape;
- Have similar size and density;
- Distinct (not overlapping).

Finally, K-means is susceptible to the presence of outliers, which can significantly shift the centroids and distort the clusters. When removing the outlier points during data cleaning is unwanted, a modification of K-means, the K-medoids algorithm, can be employed. K-medoids method uses the most centrally located point in a cluster (a medoid, which is a multidimensional extension of a median) instead of centroids, which makes it non-susceptible to occasional outliers.

The K-means algorithm is a powerful tool for partitional clustering, particularly effective for large datasets. However, its performance is contingent on the data's nature and the chosen parameters' appropriateness. Despite multiple limitations, the K-means algorithm together with its modifications, K-median and K-modes, is the most frequently used tool for cluster analysis due to its simplicity and efficiency. Meanwhile, alternative clustering methods, such as DBSCAN, offer additional flexibility in identifying more complex cluster structures.

K-means clustering

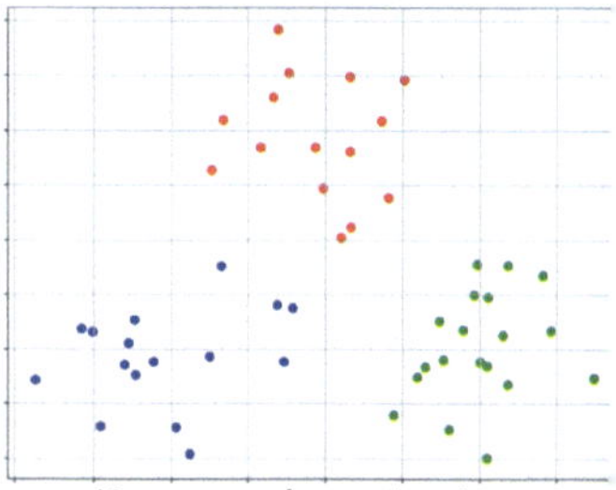

K-means clustering is optimized to search for clusters neatly organized around their centers.

5.6 Density-Based Partitioning: DBSCAN

5.6.1 The Algorithm

As a centroid-based clustering method, the K-means algorithm is optimized to search for clusters neatly organized around their centers. In contrast, the DBSCAN (Density-Based Spatial Clustering of Applications with Noise) algorithm searches for densely packed clusters of data points. The idea of the DBSCAN method is to search for dense regions in the data space, separated by regions of lower density of data points.

DBSCAN identifies clusters based on two parameters: epsilon (ε), the maximum radius of the neighborhood, and minPts, the minimum number of points required to form a dense region. A point is considered a core point if there are at least minPts within ε distance. Border points are within ε distance of a core point but do not have minPts within their ε distance. All other points are considered noise.

DBSCAN clustering

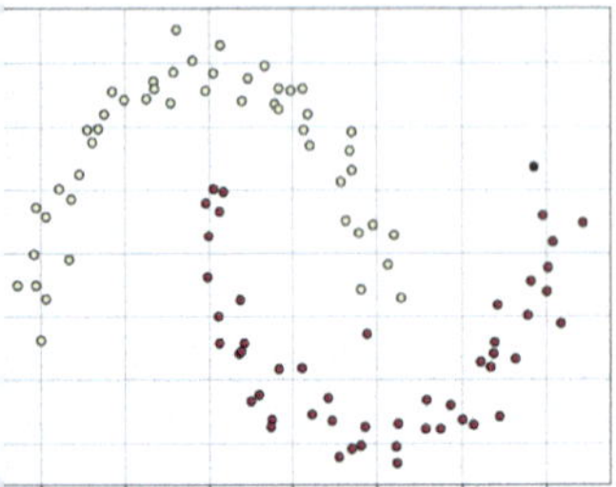

DBSCAN clustering is optimized to search for clusters of densely located data points.

To understand how DBSCAN works, we start with differentiating among three kinds of data points based on their location in multidimensional space:

1. Core point: a data point with multiple other data points in its neighborhood;
2. Border point: a data point with few or no other data points in its neighborhood, but located close to a core point; and
3. Noise point: a data point located far from all core points.

To formalize this definition, DBSCAN introduces two parameters that operationalize the constructs "multiple" and "neighborhood" in this list:

- ε is the radius of the neighborhood: all points within the ε distance from a potential core point A are considered to be *directly reachable* from point A;
- minPts is an operationalization of "multiple": the minimum number of data points in an ε-neighborhood of a point A required for point A to be a core point.

Consider points B and C. Point B is directly reachable from point A, and point C is directly reachable from point B but not from point A. The point B forms a path between the points A and C. All points that are directly reachable from the same core points are called *reachable.*

The algorithm works as follows:

1. All points are classified as core (those with at least minPts other points in their ε-neighborhood), border (those which are not core, but with at least one core point in their ε-neighborhood), and noise (all others);
2. Noise points are marked as outliers and eliminated from further consideration. These points will not be assigned to any cluster;
3. A random core point is selected and all points reachable from this core point are assigned to the same cluster.
4. A new core point is selected from those not yet assigned to any cluster, and the process repeats until all core and border points are assigned to clusters.

5.6.2 DBSCAN Data Considerations and Limitations

DBSCAN data considerations are very similar to those used in K-Means algorithm. First, DBSCAN requires an absence of data points with no data values, which can be resolved either by assigning the missing attribute a value based on the values of other records or by removing the entire record.

Second, DBSCAN requires all attributes of the dataset to have a similar scale. Otherwise, attributes with larger numerical ranges would disproportionately influence the distance calculations, leading to biased clusters.

Third, DBSCAN is designed to operate with interval or ratio data. The ordinal data, which are frequent in survey analysis, are usually treated the same way as the interval measures, but one should ensure that the "Other", and "No answer" categories are excluded from the analysis. Finally, the categorical data require special approaches similar to those discussed in ▶ Sect. 5.5.1.

Finally, DBSCAN requires careful consideration of metrics used for measuring the distance between data points. The Euclidean distance is usually appropriate for interval or ratio data and is set by default in many DBSCAN algorithms. The presence of categorical data usually requires different metrics, such as Gower's or Hamming distance.

By design, the shape of clusters is irrelevant to DBSCAN's performance, which differentiates it from K-means: DBSCAN can discover clusters of arbitrary shape, while K-means is limited to globular clusters. On the other hand, DBSCAN does not perform well when clusters have varying densities.

Unlike K-means, DBSCAN does not require the specification of the number of clusters in advance. However, its results are heavily dependent on the selection of ε and minds. Clustering of the same dataset with an inappropriate choice of parameters can easily lead to a large number of clusters, too few clusters, no clusters, or excessive noise.

Compared with K-means, DBSCAN is robust against a moderate presence of outliers and data noise, as it does not assign the respective records to any cluster. On the other hand, excessive noise may lead to the creation of spurious clusters of noisy points. Because of that, it

Clustering method selection

Multiple other clustering methods can address problems from analyzing different data types and distributions. Selecting an appropriate clustering algorithm, its parameters, data cleaning, scaling methods, and distance metrics that reflect the measured constructs requires theoretical knowledge and expert understanding of the phenomena expressed in the analyzed data.

is still recommended to remove noise and outliers during the data cleaning step.

5.7 Review Questions

1. Why might manual cluster recognition be misleading for humans?
2. What are the main goals of cluster analysis?
3. Explain the basic steps of the K-means algorithm.
4. What is the "initial centroid problem" in K-means, and how can it be mitigated?
5. How is the Sum of Squared Errors (SSE) used to evaluate the quality of clusters?
6. Describe three popular ways of measuring distance in spatial clustering.
7. Describe the elbow method for determining the optimal number of clusters.
8. Describe the silhouette method for determining the optimal number of clusters.
9. What are the average silhouette score values that indicate weak, medium, and good clusters?
10. How does DBSCAN differ from K-means in terms of identifying clusters?
11. What are the roles of the parameters epsilon and minPts in the DBSCAN algorithm?
12. Why is it important for all attributes to have a similar scale in clustering?
13. What are important considerations when clustering ordinal and categorical data?
14. What are data cleaning considerations in K-Means and DBSCAN clustering?

5.8 Lab

Objectives

1. Apply K-means cluster analysis to cluster hotel markets
2. Evaluate the optimal number of clusters
3. Practice cluster visualization

Data

The data for this chapter from the hotel industry was made available by STR, Inc. STR (▶ https://str.com/) is a leading global hotel industry market data provider. The dataset represents the averaged daily hotel industry performance for January 2017 for Florida submarkets. The dataset presents a generalized sample of data used for analysis of hurricanes' impact on travelers' evacuation. The data is stored as ▶ https://github.com/andykir/DataMiningLectures/raw/main/STR_Florida_Submarkets_January2017.xlsx.

Variables

Variable	Explanation	Notes
Location	Market name	Text variable
RoomsOCC	Room occupancy	Mean percentage of occupancy in hotel rooms
RoomsADR	Average Daily Rate	average rental income per paid occupied room
RoomsRevPAR	Average revenue per available room	Hotel's AverageADR multiplied by the occupancy rate
RoomsSupply	Room supply	Average number of rooms available for sale
RoomsDemand	Room demand	Average number of rooms sold, excluding the complimentary stays
RoomsRevenue	Daily rooms revenue	Average revenue per room, excluding taxes
CensusPropCount	Number of properties	Total number of hotels in the market
CensusRoomCount	Number of rooms	Total number of hotel rooms in the market

Part 1. Read and Clean the Data

Step 1. Read and Describe the Data

Every lab starts with identical steps of reading and describing the data. Instead of asking the Assistant to help with coding, we can reuse the code from the previous lab.

The only required change is to correct the URL for cloud data storage.

Open a new Anaconda notebook. Copy the code from the first cell of the Lab 4 notebook to the first cell of the new notebook. Then, edit the URL address. This is how the top cell of the new notebook should look like:

```
import pandas as pd

# Read the data from the URL
url = "https://github.com/andykir/DataMiningLectures/raw/main/\
STR_Florida_Submarkets_January2017.xlsx"
data = pd.read_excel(url)

# Display information on data
print(data.info())
```


Notice the backslash "\" in the URL address. The backslash is a line continuation symbol, which shows that a line of code is continued on the next line. It is used to break long lines of code for better readability.

Run the cell.

Result:

```
RangeIndex: 48 entries, 0 to 47
Data columns (total 9 columns):
 #   Column            Non-Null Count   Dtype
---  ------            --------------   -----
 0   Submarket         48 non-null      object
 1   RoomsOCC          48 non-null      float64
 2   RoomsADR          48 non-null      float64
 3   RoomsRevPAR       48 non-null      float64
 4   RoomsSupply       48 non-null      float64
 5   RoomsDemand       48 non-null      float64
 6   RoomsRevenue      48 non-null      float64
 7   CensusPropCount   48 non-null      int64
 8   CensusRoomCount   48 non-null      float64
dtypes: float64(7), int64(1), object(1)
memory usage: 3.5+ KB
None
```

Notice that there are no missing data, which helps with the analysis. Regarding data types, the attribute "Submarket" type is "object." Recall that the "object" type comprises many different types of data. We can look up a few top lines of the dataset to know more about this attribute. Add to the cell the code:

```
print(data.head())
```

Result:

```
                       Submarket   RoomsOCC    RoomsADR  RoomsRevPAR  \
0                Boca Raton, FL  73.845140  198.241223   146.825764
1        Bradenton/Airport, FL  73.392482  102.461737    75.704395
2                Clearwater, FL  69.779782  138.459483    97.520294
3  Coral Springs/Pompano Beach, FL  78.526420  143.524679   112.996542
4              Daytona Area, FL  58.392798   95.931154    58.442029

   RoomsSupply  RoomsDemand  RoomsRevenue  CensusPropCount  CensusRoomCount
0       7194.0  5312.419355  1.056265e+06               75           7194.0
1       4043.0  2967.258065  3.060729e+05               56           4043.0
2       8225.0  5739.387097  8.021044e+05              105           8225.0
3       6865.0  5390.838710  7.757213e+05              111           6865.0
4       5375.0  3138.612903  3.141259e+05               94           5375.0
```

The *Submarket* attribute is a string and is used as a unique record identifier. Other attributes have float data types, except for the *CensusPropCount* (the total number of hotels in the submarket), which is an integer. Interestingly, the *CensusRoomCount* attribute is a float, which hints that the number of rooms has changed during January.

Step 2. Analyze Data Distribution

We can start with descriptive statistics. Add a new cell in the notebook and enter the code:

```
print(data.describe())
```

Result:

```
        RoomsOCC    RoomsADR  RoomsRevPAR   RoomsSupply   RoomsDemand
count  48.000000   48.000000    48.000000     48.000000     48.000000
mean   69.271686  134.479786    96.843170   9104.226478   6447.801075
std    10.086676   58.094215    49.618386   7212.221366   5701.869062
min    36.359739   73.176363    30.658640   3646.000000   2502.516129
25%    65.099393   90.232835    60.127817   5982.750000   3649.927419
50%    69.601934  108.650090    79.177223   7282.500000   4789.967742
75%    76.351213  153.172778   116.534176   8940.500000   7025.887097
max    85.095164  301.209355   257.930457  42159.000000  30889.612903

       RoomsRevenue  CensusPropCount  CensusRoomCount
count  4.800000e+01        48.000000        48.000000
mean   9.564357e+05        87.958333      9104.226478
std    1.067499e+06        35.928391      7212.221366
min    2.367301e+05        30.000000      3646.000000
25%    3.501621e+05        61.000000      5982.750000
50%    5.886199e+05        87.500000      7282.500000
75%    1.078143e+06       105.000000      8940.500000
max    5.347827e+06       236.000000     42159.000000
```

Confirm that all numerical variables are interval or ratio types of measurement. That means that clustering on each combination of these variables can be done using a regular K-means or DBSCAN algorithm using a default Euclidean distance measure.

To understand data distribution in detail, we will use histograms. There are eight numerical attributes; to improve the visualization, arranging the histograms into a 2×4 grid makes sense. We can ask the Assistant for help with that.

USER:

Use the DataFrame "data". Plot the histograms for distribution of the attributes RoomsOCC, RoomsADR, RoomsRevPAR, RoomsSupply, RoomsDemand, RoomsRevenue, CensusPropCount, and CensusRoomCount. The histograms should be arranged in a 2×4 grid.

Assistant:

```
import matplotlib.pyplot as plt
import pandas as pd

# Selecting the columns
selected_columns = ['RoomsOCC', 'RoomsADR', 'RoomsRevPAR',
'RoomsSupply', 'RoomsDemand', 'RoomsRevenue', 'CensusPropCount',
'CensusRoomCount']

# Plotting the histograms in a 4x2 grid
fig, axes = plt.subplots(nrows=2, ncols=4, figsize=(12, 12))

for i, col in enumerate(selected_columns):
    row = i // 4
    col = i % 4
    data[col].hist(ax=axes[row, col])
    axes[row, col].set_title(selected_columns[i])

plt.tight_layout()
plt.show()
```

Explanation:

In this code, we first import the necessary packages, then select the required columns from the DataFrame, and finally plot the histograms in a 4×2 grid using a loop. The *matplotlib.pyplot* package is used for plotting, and the pandas package is used for data manipulation.

Running the code returns an error.

`KeyError: 0`

To correct this error, click "debug the error".

Assistant:

The issue in the code is that the variable *col* is being overwritten within the loop, causing the KeyError when trying to access the DataFrame. To fix this error, you can use a different variable name for the column index in the loop.

<the corrected code is provided>

After running the corrected code, you can see the histograms arranged in a grid. You can further edit the code to improve visualization. For instance:

```
import matplotlib.pyplot as plt
import pandas as pd

# Selecting the columns
selected_columns = ['RoomsOCC', 'RoomsADR', 'RoomsRevPAR',
'RoomsSupply', 'RoomsDemand', 'RoomsRevenue', 'CensusPropCount',
'CensusRoomCount']

# Plotting the histograms in a 4x2 grid
fig, axes = plt.subplots(nrows=2, ncols=4, figsize=(12, 6))

for i, col in enumerate(selected_columns):
    row = i // 4
    col_index = i % 4
    data[col].hist(ax=axes[row, col_index])
    axes[row, col_index].set_title(selected_columns[i])

plt.tight_layout()
plt.show()
```

Results:

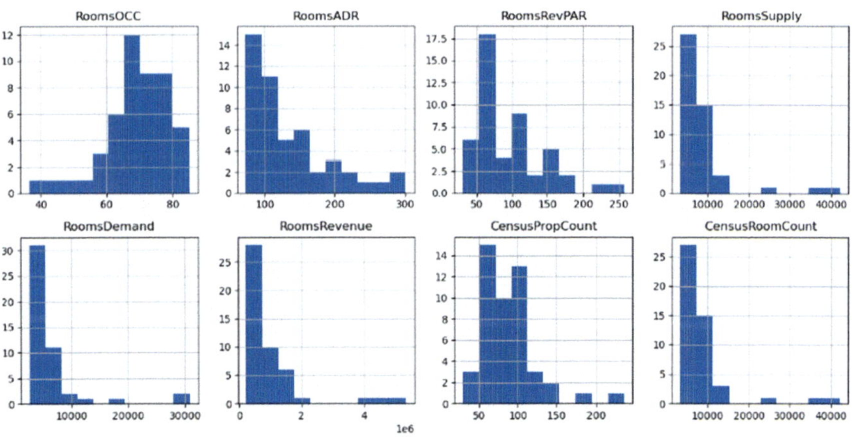

Note that data contains a potential outlier with very high room demand (variable RoomsDemand). Let us identify this submarket.

USER:

Use the DataFrame "data". Print the row with the maximum value of the column RoomsDemand.

Assistant:

5

```
# Find the index of the row with the maximum value in the
'RoomsDemand' column
max_demand_index = data['RoomsDemand'].idxmax()

# Print the row with the maximum value of 'RoomsDemand'
print(data.loc[max_demand_index])
```

Explanation:

In this code, we use the *idxmax*() function to find the index of the row with the maximum value in the *RoomsDemand* column. Then, we use the *loc* function to print the entire row with the maximum value of *RoomsDemand.*

Result:

```
Submarket          Orlando International Drive, FL
RoomsOCC                                 73.269321
RoomsADR                                141.941846
RoomsRevPAR                             105.303843
RoomsSupply                                42159.0
RoomsDemand                           30889.612903
RoomsRevenue                        4439504.709677
CensusPropCount                                122
CensusRoomCount                            42159.0
```

The potential outlier record is the International Drive submarket in Orlando, Florida. This location has a high concentration of large hotels catering to Disney World, Universal Studios, and other amusement park visitors. Other histogram plots suggest that there may be one or two other outliers. The data may require additional cleaning to remove the outliers. However, these outliers may form a separate cluster of major tourist submarkets.

Step 3. Transform Data

Clustering algorithms require all attributes to be rescaled to the same scale. The two most frequent rescaling methods are normalization and standardization. Normalization rescales data to the [-1, 1] range with zero value corresponding to the mean value. Generally, normalization is preferred when data has no outliers since outlier presence has a significant effect on the transformed values, shifting the majority of data points to an opposite edge of the [-1, 1] range. In contrast, standardization does not have a bounding edge and is more robust in outlier presence. Since our data has potential outliers, which we are unwilling to remove at this point, we will apply standardization to all numerical attributes.

USER:

Use the DataFrame "data". Use standardization to rescale columns RoomsOCC, RoomsADR, RoomsRevPAR, RoomsSupply, RoomsDemand, RoomsRevenue, CensusPropCount, and CensusRoomCount. Save results in columns stdRoomsOCC, stdRoomsADR, stdRoomsRevPAR, stdRoomsSupply, stdRoomsDemand, stdRoomsRevenue, stdCensusPropCount, and stdCensusRoomCount.

Assistant:

```
from sklearn.preprocessing import StandardScaler

# Selecting the columns to be standardized
columns_to_standardize = ['RoomsOCC', 'RoomsADR', 'RoomsRevPAR',
'RoomsSupply', 'RoomsDemand', 'RoomsRevenue', 'CensusPropCount',
'CensusRoomCount']

# Initialize the StandardScaler
scaler = StandardScaler()

# Standardize the selected columns and save the results in new columns
data[['stdRoomsOCC', 'stdRoomsADR', 'stdRoomsRevPAR',
'stdRoomsSupply', 'stdRoomsDemand', 'stdRoomsRevenue',
'stdCensusPropCount', 'stdCensusRoomCount']] =
scaler.fit_transform(data[columns_to_standardize])

# Display the updated DataFrame with standardized columns
print(data)
```

5

Explanation:

In this code, we first import the *StandardScaler* from the *sklearn.preprocessing* library. Then, we select the columns to be standardized and initialize the *StandardScaler*. We then standardize the selected columns and save the results in new columns using the *fit_transform* method. Finally, we display the updated DataFrame *data* with the standardized columns.

To visually inspect standardization outcomes, we can reuse the histogram code. Copy the code into a new cell and replace the column names with the standardized ones.

```
import matplotlib.pyplot as plt
import pandas as pd

# Selecting the columns
selected_columns = ['stdRoomsOCC', 'stdRoomsADR', 'stdRoomsRevPAR',
'stdRoomsSupply', 'stdRoomsDemand', 'stdRoomsRevenue',
'stdCensusPropCount', 'stdCensusRoomCount']

# Plotting the histograms in a 4x2 grid
fig, axes = plt.subplots(nrows=2, ncols=4, figsize=(12, 6))

for i, col in enumerate(selected_columns):
    row = i // 4
    col_index = i % 4  # Use a different variable name to avoid
overwriting 'col'
    data[col].hist(ax=axes[row, col_index])
    axes[row, col_index].set_title(selected_columns[i])

plt.tight_layout()
plt.show()
```

Result:

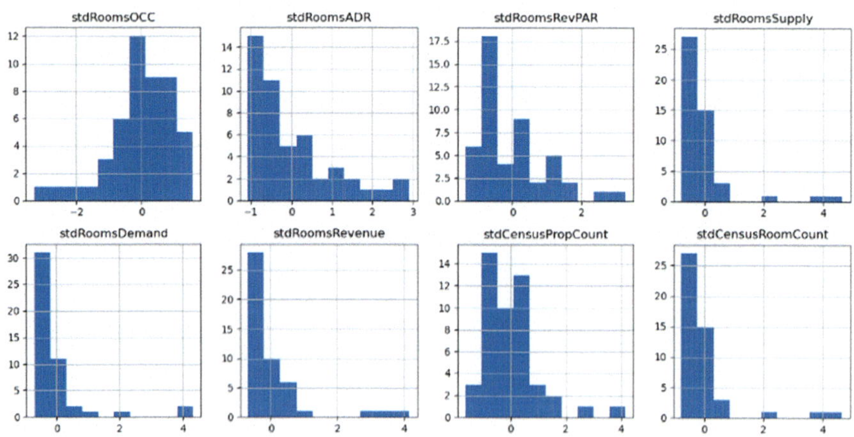

Notice that there are a few outlier instances for which room supply, demand, and revenue are four standard deviations from the mean.

Step 4. Investigate the Correlation Between the Attributes

The presence of highly correlated attributes negatively affects K-means and DBSCAN clustering performance. Therefore, it makes sense to inspect the bivariate correlations between the attributes.

USER:

Use the DataFrame "data". Print bivariate correlations between columns stdRoomsOCC, stdRoomsADR, stdRoomsRevPAR, stdRoomsSupply, stdRoomsDemand, stdRoomsRevenue, stdCensusPropCount, and stdCensusRoomCount.

Assistant:

```
# Selecting the columns for bivariate correlations
selected_columns = ['stdRoomsOCC', 'stdRoomsADR', 'stdRoomsRevPAR',
'stdRoomsSupply', 'stdRoomsDemand', 'stdRoomsRevenue',
'stdCensusPropCount', 'stdCensusRoomCount']

# Calculate the bivariate correlations
correlations = data[selected_columns].corr()

# Display the bivariate correlations
print(correlations)
```

Explanation:

In this code, we use the *corr* method to calculate the bivariate correlations between the selected columns. The resulting correlation matrix will show the relationships between the variables.

Result:

```
                    stdRoomsOCC  stdRoomsADR  stdRoomsRevPAR  stdRoomsSupply
stdRoomsOCC            1.000000     0.506454        0.646554        0.198150
stdRoomsADR            0.506454     1.000000        0.980718        0.194824
stdRoomsRevPAR         0.646554     0.980718        1.000000        0.213337
stdRoomsSupply         0.198150     0.194824        0.213337        1.000000
stdRoomsDemand         0.343662     0.252686        0.292898        0.986743
stdRoomsRevenue        0.408081     0.557425        0.573200        0.884998
stdCensusPropCount    -0.063598     0.333733        0.284566        0.317081
stdCensusRoomCount     0.198150     0.194824        0.213337        1.000000

                    stdRoomsDemand  stdRoomsRevenue  stdCensusPropCount  \
stdRoomsOCC               0.343662         0.408081           -0.063598
stdRoomsADR               0.252686         0.557425            0.333733
stdRoomsRevPAR            0.292898         0.573200            0.284566
stdRoomsSupply            0.986743         0.884998            0.317081
stdRoomsDemand            1.000000         0.907262            0.276752
stdRoomsRevenue           0.907262         1.000000            0.452659
stdCensusPropCount        0.276752         0.452659            1.000000
stdCensusRoomCount        0.986743         0.884998            0.317081

                    stdCensusRoomCount
stdRoomsOCC                   0.198150
stdRoomsADR                   0.194824
stdRoomsRevPAR                0.213337
stdRoomsSupply                1.000000
stdRoomsDemand                0.986743
stdRoomsRevenue               0.884998
stdCensusPropCount            0.317081
stdCensusRoomCount            1.000000
```

There is a very high correlation between the following attributes:

» *stdRoomsADR* and *stdRoomsRevPAR*;
stdRoomsSupply, stdRoomsDemand, stdRoomsRevenue, and *stdCensusRoomCount*.

The highly correlated attributes can be excluded, merged, or used with clustering methods specifically designed for dealing with correlated variables. For simplicity, we will leave only one variable from each highly correlated set. That leaves four attributes: *stdRoomsOCC, stdRoomsRevPAR, stdCensusPropCount,* and *stdCensusRoomCount*. To remove columns, we can use pandas method *drop*().

```
data=data.drop(columns=['stdRoomsRevPAR', 'stdRoomsSupply',
'stdRoomsDemand', 'stdRoomsRevenue'])
print(data.info())
```

Result:

```
RangeIndex: 48 entries, 0 to 47
Data columns (total 13 columns):
 #   Column              Non-Null Count  Dtype
---  ------              --------------  -----
 0   Submarket           48 non-null     object
 1   RoomsOCC            48 non-null     float64
 2   RoomsADR            48 non-null     float64
 3   RoomsRevPAR         48 non-null     float64
 4   RoomsSupply         48 non-null     float64
 5   RoomsDemand         48 non-null     float64
 6   RoomsRevenue        48 non-null     float64
 7   CensusPropCount     48 non-null     int64
 8   CensusRoomCount     48 non-null     float64
 9   stdRoomsOCC         48 non-null     float64
 10  stdRoomsADR         48 non-null     float64
 11  stdCensusPropCount  48 non-null     float64
 12  stdCensusRoomCount  48 non-null     float64
```

Part 2. Apply K-means to Cluster the Properties on Revenue Per Room and Occupancy Rate

Step 1. Use K-means with Default Parameters

USER:

Use DataFrame "data". Cluster "data" using the columns stdRoomsOCC and stdRoomsADR.

Assistant:

```
from sklearn.cluster import KMeans
import matplotlib.pyplot as plt

# Selecting the columns for clustering
selected_columns = ['stdRoomsOCC', 'stdRoomsADR']

# Extracting the selected columns for clustering
X = data[selected_columns]

# Initialize the KMeans model with the desired number of clusters
kmeans = KMeans(n_clusters=3, random_state=42)

# Fit the model to the data
kmeans.fit(X)

# Add the cluster labels to the DataFrame
data['cluster'] = kmeans.labels_

# Visualize the clusters
plt.scatter(X['stdRoomsOCC'], X['stdRoomsADR'], c=kmeans.labels_,
cmap='viridis')
plt.scatter(kmeans.cluster_centers_[:, 0], kmeans.cluster_centers_[:,
1], s=300, c='red', marker='x', label='Centroids')
plt.xlabel('stdRoomsOCC')
plt.ylabel('stdRoomsADR')
plt.title('Clustering based on stdRoomsOCC and stdRoomsADR')
plt.show()
```


Explanation:

We use the KMeans clustering algorithm from the *scikit-learn* library in Python. In this code, we first import the necessary packages and then select the columns for clustering. We then initialize the KMeans model with the desired number of clusters, fit the model to the data, and add the cluster labels to the DataFrame *data*. Finally, we visualize the clusters using a scatter plot.

Result:

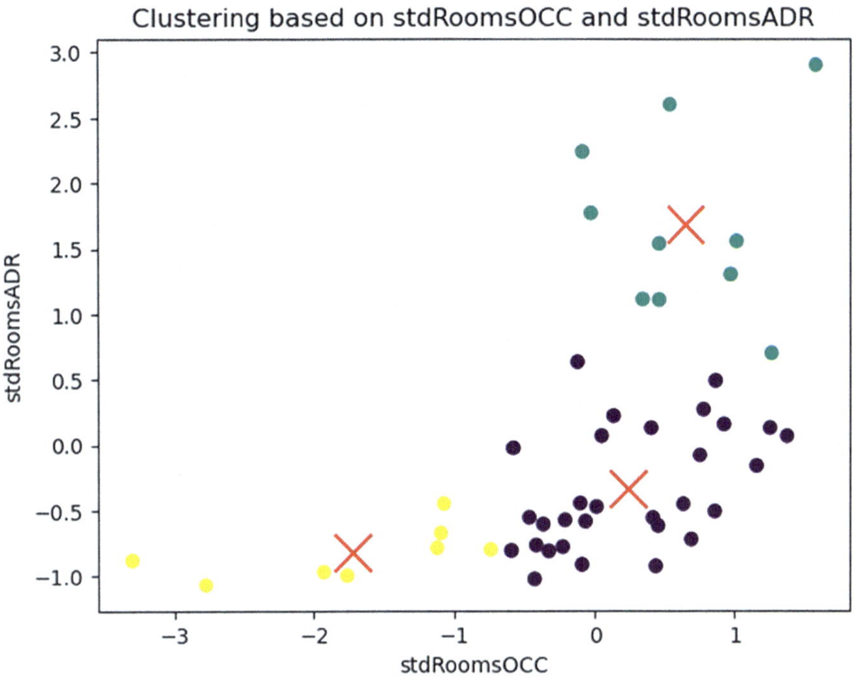

By default, K-means creates three clusters. These clusters can be interpreted as:

- Yellow: high occupancy, low room rate,
- Violette: low occupancy, low to medium room rate, and
- Cyan: high occupancy, high room rate.

In the top right corner, there is one potential outlier submarket with very high room prices and very high room occupancy rates, which are three standard deviations above the mean. Cluster membership is added to the DataFrame 'data' as a new column 'cluster':

```
data.info()
```

Result:

```
#   Column              Non-Null Count  Dtype
---  ------              --------------  -----
 0   Submarket           48 non-null     object
 1   RoomsOCC            48 non-null     float64
 2   RoomsADR            48 non-null     float64
 3   RoomsRevPAR         48 non-null     float64
 4   RoomsSupply         48 non-null     float64
 5   RoomsDemand         48 non-null     float64
 6   RoomsRevenue        48 non-null     float64
 7   CensusPropCount     48 non-null     int64
 8   CensusRoomCount     48 non-null     float64
 9   stdRoomsOCC         48 non-null     float64
 10  stdRoomsADR         48 non-null     float64
 11  stdCensusPropCount  48 non-null     float64
 12  stdCensusRoomCount  48 non-null     float64
 13  cluster             48 non-null     int32
```

To learn which submarket belongs to which cluster, we can print the relevant columns:

USER:

Use DataFrame "data". Print columns Submarket, RoomsOCC, RoomsADR, and cluster, sorted ascending by cluster and by RoomsADR.

Assistant:

```
# Print columns 'Submarket', 'RoomsOCC', 'RoomsADR', and 'cluster',
sorted ascending by 'cluster' and by 'RoomsADR'
print(data[['Submarket', 'RoomsOCC', 'RoomsADR',
'cluster']].sort_values(by=['cluster', 'RoomsADR']))
```

Explanation:

In this code, we use the *sort_values* method to sort the DataFrame *data* first by the cluster column in ascending order and then by the *RoomsADR* column. Finally, we select and print the specified columns.

Result (Abridged):

```
                          Submarket   RoomsOCC    RoomsADR  cluster
17   Jacksonville Other Areas, FL  65.012801   75.657411        0
30                      Ocala, FL  73.600835   81.325136        0
.....................................
26               Miami Beach, FL  74.615649  283.787030        1
19                  Key West, FL  85.095164  301.209355        1
.....................................

5              Daytona Beach, FL  58.607027  108.859613        2
```

Notice that the "Cyan" cluster on the plot is represented by premier beach locations. The outlier submarket we noticed earlier is Key West, which is the southmost city in the continental USA. This premium location demands the record mean daily room rate of $301 and the highest room occupancy of 85%. Inland locations mostly represent the least occupancy "Yellow" cluster. Finally, the intermediate "Violet" cluster is mostly taken by budget beach locations such as Daytona Beach and Florida Panhandle.

Step 2. Use the Elbow Method to Find the Optimum Number of Clusters

USER:

Use DataFrame "data". Your task is to cluster dataframe on columns stdRoomsOCC and stdRoomsADR. Use the elbow method to plot WCSS vs. the number of clusters. The number of clusters should vary from 2 to 10.

Assistant:

```
from sklearn.cluster import KMeans
import matplotlib.pyplot as plt

# Selecting the columns for clustering
selected_columns = ['stdRoomsOCC', 'stdRoomsADR']

# Extracting the selected columns for clustering
X = data[selected_columns]

# Initialize a list to store the WCSS values
wcss = []

# Determine the optimal number of clusters using the elbow method
for i in range(2, 11):
    kmeans = KMeans(n_clusters=i, init='k-means++', max_iter=300,
n_init=10, random_state=0)
    kmeans.fit(X)
    wcss.append(kmeans.inertia_)

# Plotting the WCSS vs. the number of clusters
plt.plot(range(2, 11), wcss)
plt.title('Elbow Method')
plt.xlabel('Number of clusters')
plt.ylabel('WCSS')
plt.show()
```

5

Explanation:

To use the elbow method to determine the optimal number of clusters for the DataFrame "data" based on the columns *stdRoomsOCC* and *stdRoomsADR*, and make a plot of WCSS (Within-Cluster Sum of Squares) vs. the number of clusters, we first import the necessary packages and then select the columns for clustering. We then use the KMeans algorithm to fit the data for different numbers of clusters (ranging from 2 to 10) and calculate the WCSS for each. Finally, we plot the WCSS vs. the number of clusters to identify the elbow point, which indicates the optimal number of clusters.

Result:

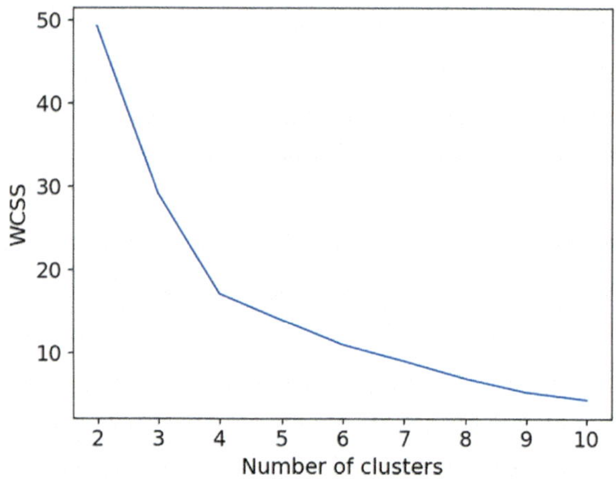

The elbow method suggests the optimal number of clusters of four.

Step 3. Use the silhouette method to find the optimum number of clusters.

USER:

Use DataFrame "data". Your task is to cluster data based on the columns stdRoomsOCC and stdRoomsADR. Compute the silhouette scores. The number of clusters should vary from 2 to 10.

Assistant:

```
from sklearn.cluster import KMeans
from sklearn.metrics import silhouette_score
import numpy as np

# Selecting the columns for clustering
selected_columns = ['stdRoomsOCC', 'stdRoomsADR']

# Extracting the selected columns for clustering
X = data[selected_columns]

# Initialize a list to store the silhouette scores
silhouette_scores = []

# Compute the silhouette scores for different numbers of clusters
for n_clusters in range(2, 11):
    kmeans = KMeans(n_clusters=n_clusters, random_state=42)
    cluster_labels = kmeans.fit_predict(X)
    silhouette_avg = silhouette_score(X, cluster_labels)
    silhouette_scores.append(silhouette_avg)

# Print the silhouette scores for each number of clusters
for i, score in enumerate(silhouette_scores):
    print("Silhouette score for {} clusters: {}".format(i+2, score))
```

Explanation:

We use the KMeans clustering algorithm and the *silhouette_score* method from the *scikit-learn* library. In this code, we first import the necessary packages and then select the columns for clustering. We then use the KMeans algorithm to fit the data for different numbers of clusters (ranging from 2 to 10) and calculate the silhouette scores for each. Finally, we print the silhouette scores for each number of clusters.

Result:

```
Silhouette score for 2 clusters: 0.43416518766799905
Silhouette score for 3 clusters: 0.43173298749161954
Silhouette score for 4 clusters: 0.45812841463254644
Silhouette score for 5 clusters: 0.42099598288227175
Silhouette score for 6 clusters: 0.3843489742622739
Silhouette score for 7 clusters: 0.4404824712164917
Silhouette score for 8 clusters: 0.4610757468979679
Silhouette score for 9 clusters: 0.45588189027014375
Silhouette score for 10 clusters: 0.4442797515166023
```

There is a local maximum of the silhouette score when the number of clusters is four and another one at eight clusters. Considering that the elbow method's estimation was four, we accept four as the optimum number of clusters regarding how tight the data points are organized around cluster centroids.

Step 3. Use K-means with the Number of Clusters Set at Four. Compute WCSS and Draw a Silhouette Plot

USER:

Use DataFrame "data". Cluster the DataFrame on columns stdRoomsOCC and stdRoomsADR. Set the number of clusters to 4. Save cluster labels in column cluster. Compute WCSS. Make a silhouette plot.

Assistant:

```
from sklearn.cluster import KMeans
from sklearn.metrics import silhouette_samples, silhouette_score
import matplotlib.pyplot as plt
import numpy as np

# Selecting the columns for clustering
selected_columns = ['stdRoomsOCC', 'stdRoomsADR']

# Extracting the selected columns for clustering
X = data[selected_columns]

# Initialize the KMeans model with 4 clusters
kmeans = KMeans(n_clusters=4, random_state=42)
data['cluster'] = kmeans.fit_predict(X)

# Compute the WCSS
wcss = kmeans.inertia_
print("WCSS:", wcss)

# Compute the silhouette scores
silhouette_avg = silhouette_score(X, data['cluster'])
print("Silhouette score:", silhouette_avg)

# Create a silhouette plot
sample_silhouette_values = silhouette_samples(X, data['cluster'])
y_lower = 10
for i in range(4):
    ith_cluster_silhouette_values =
sample_silhouette_values[data['cluster'] == i]
    ith_cluster_silhouette_values.sort()
    size_cluster_i = ith_cluster_silhouette_values.shape[0]
    y_upper = y_lower + size_cluster_i
    color = plt.cm.nipy_spectral(float(i) / 4)
    plt.fill_betweenx(np.arange(y_lower, y_upper), 0,
ith_cluster_silhouette_values, facecolor=color, edgecolor=color,
alpha=0.7)
    y_lower = y_upper + 10

plt.title("Silhouette plot for 4 clusters")
plt.xlabel("Silhouette coefficient values")
plt.ylabel("Cluster label")
plt.axvline(x=silhouette_avg, color="red", linestyle="--")
plt.yticks([])
plt.show()
```

Explanation:

In this code, we first import the necessary packages and then select the columns for clustering. We then use the KMeans algorithm to fit the data with 4 clusters, save the cluster labels in a new column *cluster*, compute the WCSS, and calculate the silhouette scores. Finally, we create a silhouette plot to visualize the silhouette coefficients for each sample and the average silhouette score.

Result:

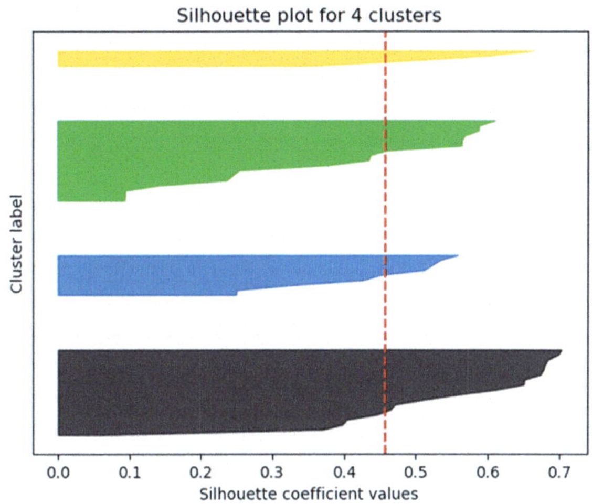

There are one small and three larger clusters. The silhouette score of 0.46 indicates poor to moderate clustering.

To investigate cluster distribution and membership, we can reuse the code from Step 1 or ask the Assistant. First, we visualize the clusters.

USER:

Visualize the clusters.

Assistant:

```
# Visualize the clusters
plt.scatter(X['stdRoomsOCC'], X['stdRoomsADR'], c=cluster_labels,
cmap='viridis')
plt.scatter(kmeans.cluster_centers_[:, 0], kmeans.cluster_centers_[:,
1], s=300, c='red', marker='x', label='Centroids')
plt.xlabel('stdRoomsOCC')
plt.ylabel('stdRoomsADR')
plt.title('Clustering based on stdRoomsOCC and stdRoomsADR')
plt.show()
```

Explanation:

In this code, we use the *scatter* method from *matplotlib* to create a scatter plot of the clusters. The points are colored based on their assigned cluster labels, and the centroids of the clusters are marked in red. This visualization helps to understand how the data points are grouped into clusters based on the specified columns.

Result:

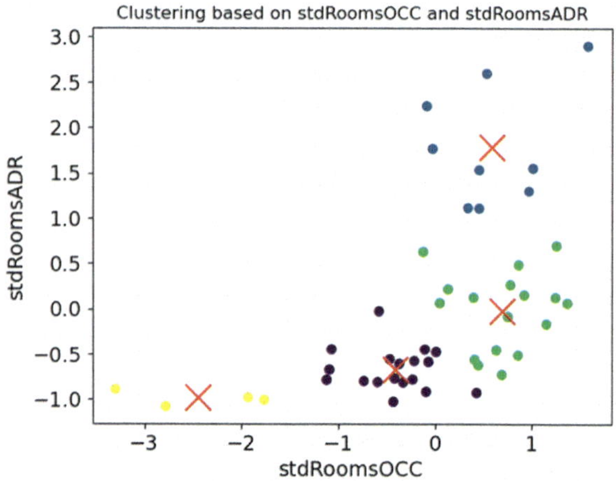

User:

Use DataFrame "data". Print columns Submarket, RoomsOCC, RoomsADR, and cluster, sorted ascending by cluster and by RoomsADR.

Assistant:

```
# Print columns 'Submarket', 'RoomsOCC', 'RoomsADR', and 'cluster',
sorted ascending by 'cluster' and by 'RoomsADR'
print(data[['Submarket', 'RoomsOCC', 'RoomsADR',
'cluster']].sort_values(by=['cluster', 'RoomsADR']))
```

Explanation:

In this code, we use the *sort_values* method to sort the DataFrame first by the *cluster* column in ascending order and then by the *RoomsADR* column. Finally, we select and print the specified columns.

Result (transformed to fit plot colors; colors represent four clusters):

Submarket	OCC	ADR	Submarket	OCC	ADR
Panama City	42	73	Orlando North	76	93
Pensacola	52	77	Melbourne/Palm Bay	74	99
Florida Panhandle Area	50	79	Bradenton/Airport	73	102
Fort Walton Beach	36	84	Fort Myers/Bonita Springs	78	105
Jacksonville Other Areas	65	76	Tampa East	76	108
Ocala	74	81	West Broward/Plantation	81	125
Jacksonville Bay	68	82	Orlando South	77	130
Kissimmee East	66	88	Miami Airport/Civic Center	83	138
Kissimmee West	63	88	Clearwater	70	138
Lakeland/Winter Haven	62	89	Orlando International Dr	73	142
Tallahassee	58	89	Lake Buena Vista	82	142
Jacksonville/Airport	67	90	Coral Springs/Pompano B	79	144
Gainesville	65	90	St Petersburg	71	147
Daytona Area	58	96	Miami South	77	150
Florida Central Area	66	100	Tampa CBD/Airport	78	163
Tarpon Springs	69	101	Sarasota/Beaches	68	171
Tampa North	67	101	Hollywood/Airport	82	175
Titusville/Cocoa Beach	65	103	Boca Raton	74	198
Fort Pierce/Port St Lucie	69	107	Fort Myers Beach	73	198
Orlando Central	68	109	West Palm Beach	79	209
Daytona Beach	59	109	Miami CBD/North	74	223
Jacksonville Beaches	64	133	Fort Lauderdale/Beach	79	224
			Upper Florida Keys	69	236
			Naples	68	263
			Miami Beach	75	284
			Key West	85	301

Compared to the initial clustering, the markets are better separated. The Florida Panhandle coastal destination entirely takes a new, very small "yellow" cluster. This is a budget beach holiday location. The "violet" cluster is mostly represented by inland locations, more likely to serve business rather than holiday travelers. A few misclassifications, such as Daytona Beach and Jacksonville Beaches, are on the border with the next cluster.

The "cyan" cluster mostly includes large urban areas such as Orlando, Miami, Tampa/St. Petersburg and Ft. Myers, catering for holiday and business travelers. Similarly to the "violet" cluster, there are a few possible misclassifications: Coral Springs and Sarasota. Finally, the "blue" cluster is mostly represented by premier beach locations such as Key West, Naples, and Miami Beach. Successful interpretation allows us to accept the 4-cluster solution finally.

Part 3. On Your Own

We used only two attributes for clustering. Repeat the clustering using all four numerical attributes. Determine the optimal number of clusters. For plotting the clusters, ask the Assistant to plot cluster assignments for all combinations of columns *stdRoomsRevPAR, stdRoomsSupply, stdRoomsDemand,* and *stdRoomsRevenue*. Interpret clustering results.

Improve Python Skills

Chapters 10–12 of the free book Learn Python with Jupyter by Serena Bonaretti. ▶ https://www.learnpythonwithjupyter.com/

Additional Reading

Tan, P.N., Steinbach, M., & Kumar, V. (2019). Chapter 7. Cluster analysis: Basic concepts and algorithms. In *Introduction to data mining* (2nd ed.). Pearson. ▶ https://www-users.cse.umn.edu/~kumar001/dmbook/ch7_clustering.pdf

6 Hierarchical Clustering

A. P. Kirilenko, *Practical Data Mining with AI for Social Scientists*,
Springer Texts in Social Sciences, https://doi.org/10.1007/978-3-031-89689-7_6

6.1 Introduction to Hierarchical Clustering

In the previous chapter, we discussed techniques for finding a pattern in data point attributes and using these findings to place data points in distinctive *flat* clusters. The word "flat" means no structure relating one cluster to another. Sometimes, however, we want more details on how clusters are structured internally and how they relate. This is where the hierarchical clusters will help.

Consider the Florida submarkets problem from the ▶ Chap. 5 lab. In this lab, we used K-means clustering to split Florida hotel industry submarkets into four large groups according to the average room price and room occupation rate (the daily average percentage of available rooms that are actually occupied):

Budget destination in Florida Panhandle

Budget Florida Panhandle beach	**Inland non-holiday + economy beach**	**Mostly large urban areas and Disney**	**Premier beach locations**
Panama City	Titusville/ Cocoa Beach	Miami South	Upper Florida Keys
Pensa-cola	Orlando Central	Tampa CBD/Air-port	Naples
Florida Panhan-dle	Daytona Beach	Sarasota/ Beaches	Miami Beach
Fort Wal-ton Beach	Jacksonville Beaches	Hollywood/ Airport	Key West

Premier destination: Key West, Florida

Notice from cluster titles that cluster separation is far from perfect in this table. Indeed, there is considerable fuzziness in cluster organization. The hierarchical clustering shows exactly how far or close the markets are to one another (◘ Fig. 6.1).

Instead of a "flat" cluster designation, hierarchical clustering creates a tree of clusters, called *a dendrogram,* which visualizes the hierarchical relationship of the data objects. The vertical axis of a dendrogram represents the distance between the data points found on the horizontal line. This can be better observed using the logarithmic scale (◘ Fig. 6.2). For example, the low location of the horizontal line connecting two data points on the ex-

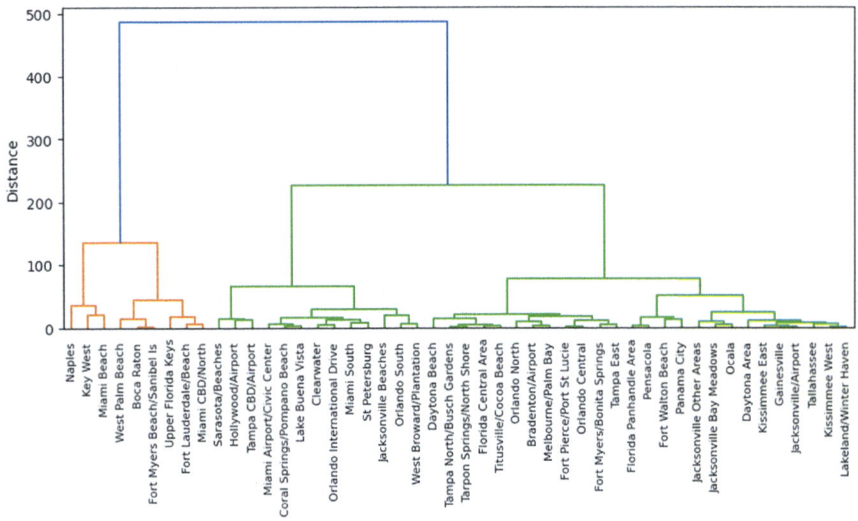

Fig. 6.1 Hierarchical clustering of Florida submarkets

treme right of the plot tells that the distance between Kissimmee West and Lakeland/Winter Haven submarkets (in terms of room occupancy and price) is just slightly about 1 unit. In comparison, the distance between the green cluster that includes both Kissimmee and Winter Haven and the cluster of premium beach locations on the extreme left of the plot (drawn on the right) is almost 500 units.

The dendrogram produced by the hierarchical clustering can be transformed into flat clusters by "cutting" the dendrogram at the desired level, as shown in Fig. 6.2. For example, cutting the dendrogram at Distance = 300 level (top dashed line) produces two large clusters visualized in orange and green. In comparison, four clusters similar to those in ▶ Chap. 5 K-means example can be produced by cutting the dendrogram at Distance = 100 level (middle dashed line). An even larger number of clusters can be produced by cutting at lower levels (bottom dashed line). Hence, the hierarchical clustering method does not require a pre-specified number of clusters, which is distinct from k-means clustering. Instead, the number of clusters is determined by cutting the dendrogram as advised by theory and your expert field knowledge.

Dendrogram

A dendrogram is a tree-like diagram that illustrates the structure of data points clusters produced by hierarchical clustering. Each branch of the tree relates to a cluster of similar data points. The vertical axis of a dendrogram represents the distance between clusters.

Flattening the clusters

The dendrogram produced by the hierarchical clustering can be transformed into flat clusters for better interpretability.

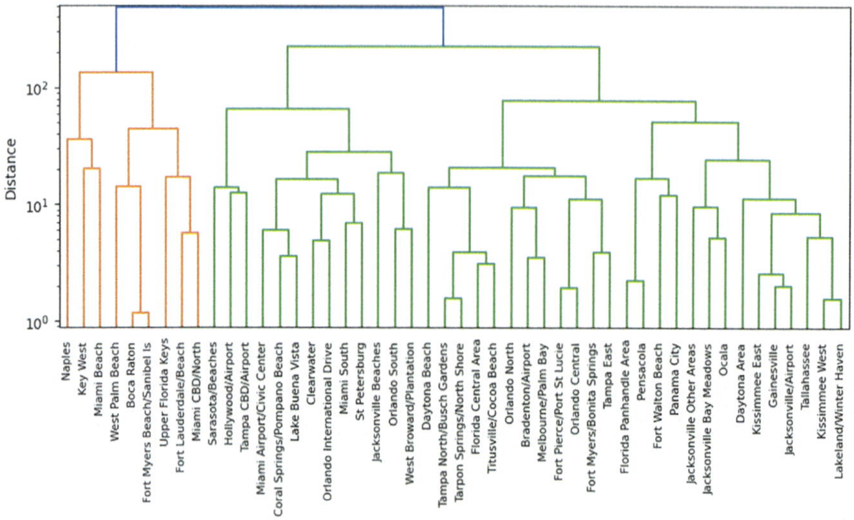

Fig. 6.2 Creating flat clusters from the dendrograms. The vertical scale is logarithmic for easier reading. The dash lines cutting the dendrogram at different distances between the clusters can create two large clusters (the top dash line), four clusters (the middle line), or seven clusters (the bottom line)

The connections between the clusters of data points, visualized by the horizontal lines on the plots, can be produced by two broad hierarchical clustering approaches, agglomerative and divisive. Agglomerative "bottom-up" clustering starts with designating each data point as a separate cluster. Then, these clusters are merged into successively larger clusters. In the end, all clusters are merged into one dendrogram. Conversely, divisive "top-down" clustering starts with all data points in one cluster and then progressively partitions them into smaller clusters. The next chapter will discuss how the clustering process works in practice and which distance measures are being used.

Bottom-up and top-down approaches

The agglomerative clustering starts with designating each data point as a separate cluster and then joins the most similar points and clusters into a single hierarchy. The divisive clustering starts with one large cluster unifying all data points and then partitions the most dissimilar parts.

6.2 Agglomerative Clustering Mechanism

In hierarchical clustering, the agglomerative approach is more popular than the divisive one, primarily due to its computational efficiency. It starts with each data point as an individual cluster. Then, the algorithm searches for two clusters separated by the shortest distance and merges them into a larger cluster. On a dendrogram, these two clusters will be connected with the lowest-positioned horizontal line. In ◘ Fig. 6.2, they are Boca Raton and Fort Myers Beach. Then, the next two closest clusters are found and connected (Tampa North and Tarpon Springs), and so on, until everything is connected into one large cluster.

Notice that since we started with clusters containing only a single data point, the similarity between these one-point clusters can be measured using distance measures identical to those used in ▶ Chap. 5 for flat clustering. However, as the algorithm progresses, the multi-point clusters are being merged. This raises the question of how to define the distance between complex, multi-point clusters. There are multiple ways of doing that.

6.2.1 Cluster Distance

The most common cluster distance measures (frequently referred to as linkage criteria) include single linkage, complete linkage, average linkage, and Ward's method (◘ Fig. 6.3):

- Single linkage: the distance between the closest points in two clusters;
- Complete linkage: the distance between the farthest points;
- Average linkage: mean distance between all pairs of points in two clusters;
- Centroid linkage: the distance between the centroids of each cluster. A similar method is medoid linkage.
- Ward's linkage: squared error increase resulting from merging two clusters. Ward’s method is distinctly different from the other three and resembles the K-Means approach: the clusters are merged to minimize the clustering squared error.

Hence, hierarchical clustering adds a new level of complexity to selecting the distance measure compared to conventional flat clustering. In addition to defining the similarity between the data points, we also need to define how this distance measure is used to measure cluster similarity. Each of the five introduced ways of measuring cluster similarity is better suited to some data patterns and poorly to others. In addition, some cluster similarity measures can be used only in combination with specific point similarity measures.

Common cluster distance measures

Single linkage (Min)
Complete linkage (Max)

Average linkage (Mean)

Centroid linkage (geometrical center)

Medoid linkage (cluster point with the least distance to other cluster points)

Ward's linkage (squared error increase

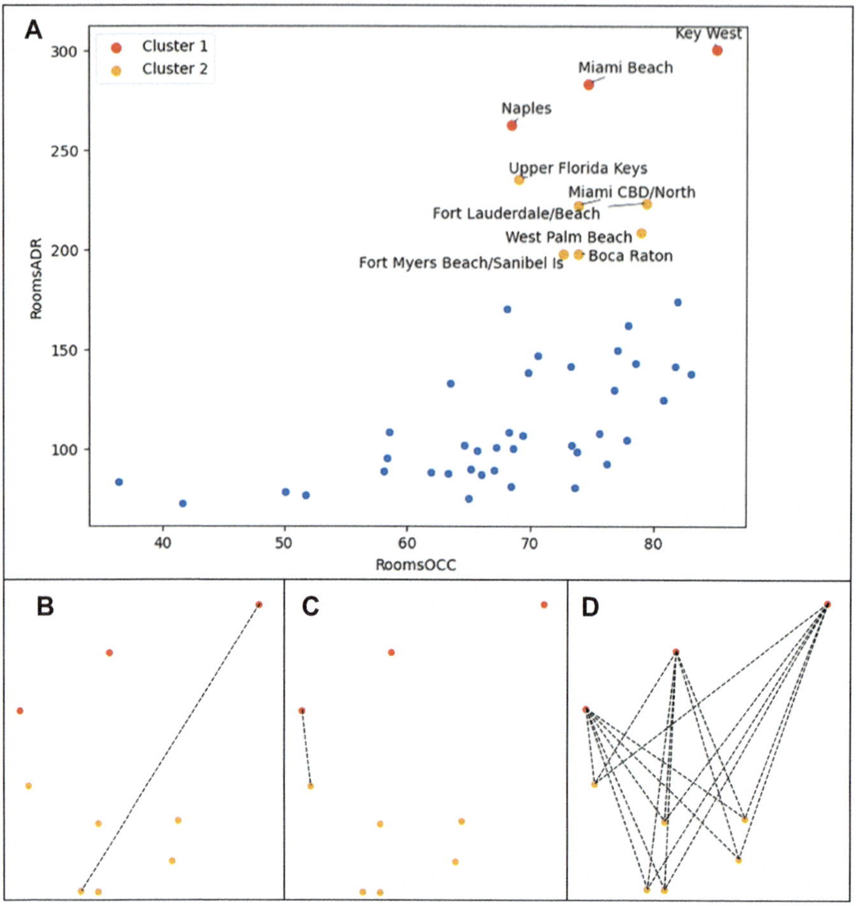

Fig. 6.3 Cluster linkage examples. **A** Two submarket clusters shown in orange on Fig. 6.2. **B** Single linkage; **C** Complete linkage; **D** Average linkage

6.2.2 Selecting Cluster Similarity Measure

The following data considerations are applied to cluster similarity measures that can be used with any point distance measures.

- The single linkage is sensitive to outliers and can result in elongated, chain-like clusters. The method minimizes the distance between the closest members of the merged clusters, while the distance between the farthest members can be high.
- Complete linkage tends to produce spherical clusters, which may or may not suit the data. The method minimizes the distance between the most remote members

of the merged clusters, but clusters' "insides" may be loose.
- Average linkage is less prone to either of these biases and is usually set as a default method in software packages. Still, the method may not work well when the data pattern is not globular in shape.

Other cluster similarity measures should be only used with Euclidean distance:
- Centroid and medoid measures are less computationally demanding than average linkage. Similarly to the average linkage, these methods are also biased towards globular clusters. In addition, inside the clusters, they tend to form "factions" of data points equidistant from the cluster's center.
- Ward's method tends to produce clusters denser towards the center with lesser density towards the clusters' periphery. This method is similar to the K-Means clustering we covered in ▶ Chap. 5.

The clusters formed using those measures may substantially differ from one another. ◻ Figure 6.4 demonstrates how different linkage measures produce distinctly different clusters. The single linkage clusters produced chains of subclusters with progressively tighter organization. The complete linkage produced "clumps" of subclusters. In comparison, the average linkage resulted in several clumps of chained subclusters. Despite this visual divergence, the most important features remain the same: one or two distinct subclusters unifying premier beach destinations and a subcluster representing economy beach holidays. Between these two poles, the cluster membership of submarkets varies, providing a degree of flexibility for an analyst to implement their vision of the hospitality industry.

6.2.3 Flattening Dendrograms

The advantage of hierarchical clustering is that it represents data as a visually clear structure of progressively more similar data points. This two-dimensional dendrogram provides a simple overview of patterns hidden in the entire dataset. However, after the exploratory analysis is finished, we frequently want a one-dimensional "flat-

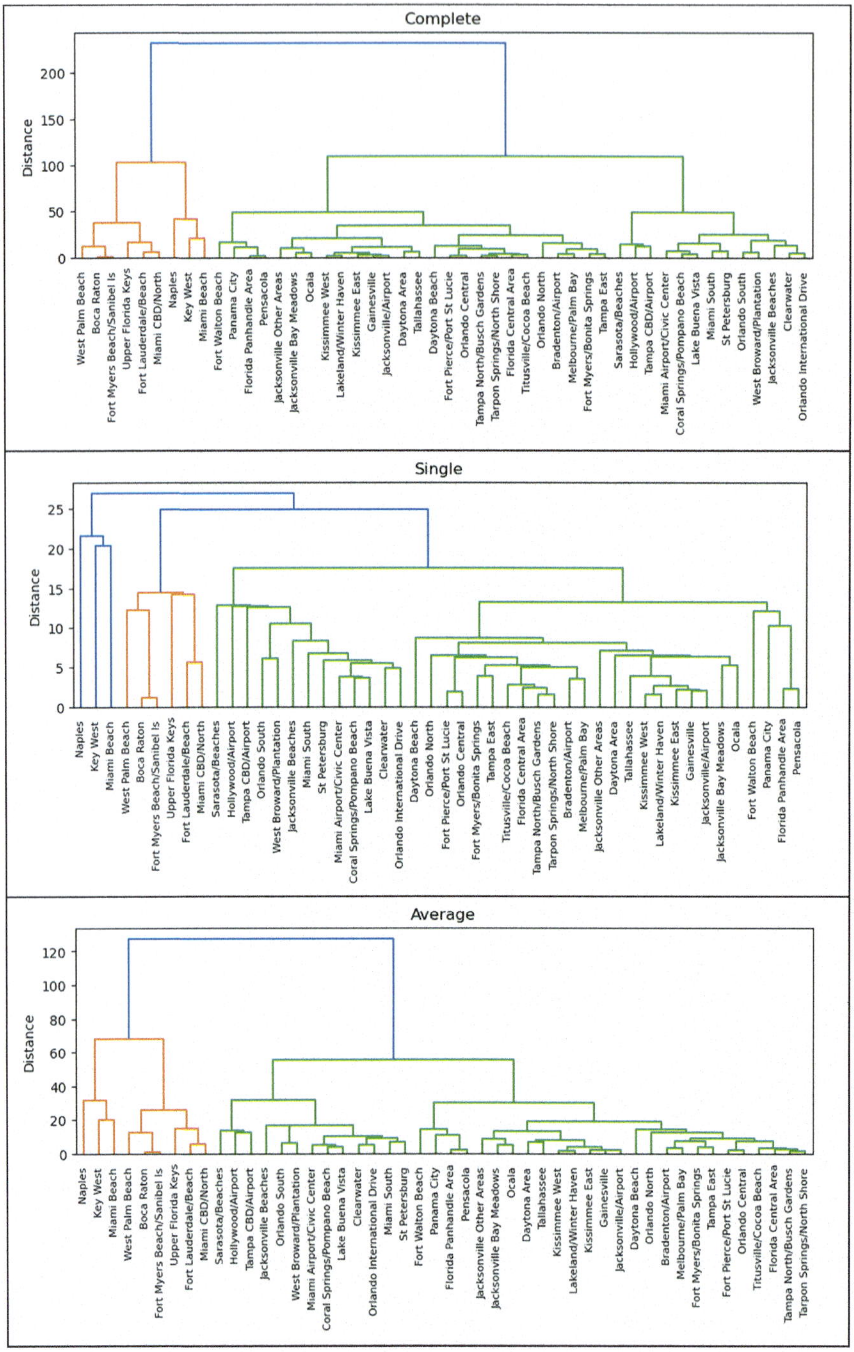

Fig. 6.4 An effect of cluster distance measures on outcomes of hierarchical clustering

tened" representation of clusters suitable for practical applications such as submarket merging.

One way of doing that is to cut the dendrogram at a straight line designated a specific distance threshold, as shown in ◘ Fig. 6.2. The subtrees below the cutting line are flattened into separate clusters. The cutting threshold can be selected to produce the required number of clusters N. Another approach is to be more flexible and find an "optimal" way of cutting the dendrogram.

We have already learned that in data mining, the word "optimal" is operationalized as minimizing (or maximizing) the value of a certain function. The role of this function called a loss or cost function, is to quantify the "penalty" of replacing the observed data points with model estimations. For example, in linear regression, the most commonly used loss function is the Mean Squared Error (MSE). The MSE quantifies the "cost" of replacing the observations with straight-line estimations of a linear regression model with coefficients that minimize the MSE.

Hence, we can formulate the problem of optimal dendrogram cutting into N sub-trees in the following way: search for N or fewer disjoint sub-trees such that the loss function is minimized. One commonly used loss function is the within-cluster sum of squares (WCSS), but other loss functions, including those discussed in ▶ Chap. 5, can also be used. Like K-Means, the number of sub-trees N may reflect your expert opinion. When N is unknown, it can be included in the optimization problem. In this chapter's lab, we will explore some methods of cutting a dendrogram using different approaches.

6.3 Feature Selection

The submarkets example uses only two independent variables (recall that independent variables are called attributes or features in data mining) to cluster the data points. This is a rare exception: commonly, we have to deal with dozens or even hundreds or thousands of features. This presents a problem for cluster analysis. From a computational point of view, a ten times increase in the number of attributes leads to a 100–1000 times increase in computational requirements, depending on the methods. Methodologically, cluster algorithms frequently don't account for the relative importance of different features, leading to potential biases towards less important attrib-

Feature selection

The process of reducing the number of features to consider in cluster analysis is called feature selection.

utes. Practically, a large number of features complicate cluster visualization and interpretation of results. This is why reducing the number of attributes before cluster analysis is advantageous.

Reducing the number of input variables is called feature selection. Basic feature selection can be done using expert knowledge. However, it is beneficial to apply a more formal feature selection process.

Feature selection can be formalized as finding a small (if possible, the smallest) subset or recombination of features that save "interesting" clusters of data points. One approach is the Principal Component Analysis (PCA) or another dimension reduction method. The PCA idea is to replace N *correlated* variables with a smaller number of new *uncorrelated* variables called principal components. Commonly, the first few principal components account for a large share of variance; hence, it is possible to replace the original attributes with the first n principal components (n << N). Even though there are methodological objections to applying PCA to feature selection for cluster analysis, it is used frequently in practice. A significant drawback, however, is that it is hard to understand the resulting clusters. Because of that, it is preferable to use a subset of the original features instead of their recombinations provided by PCA and similar dimension reduction methods.

Filter-based methods filter out features using certain criteria—for example, attributes with little variation or high correlation with others. For categorical or a mix of categorical and interval/ratio variables, ANOVA or the Chi-Squared test can be used in place of regression. Several algorithms, such as the Fast Correlation Based Filter (FCBF), consider relationships between multiple variables.

Wrapper-type methods represent a different feature selection approach. These methods "wrap" around the clustering algorithm (or another data mining method), executing it with different subsets of features. The best combination of features is then selected for the final clustering. This method is extremely computationally expensive, which explains why there are multiple algorithms realizing heuristic search strategies. The hybrid methods combine the best sides of filter- and wrapper-based feature selection methods. We will cover feature selection in greater detail later in the course since many methods apply only to supervised learning.

6.4 Methodological Considerations

Hierarchical clustering is primarily an exploratory tool that assists in understanding data structure. Its sensitivity to the selection of distance metric and linkage criterion is both a benefit and a limitation of the method. The choice of clustering parameters can significantly influence the resulting dendrogram, flattened clusters, and, consequently, their interpretation. On the one hand, the flexibility of the method allows its application to a wide variety of data patterns. Conversely, it can be easily misused when the parameters are selected to justify desired results. Importantly, the selection of clustering parameters should be guided by the data's nature and the research question's specifics. Expert field knowledge should aid this process but never replace it.

A correctly selected method …

Is suitable for attribute data types and measures

Returns robust results

Produces interpretable clusters

Not sensitive to outlier presence if outlier exclusion is unjustified

Realistic in terms of computational load

So, which considerations should be taken into account when deciding on specific methods to use? The following list constitutes some informal points to address.

- Cluster interpretability. In preparing the submarket example, I selected the methods producing easily interpretable clusters based on hotels' location (shore or inland), attraction points (beach, amusement parks, or business centers), and population size.
- Cluster shapes. The shape of the clusters should answer the project's requirements. Densely organized, well-separated clusters are frequently preferable.
- Data types and measures. Having categorical data reduces the choice of available methods. A mix of categorical and interval/ratio data radically reduces the choice of methods. Binary data require special treatment. Big data clustering needs algorithms specifically designed for working with large datasets.
- Robustness of clustering. Adding or dropping a few data points should not radically change the clusters.
- Outlier sensitivity. Many methods used in hierarchical clustering are sensitive to data outliers. Outliers distort the distance measures. This leads to subtrees that inaccurately represent the underlying data structure. On the other hand, an outlier presence is not necessarily unwanted. In the Florida hotel submarket example, the Key West hotels might be considered an outlier based on their extremely high price and room occupancy rate. Yet, they are also reflective of the state's market. When undesirable, outlier sensitivity is

ameliorated by outlier management during the data cleaning step.
- Computational requirement. Hierarchical clustering is computationally intensive, especially when a distance measure requiring similarity between multiple pairs of data points, such as the average linkage, is used. This limits the method's applicability to large datasets, such as text and image analysis. In addition, a large number of data points reduces the researcher's ability to visually analyze the dendrogram.

In conclusion, hierarchical clustering is primarily an exploratory method of data analysis widely used in a huge variety of applications to search for the underlying hierarchy of data patterns. There are many approaches and algorithms that allow customization of clustering methods to suit a particular task. We have described only a few of those approaches dealing with numerical data. Meanwhile, clustering can also be applied to text analysis, helping understand the main topics of documents, publications, social media discussions, and newspaper articles. Another important area of clustering is image segmentation and image grouping based on visual similarity. Using cluster analysis, study participants can be categorized for individual in-depth study. Despite the seeming simplicity of clustering ideas, this multitude of approaches may be a stumbling block for new data analysts.

6.5 Review Questions

1. What is the main difference between flat clustering and hierarchical clustering?
2. Explain the dendrogram's representation of clusters.
3. Describe the agglomerative approach to hierarchical clustering.
4. What is the process of merging clusters in agglomerative clustering?
5. Describe the common cluster distance measures used in hierarchical clustering.
6. How does Ward's method differ from other linkage criteria?
7. What are some data considerations for selecting a cluster similarity measure?
8. Explain how to cut a dendrogram to create flat clusters.

9. What is the role of a loss function in optimizing the cutting of a dendrogram?
10. Why is feature selection important in cluster analysis?
11. Which factors should be considered when choosing specific methods for hierarchical clustering?

6.6 Lab

Objectives

1. Apply hierarchical cluster analysis to cluster hotel markets
2. Acquire code markdown skills
3. Practice filter-based feature selection
4. Practice cluster flattening

Data

Similar to ▶ Chap. 5, the data for this chapter was made available by STR, Inc. The dataset contains averaged daily hotel industry performance data for January 2017 for Florida submarkets and is stored at ▶ https://github.com/andykir/DataMiningLectures/raw/main/STR_Florida_Submarkets_January2017.xlsx.

Variables

Variable	Explanation	Notes
Location	Market name	Text variable
RoomsOCC	Room occupancy	Mean percentage of occupancy in hotel rooms
RoomsADR	Average Daily Rate	Average rental income per paid occupied room
RoomsRevPAR	Average revenue per available room	Hotel's AverageADR multiplied by the occupancy rate
RoomsSupply	Room supply	Average number of rooms available for sale

Variable	Explanation	Notes
RoomsDemand	Room demand	Average number of rooms sold, excluding the complimentary stays.
RoomsRevenue	Daily rooms revenue	Average revenue per room, excluding taxes
CensusProp-Count	Number of properties	Total number of hotels in the market
CensusRoom-Count	Number of rooms	Total number of hotel rooms in the market

Part 1. Read and Clean the Data

Step 1. Reuse Lab 5 Code for Reading and Cleaning the Data

This step is identical to Lab 5; hence, we will reuse the code.

In the Anaconda environment, open the file browser (folder icon at the top left of the screen). Navigate to the folder where you saved your Lab 5 notebook and right-click on the code file. Select "duplicate." A new notebook with duplicated code will be created. Right-click on the notebook and select "rename." Rename the notebook as you see fit. Open the notebook.

Note: if the file browser window is closed, use the View tab of the top menu to open it.

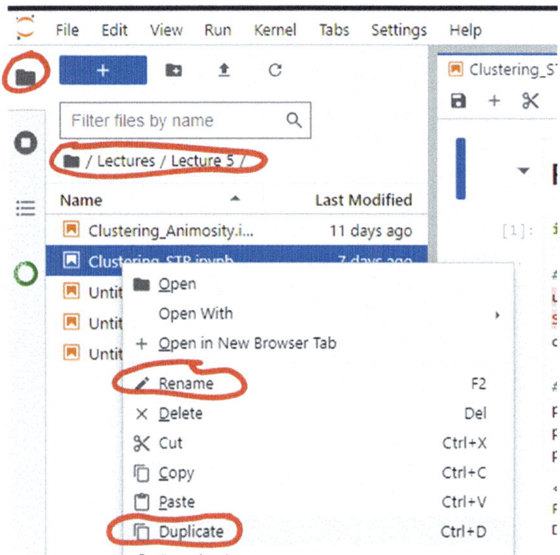

Step 2

Edit the code. Leave only the part relevant to data reading, cleaning, describing, and transformation. Recall that data should be rescaled to a unified scale before clustering; standardization or normalization are normally used. You can select the irrelevant part of the code and then delete it by pressing the key "D" twice.

You will need the following cells:

Read data

```
import pandas as pd

# Read the data from the URL
url = "https://github.com/andykir/DataMiningLectures/raw/main/\
STR_Florida_Submarkets_January2017.xlsx"
data = pd.read_excel(url)

# Display information on data
print(data.info())
print(data.head())
print(data.describe())
```

Scale data

```
from sklearn.preprocessing import StandardScaler

# Selecting the columns to be standardized
columns_to_standardize = ['RoomsOCC', 'RoomsADR', 'RoomsRevPAR',
'RoomsSupply', 'RoomsDemand', 'RoomsRevenue', 'CensusPropCount',
'CensusRoomCount']

# Initialize the StandardScaler
scaler = StandardScaler()

# Standardize the selected columns and save the results in new columns
data[['stdRoomsOCC', 'stdRoomsADR', 'stdRoomsRevPAR',
'stdRoomsSupply', 'stdRoomsDemand', 'stdRoomsRevenue',
'stdCensusPropCount', 'stdCensusRoomCount']] =
scaler.fit_transform(data[columns_to_standardize])
```


Draw a histogram of scaled data

```
import matplotlib.pyplot as plt

# Selecting the columns
selected_columns = ['stdRoomsOCC', 'stdRoomsADR', 'stdRoomsRevPAR',
'stdRoomsSupply', 'stdRoomsDemand', 'stdRoomsRevenue',
'stdCensusPropCount', 'stdCensusRoomCount']

# Plotting the histograms in a 4x2 grid
fig, axes = plt.subplots(nrows=2, ncols=4, figsize=(12, 6))

for i, col in enumerate(selected_columns):
    row = i // 4
    col_index = i % 4
    data[col].hist(ax=axes[row, col_index])
    axes[row, col_index].set_title(selected_columns[i])

plt.tight_layout()
plt.show()
```

Step 3. Execute the Code

Make sure there are no errors and that you can observe histograms of scaled data.

Part 2. Introduce the Markdown Language

The following screenshot shows what the notebook for this lab looks like on my computer screen.

Read data

```
[2]: import pandas as pd

     # Read the data from the URL
     url = "https://github.com/andykir/DataMiningLectures/raw/main/\
     STR_Florida_Submarkets_January2017.xlsx"
     data = pd.read_excel(url)

     # Display information on data
     print(data.info())
     print(data.head())
     print(data.describe())
```

```
                     Submarket   RoomsOCC    RoomsADR  RoomsRevPAR  \
0                   Boca Raton  73.845140  198.241223   146.825764
1            Bradenton/Airport  73.392482  102.461737    75.704395
2                   Clearwater  69.779782  138.459483    97.520294
3  Coral Springs/Pompano Beach  78.526420  143.524679   112.996542
4                 Daytona Area  58.392798   95.931154    58.442029

   RoomsSupply  RoomsDemand  RoomsRevenue  CensusPropCount  CensusRoomCount
0       7194.0  5312.419355  1.056265e+06               75           7194.0
1       4043.0  2967.258065  3.060729e+05               56           4043.0
2       8225.0  5739.387097  8.021044e+05              105           8225.0
3       6865.0  5390.838710  7.757213e+05              111           6865.0
4       5375.0  3138.612903  3.141259e+05               94           5375.0
        RoomsOCC    RoomsADR  RoomsRevPAR  RoomsSupply  RoomsDemand  \
count  48.000000   48.000000    48.000000    48.000000    48.000000
mean   69.271686  134.479786    96.843170  9104.226478  6447.801075
std    10.086676   58.094215    49.618386  7212.221366  5701.869062
min    36.359739   73.176363    30.658640  3646.000000  2502.516129
25%    65.099393   90.232835    60.127817  5982.750000  3649.927419
```

Scale data

```
[3]: from sklearn.preprocessing import StandardScaler

     # Assuming 'data' is the DataFrame containing the required data
```

The highlighted "Markdown" cells provide important notebook functionality for incorporating texts into computer code. This feature helps explain what parts of the code are designed to accomplish on a conceptual level. Markdown is the name of a markup language designed to make it easy to write and edit formatted texts. Compared to HTML, which is designed for Web pages, the Markdown has fewer options but is easier to use.

Anaconda notebook is a sequence of code and markdown cells. The code cells provide functionality, and the markdown cells are used for narrations explaining the computer code and its results. To add a markdown cell to the notebook, one can simply add a new cell and then change its type to Markdown. After adding text to the markdown cell, click the "Run this cell" button.

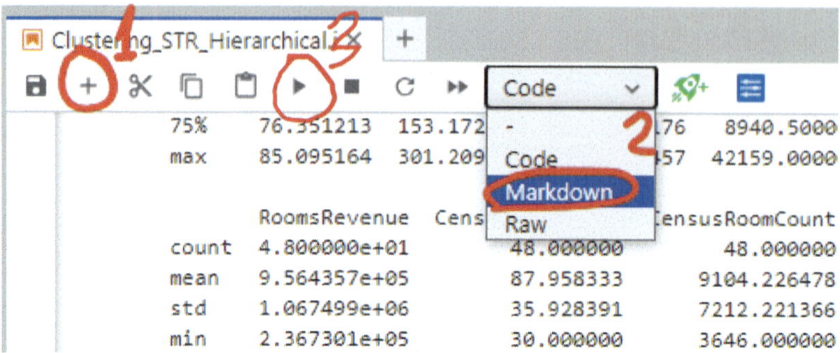

A complete guide to the Markdown language can be found at ▶ https://www.markdownguide.org/. Here, we provide some basic Markdown syntax for narrating the Python code.

Headings

The text can be marked as a header by typing a leading # symbol:

Header 1 Hierarchical clustering

Header 2Read the data ### Header 3 Modified code

Emphases

Bold text Important!

Italic *Important!*

~~Scratched~~ Important!

Lists

Ordered Unordered

1. Step 1 - Step 1

2. Step 2 - Step 2

3. Step 3 - Step 3

Add Markdown cells to the notebook and type explanations of what each code cell is supposed to accomplish. In the future, complement sections of your code with Markdown cells.

Part 3. Cluster Florida Submarkets with Two Features

This part of the lab aims to find clusters of Florida Submarkets with hierarchical clustering. We will use the same attributes as in ▶ Chap. 5.

USER:

Use DataFrame “data”. Apply hierarchical clustering to cluster “data” on columns RoomsOCC and RoomsADR. The data is already standardized.

Assistant:

```
import pandas as pd
from scipy.cluster.hierarchy import dendrogram, linkage
import seaborn as sns
import matplotlib.pyplot as plt

# If not already standardized, you can use sklearn's StandardScaler to
standardize the data

# Selecting the columns for clustering
selected_columns = data[["RoomsOCC", "RoomsADR"]]

# Calculating the distance matrix using Ward's method
linkage_matrix = linkage(selected_columns, method='ward')

# Plotting the dendrogram
plt.figure(figsize=(10, 7))
dendrogram(linkage_matrix, labels=data.index, leaf_rotation=90)
plt.title('Dendrogram')
plt.xlabel('Data points')
plt.ylabel('Distance')
plt.show()
```

Explanation:

We use the *Scipy* library to calculate the distance matrix and perform the clustering. We also use the *seaborn* library to visualize the dendrogram. In this code, we first select the columns *RoomsOCC* and *RoomsADR* from the dataframe *data*. Then, we calculate the distance matrix using Ward's method and plot the dendrogram using *seaborn* and *matplotlib*.

Result:

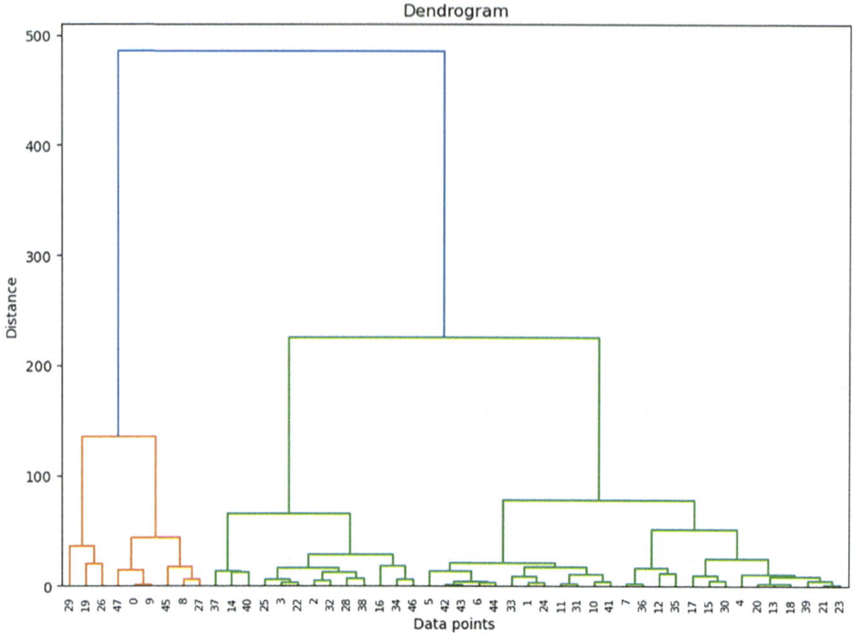

This visualization is imperfect. First, the horizontal scale is the DataFrame row number (starting with 0) instead of the submarket names. Second, the lower part of the visualization is too dense, making the distance between the clusters hard to understand. On the other hand, the top part of the histogram is sparse, simply taking space. Third, the colors used for tree brunches give an impression of one large cluster of submarkets in the right part of the plot. Finally, the titles are not informative.

We will ask the Assistant's help to deal with the first problem and then fine-tune the provided code.

USER:

Modify the code by replacing data points with data point labels. The data point labels are stored in the column Submarket.

Assistant:

```
import pandas as pd
from scipy.cluster.hierarchy import dendrogram, linkage
import seaborn as sns
import matplotlib.pyplot as plt

# Selecting the columns for clustering
selected_columns = data[["RoomsOCC", "RoomsADR"]]

# Calculating the distance matrix using Ward's method
linkage_matrix = linkage(selected_columns, method='ward')

# Setting the "Submarkets" column as the index
data_indexed = data.set_index('Submarket')

# Plotting the dendrogram with data point labels
plt.figure(figsize=(10, 7))
dendrogram(linkage_matrix, labels=data_indexed.index,
leaf_rotation=90)
plt.title('Dendrogram')
plt.xlabel('Data points')
plt.ylabel('Distance')
plt.show()
```

Explanation:

In this modified code, we set the *Submarkets* column as the index for the dataframe *data* using the *set_index* method. Then, we use the index (which now contains the data point labels) to label the dendrogram when plotting.

Result (abridged):

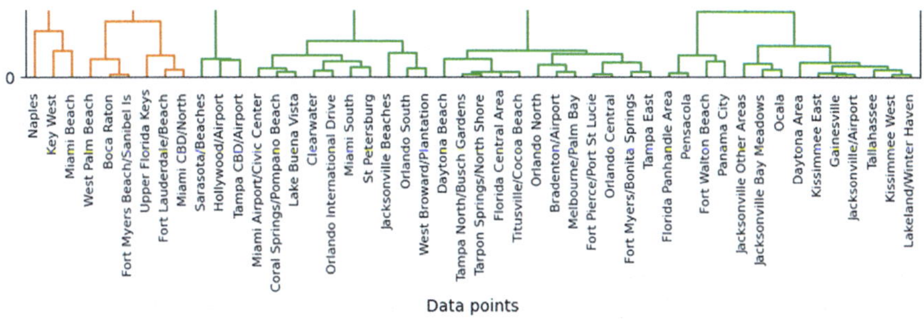

We can correct other drawbacks of this plot by applying simple modifications to the code cell. To solve the problem of overcrowded lower part, we can change the vertical scale to logarithmic. The specific instructions can be found on the Web page of the Mathplotlib project: ▶ https://matplotlib.org/stable/api/_as_gen/matplotlib.pyplot.yscale.html.

To change the scale of the "y" axis, add the following code before calling the dendrogram method.

```
plt.yscale("symlog")
```

The "symlog" scale is a modification of the logarithmic scale that accommodates non-positive values.

Run the cell.

Notice that the new dendrogram is disproportional; to improve its visual appeal, we can reduce the vertical size of the plot. The following line of code sets the horizontal and vertical plot size in inches (this is the standard unit of Mathplotlib):


```
plt.figure(figsize=(10, 7))
```

Reduce the vertical plot size to 4 inches:

```
plt.figure(figsize=(10, 4))
```

To change the color of dendrogram subclusters, add a new parameter to the dendrogram function:

```
dendrogram(linkage_matrix, labels=data_indexed.index,
leaf_rotation=90, color_threshold=70)
```

The parameter *color_threshold* sets the threshold distance, below which all descendent links have the same color as their top node below the threshold.

Next, set the x-axis name by replacing the following code:

```
plt.xlabel('Data points')
```

for

```
plt.xlabel('Submarkets')
```

Modified code:

```
import pandas as pd
from scipy.cluster.hierarchy import dendrogram, linkage
import seaborn as sns
import matplotlib.pyplot as plt

# Selecting the columns for clustering
selected_columns = data[["RoomsOCC", "RoomsADR"]]

# Calculating the distance matrix using Ward's method
linkage_matrix = linkage(selected_columns, method='ward')

# Setting the "Submarket" column as the index
data_indexed = data.set_index('Submarket')

# Plotting the dendrogram with data point labels
plt.figure(figsize=(10, 4))
plt.yscale("symlog")
dendrogram(linkage_matrix, labels=data_indexed.index,\
leaf_rotation=90, color_threshold=70)
plt.title('Dendrogram')
plt.xlabel('Submarkets')
plt.ylabel('Distance')
plt.show()
```

Run the cell.

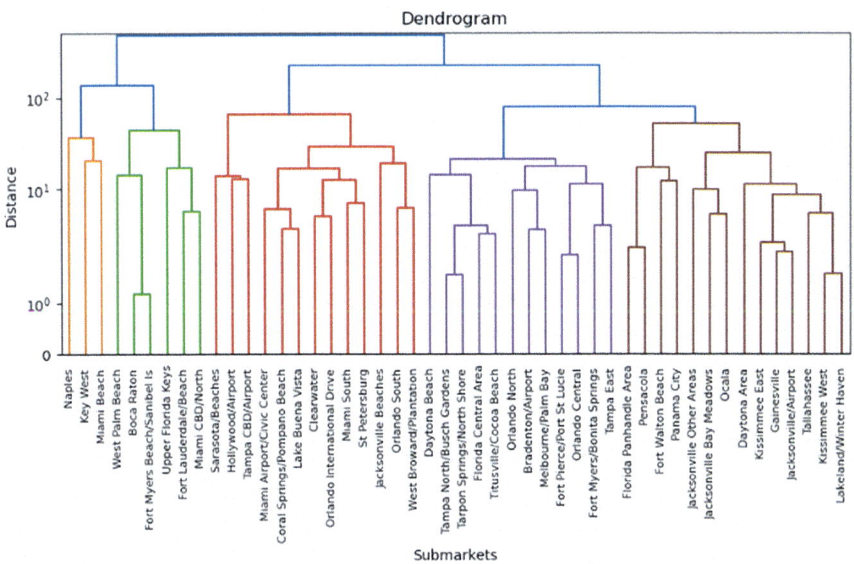

Part 4. Clustering Data Points with Many Attributes

Up to this point, we only used two attributes to cluster the submarkets. In this part, we extend our analysis to more attributes and discuss issues arising.

Step 1. Use All Numerical Attributes to Cluster Submarkets

This is a simple modification that we can accomplish by editing the code in place instead of calling the Assistant.

First, duplicate the last cell by clicking the "duplicate" icon:

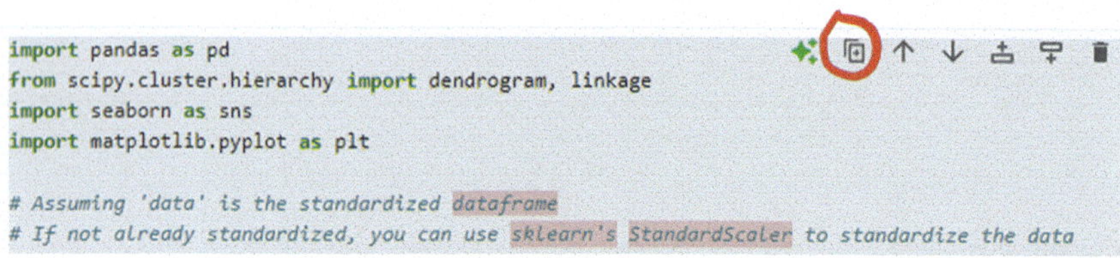

An identical cell will be inserted below the current one. Remember to add a Markdown cell explaining what the new code is doing!

In the duplicated cell, replace the following line of code:

```
selected_columns = data[["RoomsOCC", "RoomsADR"]]
```

for

```
selected_columns = data[['stdRoomsOCC', 'stdRoomsADR',
'stdRoomsRevPAR', 'stdRoomsSupply', 'stdRoomsDemand',
'stdRoomsRevenue', 'stdCensusPropCount', 'stdCensusRoomCount']]
```

Run the code.

Result:

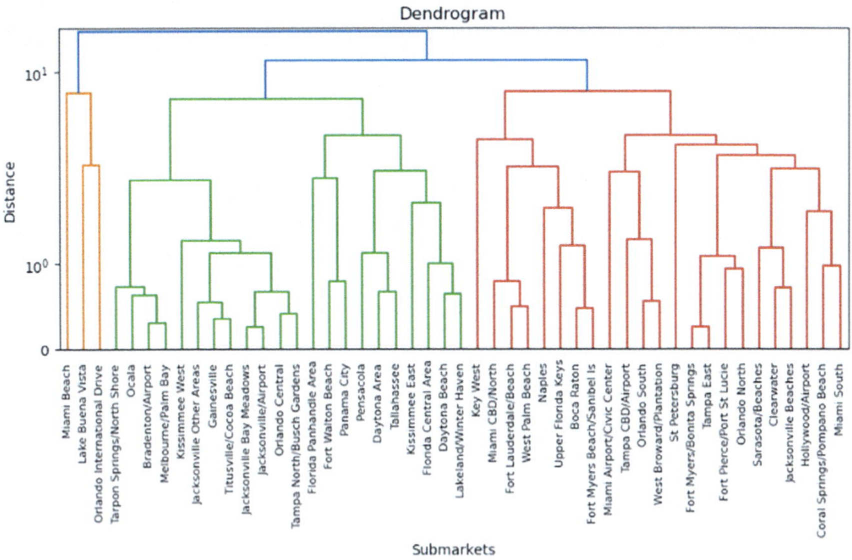

The results of clustering when all attributes are included are quite different from those when only two attributes are used. This could be due to the new data being considered when making clustering decisions. This could also be an effect of irrelevant and/or highly correlated parameters.

Step 2. Feature Selection

From the previous chapter's lab, we already know that some attributes are highly correlated, artificially inflating their importance. To resolve this problem, we will use feature selection.

In the previous five labs, we used several libraries:

- For numerical computing and data manipulation: NumPy and pandas;
- For data analysis, SciPy and scikit-learn;
- For data visualization, Matplotlib and Seaborn.

Among these libraries, the *scikit-learn* library contains several feature selection methods. However, none of these methods effectively deal with multivariate correlation of attributes in unsupervised learning. Because of that, we will use a specialized Python library for feature engineering and selection called Feature-engine. This library includes multiple feature engineering methods; among those, *SmartCorrelatedSelection* searches for groups of correlated attributes and leaves only one attribute per group. The decision on which of the attributes to leave can be based on the least number of missing values, the greatest number of unique values (highest *cardinality*), the highest variance, or the highest importance according to a user-defined estimator. Since our data has no missing values and all data are ratio, we will use the highest variance criterion.

USER:

Use the following columns of the DataFrame "data": stdRoomsOCC, stdRoomsADR, stdRoomsRevPAR, stdRoomsSupply, stdRoomsDemand, stdRoomsRevenue, stdCensusPropCount, stdCensusRoomCount. Your goal is to apply feature selection to prepare the data for clustering. Use feature_engine.selection.SmartCorrelatedSelection method.

Assistant:

```
from feature_engine.selection import SmartCorrelatedSelection
import pandas as pd

# Selecting the specified columns for feature selection
selected_columns = ['stdRoomsOCC', 'stdRoomsADR', 'stdRoomsRevPAR',
'stdRoomsSupply',
                    'stdRoomsDemand', 'stdRoomsRevenue',
'stdCensusPropCount', 'stdCensusRoomCount']

# Subsetting the data with selected columns
selected_data = data[selected_columns]

# Applying SmartCorrelatedSelection for feature selection
scs = SmartCorrelatedSelection(
    variables=None, method='Pearson', threshold=0.8,
selection_method='variance'
)

selected_data = scs.fit_transform(selected_data)

# The selected_data now contains the features selected based on
SmartCorrelatedSelection
```

Explanation:

The *feature_engine* library provides the *SmartCorrelated-Selection* method for feature selection based on correlation with the target variable and with each other. In the code, we first select the specified columns for feature selection. Then, we apply the *SmartCorrelatedSelection* method to select the features based on correlation and variance. The *selected_data* variable will contain the features selected based on the *SmartCorrelatedSelection* method.

Before running the code, do not forget to install the *feature_engine* library.

To better understand code functionality, print information on the new DataFrame *selected_data*:

```
print(selected_data.info())
```

Result:

```
RangeIndex: 48 entries, 0 to 47
Data columns (total 4 columns):
 #   Column              Non-Null Count  Dtype
---  ------              --------------  -----
 0   stdRoomsOCC         48 non-null     float64
 1   stdRoomsRevPAR      48 non-null     float64
 2   stdRoomsSupply      48 non-null     float64
 3   stdCensusPropCount  48 non-null     float64
```

The feature selection step leaves only four out of eight attributes for future processing.

Step 3. Cluster the Data Using the Attributes Left After Feature Selection

Duplicate the Step 1 code and move the duplicated cell to the end of the notebook.

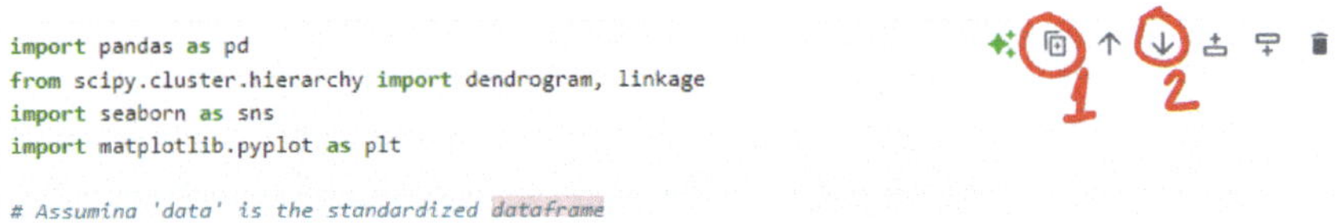

In the code, replace the list of selected columns for clustering:

```
# Selecting the columns for clustering
selected_columns = data[['stdRoomsOCC', 'stdRoomsRevPAR',
'stdRoomsSupply', 'stdCensusPropCount']]
```

Run the code.

Notice that the new dendrogram needs to adjust the link color. In addition, the premium submarkets, such as Key West, which were on the left side of the plot, are now on the left side. These issues can be resolved by adding parameters to the dendrogram method:

```
dendrogram(linkage_matrix, labels=data_indexed.index,
leaf_rotation=90, color_threshold=6,count_sort=True,
distance_sort=False)
```

Run the code.

Result:

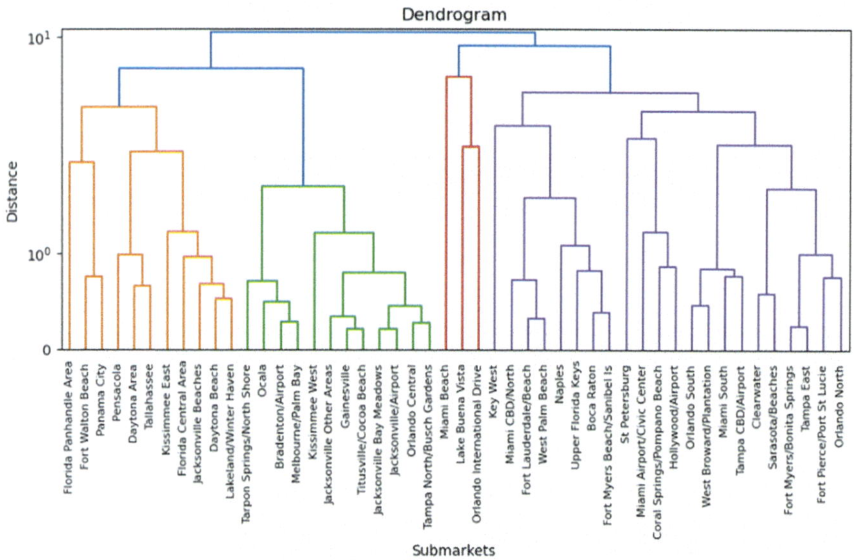

Part 5. Flatten the Dendrogram

This part aims to flatten the dendrogram and visualize the flat clusters. One approach is to cut the dendrogram along some threshold distance, for example, at distance = 7 (the same we used to color the dendrogram links).

USER:

Use the distance matrix linkage_matrix. Cut the dendrogram along the threshold distance = 7. Add results to the DataFrame “data”.

Assistant:

```
from scipy.cluster.hierarchy import fcluster

# Cutting the dendrogram along the threshold distance
threshold_distance = 7
clusters = fcluster(linkage_matrix, t=threshold_distance,
criterion='distance')

# Adding the cluster labels to the DataFrame
data['cluster_labels'] = clusters
```

Explanation:

In this code, the *fcluster* function from the *scipy.cluster.hierarchy* module is used to assign cluster labels to each data point based on the threshold distance of 7. The resulting cluster labels are then added as a new column *cluster_labels* to the DataFrame *data*.

Run the code. Confirm changes by running the code:

```
print(data.head())
```

Note a new column named *cluster_labels*.

USER:

Use the DataFrame "data". Make a scatterplot of the columns stdRoomsOCC and stdRoomsRevPAR. Use the values in column cluster_labels for data point colors.

Assistant:

```
import seaborn as sns
import matplotlib.pyplot as plt

# Create a scatter plot with cluster labels as point colors
plt.figure(figsize=(8, 6))
sns.scatterplot(data=data, x='stdRoomsOCC', y='stdRoomsRevPAR',
hue='cluster_labels', palette='viridis')
plt.title('Cluster Visualization')
plt.xlabel('Standardized Rooms OCC')
plt.ylabel('Standardized Rooms RevPAR')
plt.show()
```

Explanation:

In this code, we use *seaborn*'s scatterplot function to create a scatter plot of the columns *stdRoomsOCC* and *stdRoomsRevPAR*, with the *cluster_labels* column used to assign colors to the points. The palette *viridis* is used to specify the color scheme for the clusters.

6

Result:

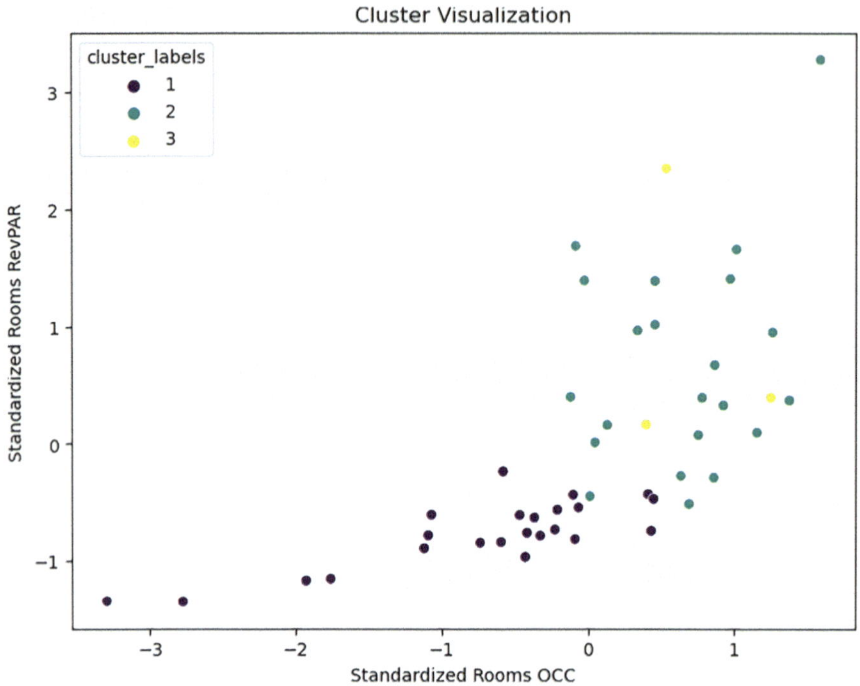

Note:

The colors used for the plot points are defined by the following parameters: *hue='cluster_labels'* and *palette='viridis'*. The *palette* parameter defines the colormap to be used. For a list of pre-defined colormaps, see ▶ https://matplotlib.org/stable/users/explain/colors/colormaps.html. The *hue* parameter designates the variable used to define which color from the palette to use for specific points on the plot.

Part 6. On Your Own

1. Add information on the submarkets from Cluster 3 to the plot. Use the values in column *Submarket* for point labels.
2. Create scatterplots for all combinations of the attributes used for clustering. Which plots are the most useful for understanding clustering decisions?

3. Modify the clustering method from Ward to single, global, and average linkage. Describe how the clustering has changed.
4. In your opinion, which of the hierarchical clusters is the most suitable for describing Florida hospitality markets?

Build Your Python Skills

Chapters 13 and 14 of the free book Learn Python with Jupyter by Serena Bonaretti. ▶ https://www.learnpythonwithjupyter.com/.

Additional Reading

Murtagh, F., & Contreras, P. (2012). Algorithms for hierarchical clustering: An overview. *Wiley Interdisciplinary Reviews: Data Mining and Knowledge Discovery, 2*(1), 86–97.

Alelyani, S., Tang, J., & Liu, H. (2018). Feature selection for clustering: A review. *Data Clustering,* 29-60. ▶ https://www.taylorfrancis.com/chapters/edit/10.1201/9781315373515-2/feature-selection-clustering-review-salem-alelyani-jiliang-tang-huan-liu

Predictive Analytics and Supervised Learning. Classification

A. P. Kirilenko, *Practical Data Mining with AI for Social Scientists*,
Springer Texts in Social Sciences, https://doi.org/10.1007/978-3-031-89689-7_7

7.1 Supervised Versus Unsupervised Learning

Loosely, the machine learning methods can be separated into supervised and unsupervised learning. Supervised learning, as the name suggests, operates under the guidance of a 'supervisor.' Practically, that means there is a dataset for which the analysis goal has already been achieved and is available to train the model. This training dataset is called labeled data. Its attributes include (1) the features representing the independent variables and (2) the label or labels (*targets*) representing the known outcomes related to the independent variables' values. Data labeling can be done manually by human annotators, which can be time-consuming and expensive, or by annotating models.

Supervised and learning

Supervised learning is a type of machine learning where the model is trained on a labeled dataset. Conversely, unsupervised learning operates without the supervision of labeled data.

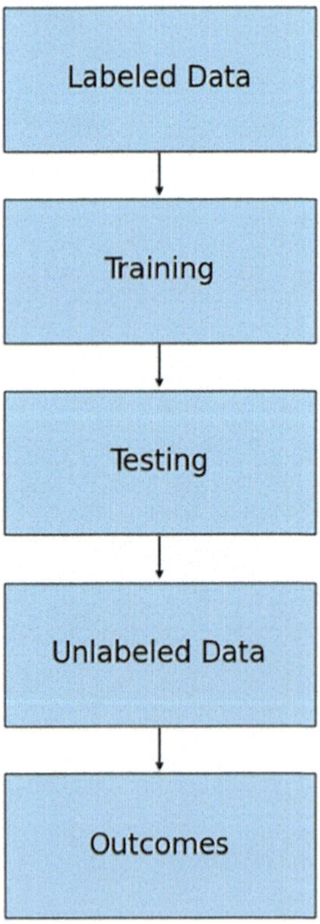

After the training and validation, the algorithm learns a function that maps inputs to desired outputs. In that sense, some machine learning algorithms are identical to inferential statistics methods you know from introductory statistics courses, such as linear regression. Then, the trained model is applied to a much larger unlabeled dataset, for which the outcomes are not available. This dataset does not contain the target attributes. Hence, the goal of supervised methods is to learn how to make predictions for an unseen set of data based on a (much smaller) set of examples.

Conversely, unsupervised learning operates without the supervision of labeled data. Instead, it seeks to identify inherent structures or patterns within the data. The algorithm is not provided with a specific target output. Rather, it tries to discover "interesting" structures in the data on its own. Among the methods we have studied so far, K-means and DBSCAN are primarily unsupervised, while Decision trees and Association rules are supervised algorithms.

This distinction between supervised and unsupervised learning is not very distinct. For instance, semi-supervised models use labeled data to guide the learning process (like in supervised learning) and unlabeled data to uncover the underlying data distribution (like in unsupervised learning). By nature, the supervised methods are more suitable for prediction tasks. For example, a time series model trained on past tourist arrivals may be used to project the next year's arrivals at a specific destination. In comparison, the unsupervised methods are more suitable for exploratory studies. On the other hand, unsupervised learning might be more suitable for exploratory data analysis, where the goal is to uncover hidden patterns or structures in the data. Still, both supervised and unsupervised methods are frequently used together in one study. For example, cluster analysis (unsupervised method), discussed in the previous two chapters, can be used to group data into clusters, which are then supplied to a supervised learning model as new features, enhancing its predictive power. The Principal Component Analysis (PCA, another unsupervised method) can be applied to data attributes to reduce dimensionality before training a predictive model. Unsupervised methods are frequently used for data anomaly detection, and so on.

This chapter primarily deals with classification methods. Classification is similar to clustering but belongs to the supervised rather than unsupervised methods. While the goal of clustering is to separate data points into naturally forming groups, classification aims to predict which pre-existing group each data point is likely to belong to. In the next two sections, we will focus on one frequently used method, the Naïve Bayes classifier. We will explore the concept of conditional probability, the Bayesian approach, and how to deal with continuous variables.

7.2 Probabilistic Approach to Classification

7.2.1 Introduction to Conditional Probability

Supervised learning can be interpreted as the process of predicting the value of an output variable based on the values of input variables. This interpretation allows the application of the conditional probability concept. Conditional probability is the probability of an event Y occurring, given that another event X has already occurred. This approach is used in many machine learning algorithms, including the Naïve Bayes classifier.

To explain the concept of conditional probability, consider a practical example from the medical diagnostics field, specifically breast cancer testing. The mammography cancer test is not without a chance of error. The following matrix represents performance measures of a mammography test of a person with cancer and a person without cancer (◘ Table 7.1).

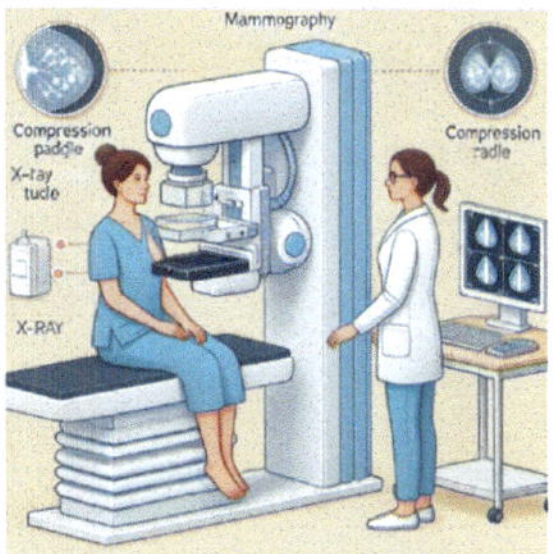

Apparently, the mammography test is able to correctly recognize existing cancer in 82% of cases, while its ability

Table 7.1 Mammography performance measures[1]

		Actual cancer presence in a patient	
		Positive: cancer	**Negatives: no cancer**
Test prediction	**Positive: cancer**	82% True positives	9% False positives
	Negative: no cancer	18% False negatives	91% True negatives

Table 7.2 Mammography test outcomes adjusted for prevalence (1,000 patients)

		Actual cancer presence in a patient		
		Positive: cancer (0.005*1000 = 5)	**Negatives: no cancer (0.995*1000 = 995)**	**Total**
Test prediction	**Positive: cancer**	0.82*5 = 4.1 True positives	0.09*995 = 89.5 False positives	93.6
	Negative: no cancer	0.18*5 = 0.9 False negatives	0.91*995 = 905.5 True negatives	906.4
	Total	5	995	1000

to predict an absence of cancer correctly is 91%. These measures are called *sensitivity* and *specificity*, respectively, and will be discussed in depth in ▶ Chap. 8.

Consider a patient taking a mammography, which returns a positive result indicating breast cancer. What is the probability that she actually has breast cancer? Looking at the "Positive: Cancer" row of the confusion matrix, one might think the answer is 82% or 91% (100%—9% of the chance of getting a false positive). This, however, is not correct. Let us do a proper statistical inference. From medical statistics, we know that only 0.5% of women coming for mammography have cancer.[2] Suppose 1000 persons are taking mammography, five of whom (0.5%) have cancer. Then, the following confusion matrix would represent a likely test outcome (Table 7.2).

1 According to the US National Center for Biotechnology Research (▶ https://www.ncbi.nlm.nih.gov/books/NBK550207/).

2 Women between ages 50 to 59. US National Center for Biotechnology Research (▶ https://www.ncbi.nlm.nih.gov/books/NBK550207/).

Then, the chance of any particular woman who tests positive for cancer actually to have cancer is only 4.4%: 4.1/93.6 = 0.044!

This is an example of the Bayesian approach to statistics. Let us answer this question using the proper statistical formulations. Denote the event of having breast cancer as C and the event of testing positive as T. We are interested in P(C|T), the probability of having breast cancer given a positive test result. While we don't have this data, we know the opposite from the laboratory data: the probability of returning a positive test result when a person has cancer P(T|C). This knowledge is enough to get the answer to our question.

P(C|T) = P(T|C) * P(C) / P(T)

According to Bayes' theorem,

$$P(C|T) = P(T|C) * P(C)/P(T)$$

Here, P(T|C) is the sensitivity of the test, P(C) is the prior probability of having breast cancer, and P(T) is the total probability of testing positive.

The prior probability P(C) is the prevalence of breast cancer in the population, which we obtained from medical statistics: P(C) = 0.005.

Key cancer test performance measures (details in Chapter 8)

Accuracy : The proportion of correctly classified instances among the total instances: (4.1 + 905.5)/1000 ~ 0.91

Precision : The proportion of correct positive predictions to the total predicted positives: 4.1/(4.1 + 89.5) ~ 0.044

Sensitivity (Recall): The proportion of correct positive predictions to actual positives: 4.1/(4.1 + 0.9) = 0.82

Specificity : The proportion of correct negative predictions to the total predicted negatives: 905.5/(905.5 + 89.5) ~ 0.91

F1-score : The harmonic mean of precision and recall: 2 × 0.044 × 0.82/(0.044 + 0.82) ~ 0.084

The total probability of testing positive P(T) can be calculated as the sum of the probabilities of testing positive with and without breast cancer, weighted by their respective prevalences: P(T) = 0.936.

Then, P(C|T) = 0.82 * 0.005/0.936 = 0.0044.

In essence, the Bayes statistics allow us to update our beliefs about the likelihood of an event (having breast cancer) based on new evidence (the test result). This example also highlights the potential pitfalls of relying solely on test sensitivity and false-positive rates without considering the prior probability of the event. This is especially important when there is a significant disbalance between the population groups. For example, the author observed this type of error in a study published during the early stages of COVID-19, which overestimated the pandemic rate at a famous holiday destination by an order of 10 due to not accounting for prior probability.

The approach we used to estimate the chances of having cancer in a woman with a positive mammogram is called Bayesian. The Bayesian approach is a method of statistical inference that integrates prior knowledge, often subjective (in our example, we used a guesstimate of cancer prevalence in the population), with empirical data to guide statistical inference. This flexibility greatly simplifies the inclusion of expert knowledge into statistical inference, which is especially important in social sciences.

7.2.2 Naïve Bayes Classifier

The Bayesian approach is the core of a very popular Naïve Bayes classification method. Consider the following example.

Researchers are interested in studying a discourse on climate change's impact on tourism appearing in mass media. Using a keyword search of the newspaper database available from LexisNexis data analytics company, the researchers secured a sample of 10,000 publications. However, the researchers are not confident that all collected publications deal with climate change.

LexisNexis provides a list of topics in each newspaper publication's text. Hence, the researchers decided to predict the relevance of each newspaper publication to their research using a supervised learning model. They started with creating a topic presence/absence matrix with rows representing the publications and columns representing

Training sample

ID	Airlines	Beaches	Carbon	Coast	Attractions	Drought	Ecotourism
103	0	0	1	0	1	0	0
437	0	1	0	1	1	0	0
2	1	0	0	1	0	1	0
1890	1	0	1	1	1	0	1

Manual sample labeling

ID	Relevant?	Airlines	Beaches	Carbon	Coast	Attractions	Drought	Ecotourism
103	1	0	0	0	0	1	0	0
437	1	0	1	0	1	1	0	0
2	0	0	0	0	1	0	0	0
1890	1	1	0	1	1	0	0	1

Estimation of probabilities

Relevant?	Airlines	Beaches	Carbon	Coast	Attractions	Drought	Ecotourism
1	0.33	0.33	0.67	0.67	1	0	0.33
0	...	...	...	...	...	...	...

Fig. 7.1 Naïve Bayes model training. The probabilities are calculated for the relevant class only

the topics. For each publication, the matrix cell contained ones for present and zeros for absent topics (Fig. 7.1).

Then, the researchers manually classified ("*labeled*") a random sample of 500 newspaper publications. Each publication discussing climate change and tourism was coded with "1" (relevant); all other publications were coded with "0" (irrelevant). Then, this coded *training* sample was used to train a Naïve Bayes classifier. The trained model was used to classify the rest of the newspaper publications ("*unlabeled*") *into* relevant and irrelevant groups.

The Naïve Bayes classifier works as follows. To be able to classify the unlabeled dataset, we need to compare two probabilities: the probability of the events "publication is relevant" given our knowledge of the values of the keyword attributes $k_1, k_2, \ldots, k_n$ (Airlines, Beaches, and so on):

$$\text{P}(\text{Relevant}|k_1, k_2, \ldots, k_n)$$

and probability of the events "publication is relevant" given our knowledge of the values of the keyword attributes $k_1, k_2, \ldots, k_n$:

$$\text{P}(\text{Irrelevant}|\, k_1, k_2, \ldots, k_n)$$

Whichever probability is greater will define our decision to select or reject a particular paper. These probabilities are evaluated from the labeled dataset in the following way.

According to Bayes' theorem,

$$\mathrm{P(Relevant}\ |k_1, k_2, \ldots, k_n) = \mathrm{P}(k_1, k_2, \ldots, k_n|\ \mathrm{Relevant)}\frac{\mathrm{P(Relevant)}}{\mathrm{P}(k_1, k_2, \ldots, k_n)}$$

Similarly,

$$\mathrm{P(Irelevant}\ |k_1, k_2, \ldots, k_n) = \mathrm{P}(k_1, k_2, \ldots, k_n|\ \mathrm{Irrelevant)}\frac{\mathrm{P(irrelevant)}}{\mathrm{P}(k_1, k_2, \ldots, k_n)}$$

Notice that the fractions have identical denominators. That means, for comparison of the "Relevant" and "Irrelevant" probabilities, we only need to compare.

$\mathrm{P}(k_1, k_2, \ldots, k_n|\ \mathrm{Relevant)P(Relevant)}$ to $\mathrm{P}(k_1, k_2, \ldots, k_n|\mathrm{IrrRelevant)P(Irrelevant)}$

Probabilities P(Relevant) and P(Irrelevant) are easy to compute: count the proportion of 1 s and 0 s in the training sample. The $\mathrm{P}(k_1, k_2, \ldots, k_n|\ \mathrm{Relevant)}$ and $\mathrm{P}(k_1, k_2, \ldots, k_n|\ \mathrm{Irrelevant)}$ are also possible to estimate by counting the proportion of each combination of the attributes among the relevant publications and doing the same with the irrelevant publications. However, it is unlikely that each combination will be present if the training sample is small. Because of that, the Naïve Bayes method makes a "naive" assumption of independence among predictors. For example, it assumes that the event "relevant publication has a keyword ecotourism" is independent of the event that it has a keyword "carbon." Then, using the joint probability formula,

$$\mathrm{P}(k_1, k_2, \ldots, k_n|\ \mathrm{Relevant)} = \mathrm{P}(k_1|\ \mathrm{Relevant)} \times \ldots \times \mathrm{P}(k_n|\ \mathrm{Relevant)}.$$

These probabilities are easy to estimate from the training dataset. For example, $\mathrm{P}(k_1|\ \mathrm{Relevant)}$ is computed as the number of "relevant" records with the keyword "airlines" divided by the total number of relevant records. The same applies to irrelevant records.

7.2.3 Advantages of the Naïve Bayes Approach

The Naïve Bayes classifier is fast and robust, which defines its popularity among the analysts. Regarding robustness, even though the "naive" assumption that all features are mutually independent (conditionally on the category) is rarely met in real-world data, this has a small impact on the classification results. Indeed, the algorithm does not require the exact computations of the probabilities for each classification group; instead, it only relies on their comparisons. The classification result is correct as long as the order of the groups according to their probabilities is not affected by deviations from the "naive" assumption.

In turn, the mutual independence assumption eliminates the need to estimate the multivariate probability distribution function for all features. By treating each feature independently, a classifier can be trained faster and with less data. Moreover, this assumption leads to better interpretable models, as the relationships between individual features and the target variable can be examined in isolation. When only a small amount of training data is available, the Naïve Bayes is one of the best classifiers to consider.

The Naïve Bayes classifier is not strongly affected by missing values. When a certain record has missing values, the respective probabilities are not included in the calculations with little impact on the outcome. Similarly, the Naïve Bayes is robust against noisy data. These characteristics relax the requirements of making arbitrary decisions on the treadment of noise or missing data during the data cleaning step. Finally, the Naïve Bayes is tolerant towards the presence of irrelevant attributes, making feature selection, discussed in the previous chapter, less important.

The algorithm's results are presented probabilistically rather than "crisp" decisions on group membership. This provides a clear confidence measure. For instance, in the text classification example from the previous section, if the researcher is interested only in publications that are almost definitely on tourism and climate change, they can set the threshold cluster membership probability at 90%. In comparison, a researcher more interested in the quantity of group membership, can set a lower threshold.

Naïve Bayes classifier

- Very fast and robust
- Handles missing data well
- Suitable for text classification
- Returns results in probabilistic form
- Assumes attribute independence conditionally on the category
- Assumes attributes normal distribution

7.2.4 Limitations

Naïve Bayes limitations

The base method is designed to work with categorical data

Gaussian NB assumes a normal distribution of interval/ratio features

Sensitivity to multiple instances of missing data

Zero-frequency problem

Highly correlated features overestimate their importance

For simplicity, we used an example where all attributes were binary. However, the next section discusses a variant of the Naïve Bayes classifier, which can be used with interval/ratio independent variables as long as the dependent variable (group membership) is categorical. Yet, in dealing with interval/ratio features the algorithm assumes that they are normally distributed. In addition, the feature independence assumption of the Naive Bayes method limits its ability to capture feature relationships.

Another limitation specific to the Naïve Bayes method is the "Zero probability" problem. Notice that in the tourism and climate change example, there are no relevant publications on Drought in the training sample. Hence, P(Drought|Relevant) = 0 (▫ Fig. 7.1). That means that all publications on Drought will be classified as Irrelevant, no matter how high the other probabilities are. A similar problem arises when dealing with very small probabilities. Those problems are resolvable, but they add complexity to the model.

7.2.5 Naïve Bayes Variants

The "classical" *categorical Naïve Bayes* classifier discussed in the previous sections requires categorical features and labels. In the newspaper publications example, the features and the labels were binominal (Features: 1 for the "present" category, 0 for "absent"; labels: 1 for the "relevant" category, 0 for "irrelevant"). Modifications of the Naïve Bayes methods widen the applicability of this method.

The *Gaussian* Naïve Bayes classifier allows interval/ratio features. In this chapter's lab, the thematic presence/absence matrix contains values corresponding to the topic relevance score. That means the features are continuous rather than categorical, and classification cannot be completed using the categorical Naïve Bayes. However, the Gaussian Naïve Bayes, requires all features to be normally distributed.

The *Multinominal* and *Bernoulli* Naïve Bayes classifiers relax another limitation of the "classical" algorithm: they allow multinominal labels. That means the data can be classified into an arbitrary number of classes instead of only two. This makes these methods suitable for processing textual data: classification, topic modeling, sentiment analysis, and similar tasks. The Bernoulli Naïve Bayes requires the presence/absence features, while the Multinominal Naïve Bayes requires the frequency features (occurrence counts).

Finally, the *Complement* Naive Bayes classifier addresses the problem of imbalanced datasets, that is, the datasets with high variation in different classes' label counts. The original method tends to discriminate against the low-frequency classes. This vulnerability is especially characteristic of the multinominal Bayes. Hence, this method provides a robust alternative to traditional Naive Bayes methods, particularly in text classification tasks involving imbalanced datasets such as customer reviews, where the 1 or 2-star negative reviews typically are a minority.

7.3 Decision Trees and Random Forest

We have already discussed the Decision Trees algorithm as an explanatory analysis tool in ► Chap. 1. There, we used it to investigate the factors influencing Titanic passengers' survival. Then, in ► Chap. 4, we used Decision Trees to understand factors contributing to a decision to travel to an adversary country. Meanwhile, this algorithm can also be used as a classifier.

Similar to the Bayesian approach, the Decision Tree classifier is first trained on a sample of data. An important caveat is that if not stopped early, the Decision Tree training process splits the data into smaller and smaller subsets, resulting in an overfitted model. Such a model exhibits excellent performance on the training data but has no skill in classifying unseen data items. Because of that, Decision Tree model testing on an independent dataset is a must. We will discuss model evaluation strategies in the next chapter. In comparison, the Naïve Bayes models, even not completely safe from overfitting, are unlikely to have this problem.

Decision Trees are also not well suitable for handling missing data. Consider the model trained in ► Chap. 4 lab to explain factors contributing to a decision to travel to an adversary country (specifically, from Russia to the United States) in ▫ Fig. 7.2.

We can use this trained model as a classifier to evaluate if a potential customer will likely buy a tourism tour

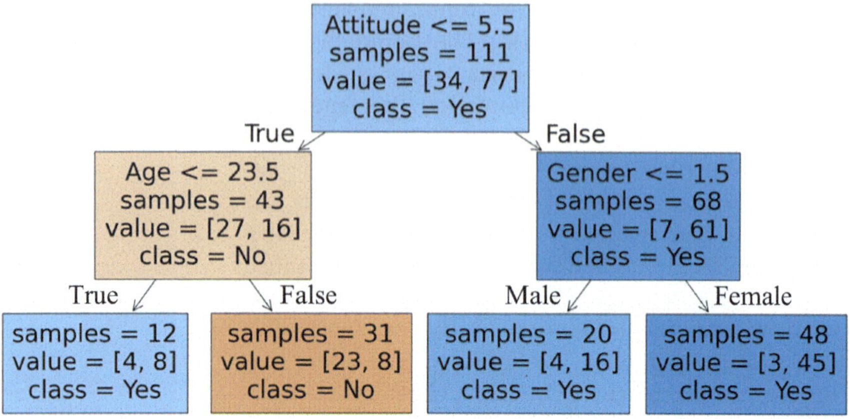

Fig. 7.2 A simple classifier for a decision-making process regarding traveling to an adversary country. Variables: Attitude towards the US, Age, and Gender (1: Male, 2: Female)

to an adversary country. However, a new customer did specify their gender. That means the decision-making process is unable to proceed past the Gender node. One popular algorithm, CART, substitutes the missing values with statistically similar ones from other attributes. For example, suppose the participant has a very high attitude towards the adversary countries, and from other data, we know that this is more typical of females. In that case, we can replace the missing gender value with "Female." Another popular approach, C4.5, would evaluate the probability distribution for the missing values. These approaches, however, would be viable when missing values are sparse.

Another limitation of the algorithm is its sensitivity to small changes in data and methods. One new data point can dramatically change the decision tree, reducing its major method's benefit of explainability. Like many other classification algorithms, Decision Trees tend to be biased towards more dominant classes. Finally, the algorithm does not handle datasets with many attributes well. Because of that, it is not well suited for the classification of textual data.

Many of these problems are solved by replacing a single Decision Tree classifier with a "*Random Forest*" ensemble of classifiers, each giving its classification outcome independently. The most popular outcome is then selected as the final one. To reduce major problems of instability against the introduction of new data and

Decision trees

- Results are visual and easy to interpret
- Unstable against new data and parameter change
- Requires special processing of missing data
- Prone to overfitting
- Random forest classifier resolves the overfitting and instability problems, but computationally expensive

changes in parameters, the classifiers are trained on random subsets of labeled data and with variations in the method's parameters. As an additional benefit of this approach, it eliminates the overfitting risk. This improvement, however, is at the expense of a radical increase in computational requirements.

7.4 Support Vector Machines

The previous section has one solution to the common problem of machine learning algorithms: overfitting. Another classification algorithm, the Support Vector Machine (SVM), can do that faster and deliver accurate results, but has its own data limitations.

The idea of the SVM is very simple: given a set of labeled data points in an N-dimensional space, draw the "best" hyperplane separating the points from different classes. Here, N is the number of features, and the label is the class to which the data point is classified. Two examples of ◘ Fig. 7.3 show how the SVM classification works for two and three features. During the training process, the "best" hyperplane (for two features, the hyperplane is a line; for three features, a plane) is drawn in such a way that it maximizes the distance to points from each class (blue and red dots on the figure). Essentially, the SVM tries to find the line separating the classes with the widest margin. Then, during the classification step, the trained model assigns a new unlabeled point to the class based on the point's position relative to the hyperplane.

Support Vector Machines SVM

- Fast and robust, but may be slow to train
- Requires binary labels

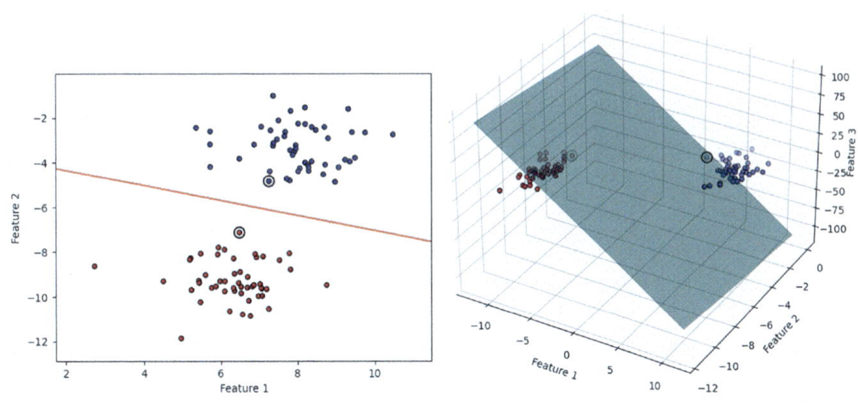

◘ **Fig. 7.3** SVM decision boundary for two features (left) and three features (right)

In the real world, the data rarely allows a clean separation, as shown in ◘ Fig. 7.3. This problem is solved with feature transformation, which in turn requires careful model parameter selection. While SVM is a fast and robust algorithm, it may be slow to train, especially for high-dimensional data such as textual data.

One fundamental limitation of the SVM is that it requires binary labels. This can be solved, e.g., by training multiple "one class against all other classes" models, which further slows down the model training process.

7.4.1 Artificial Neural Networks

The idea of the Artificial Neural Networks (ANN) is simple: to train a network of interconnected nodes (neurons) to process data received on the input neurons to obtain the desirable outcome. Structurally, the neurons are organized in layers (◘ Fig. 7.4). The input layer neurons receive data and send it to the hidden layer's neurons, where the data is transformed and sent to the neurons of the output layer, where the computations are finalized and data processing results are compared with expectations (truth). The network links are recalibrated based on the comparison outcomes, aiming at error reduction. Then, a revised output is generated, and the model training process continues.

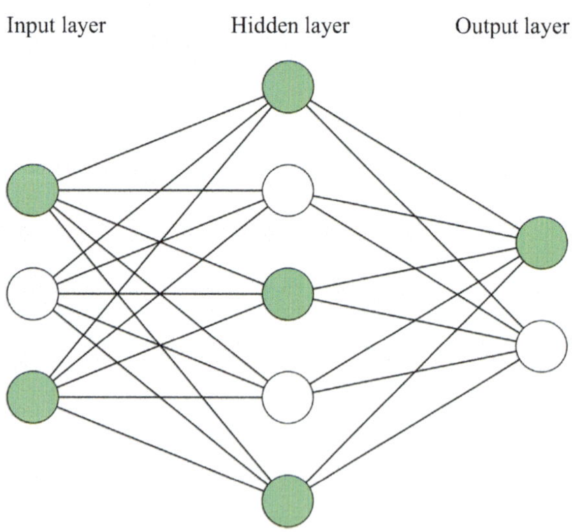

◘ **Fig. 7.4** A conceptual representation of an Artificial Neuron Network. Note that there can be multiple hidden layers

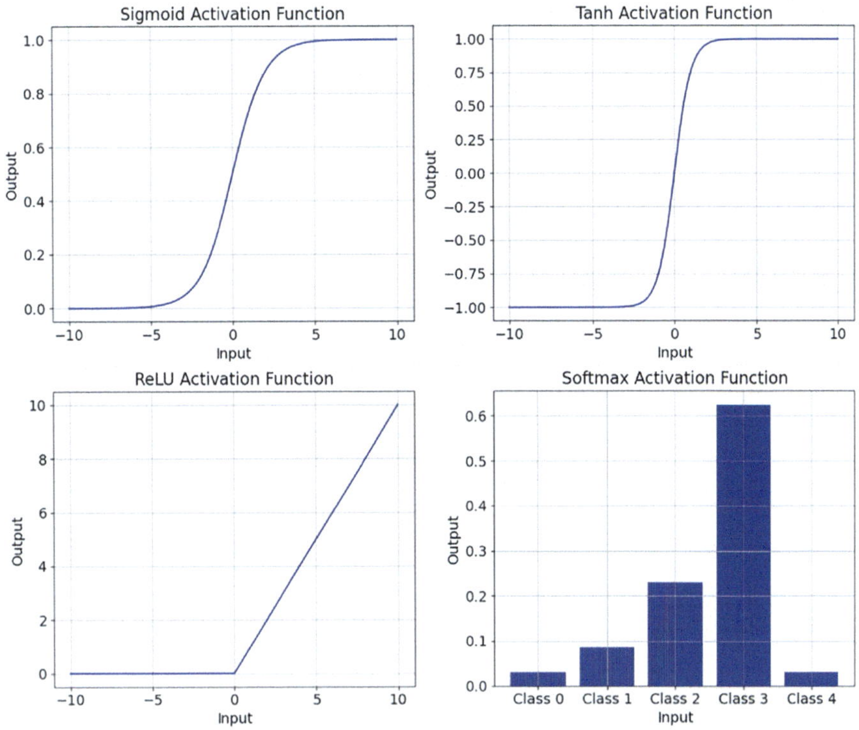

Fig. 7.5 Frequently used activation functions

Data transformation at the neuron is completed through the *activation functions*. The role of the activation function is to apply a non-linear transformation to a *weighted sum* of the neuron's input. There are many different ways of defining the activation function, as long as they are non-linear (Fig. 7.5). Among the ANNs used for classification, the softmax function is a popular choice for the output layer, as its result is interpreted in terms of class probabilities.

The weighted sum at the neuron's input uses the weights of each of the neuron's input linkages. Initially, the weights are random; during training, the weights are adjusted to reduce the error at the output layer. Importantly, this adjustment is independent at each neuron, making the ANNs very attractive for parallel computing. Still, compared to other classifiers we have discussed, the ANN training has high computational requirements. Another important limitation, shared with others but not all of the classification approaches, is the danger of overfitting.

Artificial Neural Networks ANNs

- Extremely flexible and suitable for many different data analysis problems
- There are many diverse ANN types
- In classification, returns very good results, but may be prone to overfitting
- Very slow to train

There is a huge variety of neural network types that can be used not only for classification but also for image recognition, regression, clustering, text generation, robotics control, and many other tasks. The famous ChatGPT is also a type of neural network, specifically a deep learning model based on the transformer architecture. Appropriate selection of the ANN type suitable for the intended task is a must.

7.4.2 Conclusion

There are many approaches to classification problems, each with its advantages and limitations. This chapter covered only a few of the most popular methods. The nature of data, its dimensionality, measurement types, amount of noise and missing data, intended output use, and computational requirements define the method most suitable for the project.

7.5 Review Questions

1. What is the difference between unsupervised and supervised learning?
2. How does classification differ from clustering?
3. Explain the concept of conditional probability with an example.
4. Explain Bayes' theorem
5. Explain how Bayesian inference allows for updating beliefs with new evidence.
6. What is the main assumption behind the Naïve Bayes classifier?
7. What are the main advantages and limitations of the Naïve Bayes classifier?
8. How does a Decision Tree model decide the outcome of a classification?
9. What are the main advantages and limitations of the Decision Tree classifier?
10. What problem does the Random Forest algorithm solve compared to a single Decision Tree?
11. Describe the main idea of the Support Vector Machine classifier (SVM).
12. What is a fundamental limitation of SVM, and how can it be addressed?

13. What are the main components of an Artificial Neural Network?
14. Explain how the ANN activation functions work.
15. Discuss the applications you would prefer using Naïve Bayes, Decision Trees, SVM, or ANN.

7.6 Lab

Objectives

1. Practice code markdown skills
2. Apply Naïve Bayes classifier to select newspaper publications on tourism and climate change
3. Investigate Naïve Bayes classifier options

Data

The data for this chapter comes from the LexisNexis Newspapers and Wires database of global news media (▶ https://www.lexisnexis.com/ap/academic/form_news_wires.asp). The data is stored in a Microsoft Excel file with two sheets: the LabeledData sheet contains 324 labeled records, and the UnlabeledData sheet contains 2141 unlabeled records.

The goal is to:

(1) Using the labeled dataset, train a classifier to group the publications into two classes: (1) those that are relevant for discussion of climate and tourism interaction and (2) others.
(2) Using the trained classifier, to classify the unlabeled data.

The data is stored as ▶ https://github.com/andykir/DataMiningLectures/raw/main/Newspapers.xlsx.

Variables

Variable	Explanation	Notes
LABEL	Label. YES: relevant: NO: irrelevant	Binominal. Only for labeled data
YEAR	Year of the publication	Ratio

Variable	Explanation	Notes
PUBLICATION	The outlet	String variable
TITLE	Publication title	String variable
LENGTH	Number of words	Ratio
AIRLINES … WETLANDS	41 attributes showing the presence/absence of a topic in the publication. The attribute name defines the topic	Ratio. Values: 0 (lesser than 50% chance of topic presence); 50–100 (probability of the topic presence)

Part 1. Read, Describe, and Clean the Data

Step 1. Read the Data

Every lab starts with identical steps of reading and describing the data. Instead of asking the Assistant to help with coding, we can reuse the code from the previous lab. The only required change is to correct the URL for cloud data storage.

Open a new Anaconda notebook.

Start with a new Markdown cell. Describe the purpose of the code. This is what it looks like in my notebook:

```
The code for the Chapter 7 Lab: Classification
Goal: practice Naive Bayes classification
Data: LexisNexis Newspapers and Wires database of global news
media
```

Do the same for the rest of the code.

First, read data from the labeled dataset. Since we have not read the data from multiple sheets of an Excel fine, we will use the Assistant.

USER:

Read the data from an Excel file stored at the URL ▶ https://github.com/andykir/DataMiningLectures/raw/main/Newspapers.xlsx. Read the sheet LabeledData to DataFrame labeledData. Print labeledData information.

Assistant:

```
import pandas as pd

# Read the Excel file from the URL
url =
"https://github.com/andykir/DataMiningLectures/raw/main/Newspape
rs.xlsx"
labeledData = pd.read_excel(url, sheet_name="LabeledData")

# Print information about the labeledData DataFrame
print(labeledData.info())
```

Explanation:

This code first imports the pandas library as *pd*. Then, we use the *pd.read_excel*() function to read the Excel file from the provided URL and load the *LabeledData* sheet into the DataFrame called labeledData. *Finally*, we use the DataFrame *info*() method to print information about the *labeledData* DataFrame, including attribute data types, non-null values, and memory usage.

Result:

```
RangeIndex: 48 entries, 0 to 47
RangeIndex: 324 entries, 0 to 323
Data columns (total 46 columns):
 # Column                    Non-Null Count Dtype
--- ------                   -------------- -----
 0   LABEL                                       324 non-null    object
 1   YEAR                                        324 non-null    int64
 2   PUBLICATION                                 324 non-null    object
 3   TITLE                                       324 non-null    object
 4   LENGTH                                      324 non-null    int64
 5   AIRLINES                                    324 non-null    int64
 6   BEACHES                                     324 non-null    int64
                            . . .
 45  WETLANDS                                    324 non null    int64
```

Notice that there is no missing data, which helps with the analysis. Regarding data types, the attributes *LABEL*, *PUBLICATION*, and *TITLE* have the *object* type, and the rest are integers.

Print the initial lines of the dataset. To reduce the output size (46 attributes), limit it to the initial six features.

USER:

Print the initial five lines of labeledData. Print only the first six columns.

Assistant:

To print the first five rows of the *labeledData* DataFrame, displaying only the first six columns, you can use the following Python code:

```
print(labeledData.iloc[:5, :6])
```

In this code, *labeledData.iloc*[:5,:6] selects the first five rows and the first six columns of the *labeledData* DataFrame, and then prints the resulting subset of the data.

Result:

```
  LABEL  YEAR                              PUBLICATION  \
0    NO  2016                     The Times (London)
1    NO  2016                                      DNA
2    NO  2016                      Mirror Publications
3    NO  2016                           Times of Oman
4    NO  2016  The Hamilton Spectator (Ontario Canada)

                                          TITLE  LENGTH  AIRLINES
0  Cut duty to give Edinburgh the edge says airpo...     329         0
1  Mumbais CST Agras Taj part of Swachh Bharat drive     388         0
2                                   Foot shoulder     664         0
3  Young Americans would gain a lot from seeing t...     708         0
4     Halton is big winner in Wynnes cabinet shuffle     495         0
```


Notice how we sliced the DataFrame *labeledData* using the *iloc* attribute. The code *labeledData.iloc*[:5,:6] selects the slice of rows from the beginning (row 0) to five and the columns from the beginning (column 0) to six. The following notations are equivalent[3]:

» *labeledData.iloc*[:5, :6]
labeledData.iloc[0:5, 0:6]
labeledData.iloc[0:5].*loc*['LABEL':'AIRLINES']

3 The indexing attribute *iloc* selects rows and columns by their location, and the attribute *loc* selects by index. For example, the column "AIRLINES" index is 6. When we printed the DataFrame data (*print(data.head())*), the left-most column with values from 0 to 4 represented the rows' indices.

Read and describe the unlabeled variable:

```
# Read the Excel file from the URL
url = "https://github.com/andykir/DataMiningLectures/raw/main/\
Newspapers.xlsx"
unlabeledData = pd.read_excel(url, sheet_name="UnlabeledData")

# Print information about the labeledData DataFrame
print(unlabeledData.info())
unlabeledData.iloc[0:5, 0:6]
```

Step 2. Remove Frequent Themes

USER:

Build a horizontal bar chart with the mean values of columns 4 to 44, sorted in ascending order. Label the bars with column names.

Assistant:

```
import matplotlib.pyplot as plt

# Calculate the mean values of columns 4 to 44
mean_values = labeledData.iloc[:, 5:45].mean()

# Sort the mean values in descending order
mean_values_sorted = mean_values.sort_values(ascending=False)

# Create a horizontal bar chart
plt.figure(figsize=(10, 8))
mean_values_sorted.plot(kind='barh')
plt.xlabel('Mean Value')
plt.ylabel('Column Name')
plt.title('Mean Values of Columns 5 to 44')
plt.show()
```

Explanation:

In this code, we first calculate the mean values of columns 5 to 44 using *labeledData.iloc*[:, 5:45].*mean*(). Then, we sort the mean values in descending order using *sort_values(ascending=False)*. Finally, we create a horizontal bar chart using matplotlib to visualize the mean values, with the column names used to label the bars.

Result:

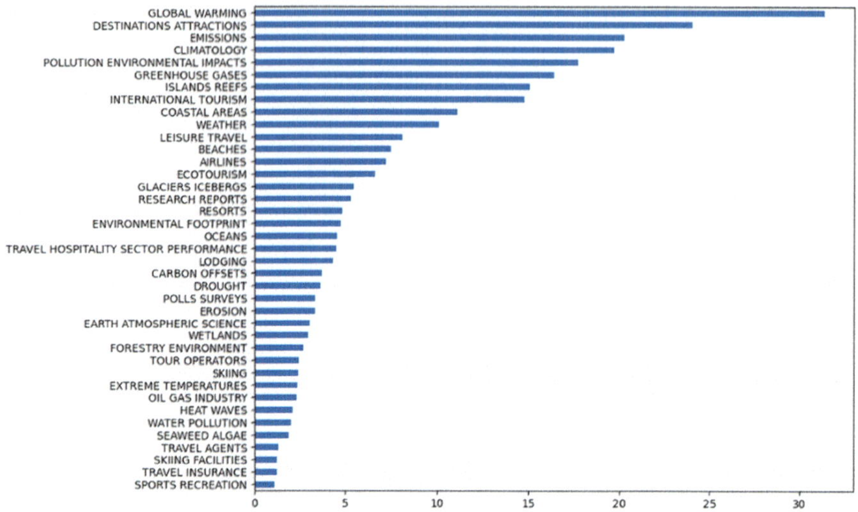

7

The chart shows that the topic "GLOBAL WARMING' is present in 32% of publications. This suggests that this topic could be pervasive and not unique to specific subsets of the dataset, reducing its usefulness for analysis. While this issue is not critical, we will remove this column to demonstrate this operation.

USER:

Remove column 'GLOBAL WARMING' from the DataFrames labeledData and unlabeledData.

Assistant:

```
# Remove columns from labeledData
labeledData = labeledData.drop([' GLOBAL WARMING '], axis=1)

# Remove columns from unlabeledData
unlabeledData = unlabeledData.drop([' GLOBAL WARMING '], axis=1)
```

Explanation:

In this code, the *drop* method removes the specified columns from the DataFrames. The *axis*=1 parameter indicates that the operation should be performed along the columns (as opposed to rows). After running this code, the 'GLOBAL WARMING' column will be removed from both the *labeledData* and *unlabeledData* DataFrames.

Run the code to confirm that the columns are removed:

```
unlabeledData.info()
labeledData.info()
```

Part 2. Train a Naïve Bayes Classifier

Step 1. Classifier Training

USER:

Use the DataFrame "labeledData" to train a Naive Bayes classifier. The labels are in column "LABEL", and the features are in columns 5–43.

Assistant:

```
from sklearn.naive_bayes import GaussianNB
from sklearn.model_selection import train_test_split

# Define features and labels
X = labeledData.iloc[:, 5:44]  # Features
y = labeledData["LABEL"]  # Labels

# Split the data into training and testing sets
X_train, X_test, y_train, y_test = train_test_split(X, y,
test_size=0.2, random_state=42)

# Initialize the Naive Bayes classifier
nb_classifier = GaussianNB()

# Train the classifier
nb_classifier.fit(X_train, y_train)
```

Explanation:

In this code, we first define the features (X) and labels (y) using the specified columns from the *labeledData* DataFrame. We then split the data into training and testing sets using the *train_test_split* function. Next, we initialize the Naive Bayes classifier using *GaussianNB*() method from the *scikit-learn* library and train the classifier using the training data with *fit(X_train, y_train)*. After running this code, the Naive Bayes classifier will be trained using the *labeledData* DataFrame.

Notice that we, once again, use the scikit-learn data mining library. Since the features measurement level is ratio, we use the Gaussian Naïve Bayes:

```
from sklearn.naive_bayes import GaussianNB
```

From the labeled data, 80% are used for training and 20% for testing:

```
X_train, X_test, y_train, y_test = train_test_split(X, y,
test_size=0.2, random_state=42)
```

Running the code returns a trained Naive Bayes classifier:

```
nb_classifier.fit(X_train, y_train)
```


Step 2. Classifier Testing

In the code, notice the variables storing testing data: *x_test* stores the features and *y_test* stores respective labels:

```
# Split the data into training and testing sets
X_train, X_test, y_train, y_test = \
train_test_split(X, y, test_size=0.2, random_state=42)
```

USER

Test performance of the Naive Bayes classifier nb_classifier using the features x_test and the labels y_test. Print the accuracy, precision, recall, and f1-score.

Assistant:

```
from sklearn.metrics import accuracy_score, precision_score,
recall_score, f1_score

# Use the trained classifier to make predictions
y_pred = nb_classifier.predict(X_test)

# Calculate evaluation metrics
accuracy = accuracy_score(y_test, y_pred)
precision = precision_score(y_test, y_pred, average='weighted')
recall = recall_score(y_test, y_pred, average='weighted')
f1 = f1_score(y_test, y_pred, average='weighted')

# Print the evaluation metrics
print("Accuracy:", accuracy)
print("Precision:", precision)
print("Recall:", recall)
print("F1-score:", f1)
```

Explanation:

In this code, we use the trained *nb_classifier* to make predictions on the test features *X_test*. We then calculate the accuracy, precision, recall, and F1-score using the *accuracy_score, precision_score, recall_score,* and *f1_score* functions from the *sklearn.metrics* module. Finally, we print the calculated evaluation metrics.

Result:

```
Accuracy: 0.7692307692307693
Precision: 0.8176941262047646
Recall: 0.7692307692307693
F1-score: 0.7571370428513287
```

The algorithm performs fairly well in discriminating between relevant and irrelevant publications. Now, we can apply it to the unlabeled dataset.

Part 3. Apply The Trained Classifier To Unlabeled Data

Step 1. Classify Unlabeled Data into "Yes" (Relevant) and "No" (Irrelevant)

User:

Apply the trained classifier nb_classifier to the unlabeled dataset unlabeledData.

Assistant:

```
# Assuming the features for prediction are in columns 4 to 42 of
the unlabeledData DataFrame
unlabeled_features = unlabeledData.iloc[:, 4:43]

# Use the trained classifier to make predictions on the
unlabeled dataset
predicted_labels = nb_classifier.predict(unlabeled_features)

# Print the predicted labels
print(predicted_labels)
```

Explanation:

In this code, we extract the features from columns 4 to 42 of the *unlabeledData* DataFrame and then use the trained *nb_classifier* to make predictions on these features. The predicted labels are then printed.

Run the code.

USER:

Print frequency distribution of predicted_labels.

Assistant:

```
import pandas as pd

# Convert the predicted labels to a pandas Series
predicted_labels_series = pd.Series(predicted_labels)

# Print the frequency distribution of the predicted labels
print(predicted_labels_series.value_counts())
```


Explanation:

In this code, we convert the predicted labels array to a pandas Series and then use the *value_counts*() method to calculate the frequency distribution of the predicted labels. Finally, we print the frequency distribution of the predicted labels.

Result:

```
NO     1494
YES     647
```

These results show that approximately 30% of the unlabeled data was classified as relevant for climate change and tourism research.

Include the predicted labels as a column to the unlabeled data. Confirm that the classifier did complete the classification of the unlabeled data. At this point, we have learned Python enough to do that without relying on the Assistant.

Code:

```
unlabeledData['LabelCategorical']=predicted_labels
unlabeledData.info()
```

Result:

```
#    Column                                   Non-Null Count  Dtype
---  ------                                   --------------  -----
 0    YEAR                                     2141 non-null   int64
 1    PUBLICATION                              2141 non-null   object
 2    TITLE                                    2141 non-null   object
 3    LENGTH                                   2141 non-null   int64
 4    AIRLINES                                 2141 non-null   int64
      . . .
44   LabelCategorical                         2141 non-null   object
```

Notice the new column *LabelCategorical,* which stores classification results.

Step 2. Estimate Classification Uncertainty

Recall that the Naïve Byas classifier can return probabilities of class assignment.

USER:

Apply the trained classifier nb_classifier to the unlabeled dataset unlabeledData. The results should represent the probability of class "YES".

Assistant:

```
# Use the trained classifier to predict the probabilities of
each class
predicted_probabilities =
nb_classifier.predict_proba(unlabeled_features)

# Extract the probability of class "YES" (assuming "YES" is the
positive class)
probability_of_yes = predicted_probabilities[:, 1]

# Print the probabilities of class "YES"
print(probability_of_yes)
```

Explanation:

In this code, we use the *predict_proba* method of the trained *nb_classifier* to obtain the predicted probabilities of each class for the unlabeled dataset. We then extract the probability of class "YES" (assuming "YES" is the positive class) and print the resulting probabilities.

Result:

```
[1.00000000e+00 1.00000000e+00 1.00000000e+00 ... 1.90664248e-56
 3.57386028e-34 5.51425104e-22]
```

To understand the relevant class probability distribution, we will draw a histogram.

USER:

Draw a histogram of variable probability_of_yes. Use 10 bins.

Assistant:

```
import matplotlib.pyplot as plt

# Create the histogram with 10 bins
plt.hist(probability_of_yes, bins=10, alpha=0.6, color='b',
edgecolor='black')

# Add labels and title
plt.xlabel('Probability of Yes')
plt.ylabel('Frequency')
plt.title('Histogram of Variable probability_of_yes
Distribution')

# Display the plot
plt.show()
```


Explanation:

In this example, we assume *probability_of_yes* is a list or array of probability values. We use *plt.hist* to create a histogram with 10 bins, and then add labels and a title to the plot. The *matplotlib* package is used to create histogram.

Result:

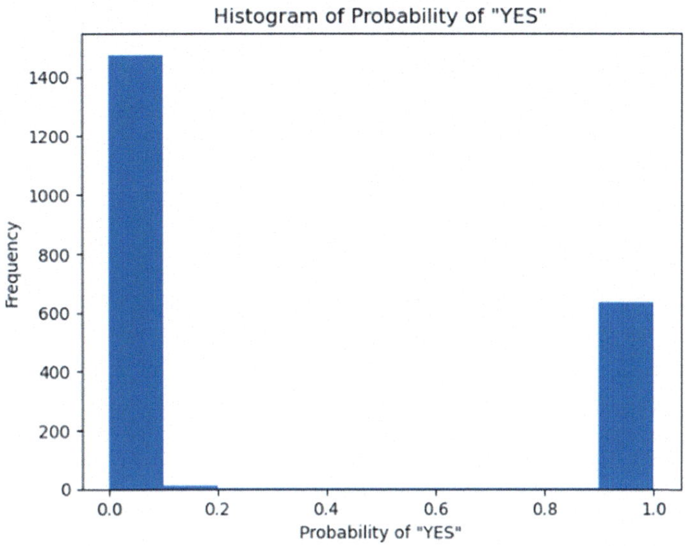

Notice that most records are labeled as "YES" with a probability close to 1 or close to 0. The zero probability of "YES" for a particular publication means it is predicted as unrelated to climate change and tourism.

Next, we will include the probability as new data columns.

Code:

```
unlabeledData['ProbabilityOfYES'] = probability_of_yes
unlabeledData['ProbabilityOfNO'] = 1-probability_of_yes
unlabeledData.info()
```

Result:

```
#   Column                                Non-Null Count   Dtype
---  ------                                --------------   -----
 0   YEAR                                  2141 non-null    int64
 1   PUBLICATION                           2141 non-null    object
 2   TITLE                                 2141 non-null    object
 3   LENGTH                                2141 non-null    int64
 4   AIRLINES                              2141 non-null    int64
    . . .
 44  LabelCategorical                      2141 non-null    object
 45  ProbabilityOfYES                      2141 non-null    float64
 46  ProbabilityOfNO                       2141 non-null    float64
```

Notice two new columns that store classification uncertainty.

Step 3. Compare the Classifier that Uses All 39 Features with a Classifier Using Only the First Five Features

We will reuse the code to compare our results with those based on fewer independent variables.

Code:

Train classifier; use features 5 through 9

```
from sklearn.naive_bayes import GaussianNB
from sklearn.model_selection import train_test_split

# Define features and labels
#X = labeledData.iloc[:, 5:44]  # Features
X = labeledData.iloc[:, 5:10]  # Features
y = labeledData["LABEL"]  # Labels

# Split the data into training and testing sets
X_train, X_test, y_train, y_test = train_test_split(X, y,
test_size=0.2, random_state=42)

# Initialize the Naive Bayes classifier
nb_classifier = GaussianNB()

# Train the classifier
nb_classifier.fit(X_train, y_train)
```

Test classifier

```
from sklearn.metrics import accuracy_score, precision_score,\
recall_score, f1_score

# Use the trained classifier to make predictions
y_pred = nb_classifier.predict(X_test)

# Calculate evaluation metrics
accuracy = accuracy_score(y_test, y_pred)
precision = precision_score(y_test, y_pred, average='weighted')
recall = recall_score(y_test, y_pred, average='weighted')
f1 = f1_score(y_test, y_pred, average='weighted')

# Print the evaluation metrics
print("Accuracy:", accuracy)
print("Precision:", precision)
print("Recall:", recall)
print("F1-score:", f1)
```

Result:

```
Accuracy: 0.6153846153846154
Precision: 0.6197755732639453
Recall: 0.6153846153846154
F1-score: 0.605206114640077
```

Model testing shows that the new algorithm's performance is worse than that of the one trained on the entire set of features.

Draw a histogram of the "Yes" class probability distribution

```
# Classify unlabeled features
unlabeled_features = unlabeledData.iloc[:, 4:9]

# Use the trained classifier to predict the probabilities of
each class
predicted_probabilities =\
nb_classifier.predict_proba(unlabeled_features)

# Extract the probability of class "YES" \
(assuming "YES" is the positive class)
probability_of_yes = predicted_probabilities[:, 1]

# Create a histogram with 10 bins
plt.hist(probability_of_yes, bins=10)
plt.xlabel('Probability of "YES"')
plt.ylabel('Frequency')
plt.title('Histogram of Probability of "YES"')
plt.show()
```

Result:

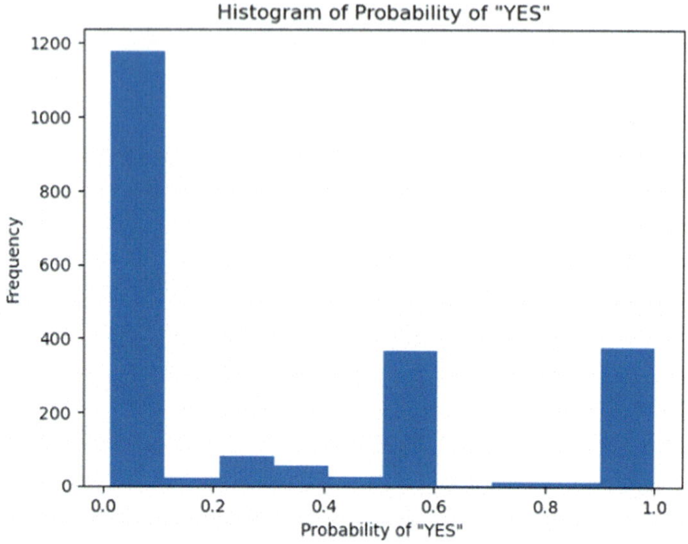

Compared with the previous histogram, many predictions with only five attributes have low confidence. This shows that having a large number of attributes is imperative for text classification.

Part 4. On Your Own

Train and test SVM and Random Forest classifiers on the labeled dataset. Based on classifier validation, which model performs best?

Build Your Python Skills

▶ Chapter 15 of the free book Learn Python with Jupyter by Serena Bonaretti. ▶ https://www.learnpythonwithjupyter.com/.

Additional Reading

Aggarwal, C. C. (2014). *Data algorithms and applications*. Chapman & Hall.

Validation and Evaluation Methods

A. P. Kirilenko, *Practical Data Mining with AI for Social Scientists*,
Springer Texts in Social Sciences, https://doi.org/10.1007/978-3-031-89689-7_8

8.1 Introduction

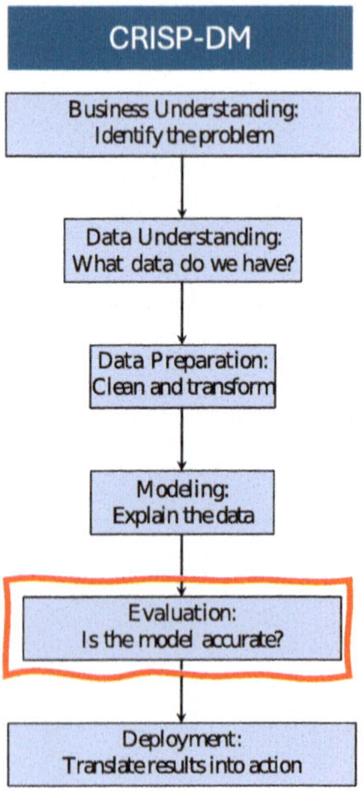

▶ Chapter 1 discussed the Cross-Industry Standard Process for Data Mining (CRISP-DM), the standard methodology for conducting data mining projects. The process is divided into six phases, and in this chapter, we will discuss Phase 5: model evaluation. Specifically, we will talk about evaluating model performance where the dependent variable is categorical (such as in classification tasks).

There are several terms related to the process of assuring that the model design is appropriate for the intended purpose, the model reflects the data, and the model output meets expectations. These terms are field-specific and frequently used interchangeably, adding to confusion. We will narrow the focus to machine learning models, especially supervised learning. In this context, model validation refers to the process of assuring that model parameters are well-selected (the model is "tuned") and the model is not overfitted. Recall that in the previous lab, we split a dataset into two parts (80% and 20% of the data), evaluating model performance on a smaller part of the data and then discussing the measures of error. This was an example of how model validation works.

Hence, *model validation* verifies that the model is well-fitted to data (but not overfitted). If model validation results are unsatisfactory, model parameters would be modified, or the model would be entirely replaced (for example, a Decision Tree classifier could be replaced with a Support Vector Machine). After the final decision on a fully validated model is made, the model is exposed to a new unseen dataset, not involved in the process of model training and testing. This step of *model evaluation* aims to obtain realistic measures of expected model performance in the real world. Note that the purpose of this step is limited to understanding the model's predictive power; the model is not to be replaced based on the evaluation results. Using the social science terminology, model validation relates to ensuring its internal consistency, while the evaluation relates to understanding the external validity of the model.

There are various validation and evaluation approaches related to different types of models. For example, the fit of a linear regression model is often assessed using metrics such as R^2, Mean Absolute Error (MAE), or Root Mean Squared Error (RMSE). Algorithms that summarize texts into a small number of main topics, such as the Latent Dirichlet Allocation discussed in ▶ Chapter 12, utilize perplexity or coherence measures. This chapter focuses on measures and approaches that are most commonly used to assess the quality of *classification tasks*. In classification, the dependent variable (the class) is nominal, which allows for the convenient presentation of model performance as a matrix displaying the count of correctly and incorrectly classified instances.

8.2 Model Performance Evaluation Metrics

The previous chapter started with discussing the performance of the mammogram-based breast cancer test. We can think of this test as a classifier that groups the patients into two classes: those with and without cancer evidence. Suppose 1000 patients pass through the mammography screening. Post-screening, the patients are re-evaluated for cancer presence with the following results (◻ Table 8.1).

In this table, the blue-shaded cells represent correct predictions: four patients were correctly classified as having cancer, while 906 patients were correctly classified as cancer-free. The red-shaded cells represent model errors: 89 patients were incorrectly classified as having cancer, while one patient was incorrectly placed in the cancer-free group. Notice that the goal of model optimization was to minimize the classifier error (the values in the red cells), but how do we measure this error? This is the topic of this subsection.

◻ **Table 8.1** Illustrative outcomes of mammography cancer screening evaluation of 1000 patients

		Actual cancer presence in a patient	
		Positive: cancer	**Negatives: no cancer**
Test prediction	**Positive: cancer**	4 True positives	89 False positives
	Negative: no cancer	1 False negatives	906 True negatives

Table 8.2 Confusion matrix for a binary classifier

		Actual values	
		Positive	**Negative**
Predicted values	**Positive**	a True positives	b False positives: Type I error
	Negative	c False negatives: Type II error	d True negatives

8.2.1 Confusion Matrix

Confusion matrix

A confusion matrix (also known as a contingency table, cross-tabulation, or a crosstab) provides a summary of prediction results by comparing the predicted labels with the actual labels. The diagonal cells of a confusion matrix show correct predictions, while the off-diagonal cells show misclassifications.

8

Table 8.1 shows the frequency distribution of two categorical variables: cancer prediction and presence variables. In statistics, this table is called a contingency table, cross-tabulation, or a crosstab. In the machine learning realm, however, this table is called a confusion matrix. We will start our discussion with a confusion matrix for binary variables, like in the mammography example. Let us consider a binary classification problem with two classes: Positive (P) and Negative (N). The following confusion matrix (Table 8.2) represents the classification performance:

Notice two types of error. Type I error happens when a model predicts a positive test result incorrectly. Type II error is the opposite: predicting a negative result incorrectly. In terms of hypothesis testing, the null hypothesis is that a person is cancer-free. Type I error happens when we reject the null hypothesis in error; type II error is failing to reject the null hypothesis in error. Which of the errors is more important depends on the context.

In many “alarm” type classifiers, the Type II error is more critical. For example, missing an emerging cancer in a patient (Type II error) is significantly more dangerous than falsely alarming a health patient (Type I error). This is one of the reasons the top right cell of Table 8.1 has many more patients than the bottom left cell (the other reason is class imbalance, discussed later in the chapter). In this context, the Type I error is annoying, while the Type II error has serious consequences and should be minimized.

In other situations, the relative importance of errors may be equal or reversed. Think of a jury trial. The null hypothesis is that the suspect is innocent. A Type I error means an innocent person goes to prison, while a Type II error means a guilty suspect walks free. In this context, Type II error is unfortunate. In contrast, the Type I error is immoral and should be as low as possible. To reach this effect, the "beyond a reasonable doubt" legal burden of proof is applied to affirm a criminal conviction.

8.2.2 Common Measures of Model Performance

There are four common measures of model performance: accuracy, precision, recall, and the F-measure.

Accuracy is the proportion of *total* predictions that a model makes *correctly*:

$$\text{accuracy} = \frac{a+d}{a+b+c+d}$$

This measure seems reasonable and comprehensive: it shows how good the model is in making successful guesses. Still, accuracy is *not enough* to understand the model's performance. Take, for instance, the cancer example. A naïve classifier that classifies all patients into the "negative" (that is, cancer-free: Table 8.3) category would be very accurate:

$$\text{accuracy} = \frac{950}{1000} = 0.95$$

Model performance measures

The most commonly used measures are accuracy, precision, sensitivity (also known as recall), f1 measure, and specificity.

Table 8.3 Illustrative outcomes of a naïve classifier that classifies all mammography tests into the Negative group The values reflect the data showing that 0.5% of women coming for mammography have cancer[1]

		Actual cancer presence in a patient	
		Positive: cancer	**Negatives: no cance**
prediction	**Positive: cancer**	0 True positives	0 False positives
	Negative: no cancer	50 False negatives	950 True negatives

1 Women between ages 50–59. US National Center for Biotechnology Research (► https://www.ncbi.nlm.nih.gov/books/NBK550207/).

In fact, this naïve classifier is even more accurate than the real-world mammography test (Table 8.1):

$$\text{accuracy} = \frac{906 + 4}{1000} = 0.91$$

We keep the less-accurate classifier instead of using the naïve classifier that tells all patients they are cancer-free because, in this particular case, we are not very interested in overall accuracy. Rather, we want the test to emphasize the cancerous patients, that is, cell a of the confusion matrix (Table 8.2). The two measures that stress the importance of specific cells of the confusion matrix are precision and sensitivity.

Precision is the proportion of true positive predictions among all positive predictions. It is a measure of a model's ability to identify positive instances correctly:

$$\text{precision} = \frac{a}{a + b}$$

Sensitivity (also called **recall**) is the proportion of true positives the model identifies correctly. It measures the model's capacity to detect all positive instances.

$$\text{sensitivity} = \frac{a}{a + c}$$

Harmonic mean

The harmonic mean is useful when dealing with rates or ratios. Compared to the "regular average", the harmonic mean gives more weight to smaller numbers. This is why the f1 measure is sensitive to low values of both precision and recall.

For the mammography example, the important measure is the test's sensitivity, as it measures the ability to detect all patients with cancer. Note that, once again, this test is not fully sufficient: another naïve classifier that puts all mammography results into the cancer-positive group would have a 100% sensitivity, but its precision would be only 0.5%. **The F1 measure** is a *harmonic mean* of precision and sensitivity:

$$F1 = \frac{2 \times precision \times recall}{precision + recall}$$

Finally, we may also be interested in the model's ability to identify negatives. For instance, a criminal justice researcher would be interested in the court system's ability to identify innocent people correctly. The **specificity measure** is the proportion of actual negatives that were correctly identified.

$$\text{specificity} = \frac{d}{b + d}$$

Table 8.4 Comparison of the main performance measures of the real and "naïve" mammography tests. The "N/A" exemplifies division by 0 calculation failures

Test	Accuracy	Precision	Sensitivity	F1	Specificity
Real	0.91	0.04	0.80	0.08	0.91
"Naïve"	0.95	N/A	0	N/A	1

Table 8.4 compares the performance of the real and "naïve" mammography tests. Notice that the real-world test has high sensitivity while being reasonably good in terms of its accuracy and specificity. In comparison, while the "naïve" test is superior in terms of its accuracy and specificity, it fails in the most important measure: the sensitivity.

8.3 Performance Metrics for Multi-class Classifiers

8.3.1 Accuracy, Precision, Sensitivity, Specificity, and F1

The previous section, 8.2, dealt with binary dependent variables. This section extends the model performance measures to any categorical dependent variable. Table 8.5 provides a generic result of a multi-class classifier. The cells highlighted in blue represent correct predictions; the red cells are misclassifications.

The following measures can be used to understand the performance of this model.

The overall accuracy can be computed as the sum of diagonal values divided by the overall sum of all elements of the confusion matrix:

Table 8.5 Generalized confusion matrix for a multi-class classifier

		Actual values			
		Class 1	Class 2	…	Class N
Predicted values	Class 1	a_{11}	a_{12}	…	a_{13}
	Class 2	a_{21}	a_{22}	…	a_{23}
	…	…	…	…	…
	Class N	a_{N1}	a_{N2}	…	a_{NN}

Table 8.6 True positives (tp), true negatives (tn), false positives (fp), and false negatives (fn) for Class 1 prediction of a multi-class classifier

<table>
<tr><td colspan="2" rowspan="2"></td><td colspan="4">Actual values</td></tr>
<tr><td>Class 1</td><td>Class 2</td><td>…</td><td>Class N</td></tr>
<tr><td rowspan="4">Predicted values</td><td>Class 1</td><td>$tp_1 = a_{11}$</td><td colspan="3">$fp_1 = \sum_{j=2}^{N} a_{1j}$</td></tr>
<tr><td>Class 2</td><td rowspan="3">$fn_1 = \sum_{i=2}^{N} a_{1i}$</td><td colspan="3" rowspan="3">$tn_1 = \sum_{i=2}^{N} a_{ii}$</td></tr>
<tr><td>…</td></tr>
<tr><td>Class N</td></tr>
</table>

$$\text{accuracy} = \frac{\Sigma a_{ii}}{\Sigma a_{ij}}$$

But this measure is not very interesting as the measure is biased towards the large classes. Instead, the average accuracy can be computed by averaging the classification accuracy for each class. Consider classification for class 1. Table 8.6 shows the respective calculations of the number of true positives (the model correctly predicted class 1), true negatives (the model correctly predicted that the data point does not belong to class 1), false positives, and false negatives. Similarly, we can calculate classification accuracies for other classes. The average classification accuracy can then be calculated as the mean accuracies for all N classes.

$$\text{Average accuracy} = \frac{1}{N}\sum_{i=1}^{N} \frac{tp_i + tn_i}{tp_i + tn_i + fp_i + fn_i}$$

Similarly, we can compute precision, sensitivity, specificity, and F1 score by averaging the respective scores for each class. There are many more similar measures of classifier performance; refer to the additional reading for a comprehensive review.

8.3.2 Accounting for By-Chance Agreement: Cohen's Kappa

We have observed that accuracy is a poor measure of model performance when there is a severe imbalance in class distribution, such as in testing for cancer. Cohen's kappa is somehow similar to the accuracy measure in its intent, yet it takes class imbalance into account, hence returning a more objective evaluation of model error.

Cohen's kappa is a statistical measure that compares trained model performance with the performance of a "random guess" model. That is, it compares the *observed* accuracy of a classification model to the accuracy that would be *expected* if the classifications were made randomly:

$$\kappa = \frac{P_o - P_E}{1 - P_E}$$

where P_o is the observed and P_E is the expected agreement by chance alone. The observed agreement is simply the accuracy measure. The expected agreement is calculated from the marginal probabilities and is a bit more involved. As an example, for binary classification (Table 8.2)

$$P_E = \frac{a+b}{S} \times \frac{a+c}{S} + \frac{c+d}{S} \times \frac{b+d}{S}$$

where S = a + b + c + d is the sum of all elements of the matrix. The value of kappa is distributed from -1 to 1, with the following interpretation[2]:

- 0.01–0.20: Slight agreement.
- 0.21–0.40: Fair agreement.
- 0.41–0.60: Moderate agreement.
- 0.61–0.80: Substantial agreement.
- 0.81–1.00: Almost perfect agreement.

To illustrate Cohen's kappa, let us compute the expected agreement for the mammography example (Table 8.1). Out of 1000 patients, there are five patients with and 995 patients without cancer. That is, the probability that any new patient has cancer is 5/1000 = 0.005. Similarly, the probability that the patient is cancer-free is 995/1000 = 0.995.

Consider a random classifier that provides *random* cancer diagnoses while keeping the same proportion of positive and negative predictions as the original mammography test, that is, 93/907. The probability that a

Cohen's kappa

Cohen's kappa compares trained model performance with the performance of a "random guess" model. The assumption is, a successfully train model should perform better than a random guess classifier.

2 Landis, J. R., & Koch, G. G. (1977). The measurement of observer agreement for categorical data. *Biometrics, 33*(1), 159–174.

patient will be diagnosed by the random classifier with cancer is 93/1000 = 0.093, while the probability of cancer-free diagnosis is 907/1000 = 0.907.

To compute the chances of a random classifier making a correct prediction, we can use the probability multiplication rule. The chance that a new patient will be correctly diagnosed with cancer is $0.005 \times 0.093 = 0.000465$. Here, 0.005 is the probability that the patient indeed has cancer, and 0.093 is the probability that the random classifier returns the cancer diagnosis. Similarly, the chance that a new patient will be correctly diagnosed as cancer-free is $0.995 \times 0.907 = 0.902465$. Hence, the chance of the random classifier to return the correct answer is:

$$P_E = 0.000465 + 0.902465 \sim 0.903$$

This means that if the classifier was making random predictions based only on the class distribution, we would expect it to agree with the actual labels 90.3% of the time just by chance. Compare that with accuracy of the real mammography test computed at the beginning of this section:

$$P_O = 0.91.$$

Using the kappa statistics formula,

$$\kappa = \frac{P_o - P_E}{1 - P_E} = \frac{0.91 - 0.903}{1 - 0.903} \sim 0.07.$$

The low value of the kappa shows that the real mammography test performs only slightly better than a random guess. This seemingly deficient outcome is due to the fact that mammography targets high specificity (not missing the cancer) rather than accuracy. Comparably, for the "naïve" cancer test $\kappa = 0$.

8.4 Performance Evaluation Procedures

8.4.1 Holdout Method

The previous sections introduced common measures of model performance. In this section, we cover strategies for the actual model evaluation. Recall that model evaluation *should not* be done on the same data used for model training. For this reason, in the labs, we divided the labeled dataset into two parts, with subsequent use of the larger part for model training and the smaller part for model testing.

This approach is called the *holdout* method. The holdout evaluation procedure works as follows. First, classification model parameters are adjusted to optimize the model performance using a performance measure of choice, for instance, maximizing accuracy, maximizing sensitivity, or maximizing accuracy while keeping sensitivity above 0.8. Then, the trained model is exposed to the testing part of the dataset, and its performance is evaluated again. Only the latter performance measures are relevant and should be reported.

Sometimes, machine learning involves an extra step. For instance, a researcher wants to compare the performance of multiple model types with the goal of selecting the best performer for further data processing, such as selecting between a decision tree and a naïve Bayes classifiers. In another example, an analyst wants to fine-tune a trained model, adjusting its hyperparameters. For instance, the analyst would compare decision tree models that use Gini impurity and Information gain splitting decisions and select the best performing one. This, however, would require a three-part rather than two-part data split into the training (used for training the models), validation (used for final model selection), and testing (used for a final performance test of the selected model).

A common split ratio in the two-part split holdout method is 80% training and 20% testing, or 70% training and 30% testing. In the three-part split, the training set is commonly 60–70% of the data, while validation is completed on 10–20% and testing on 20% of the dataset.

8.4.2 Cross-Validation

The holdout procedure has a fundamental problem of "wasting" labeled data. Labeled datasets are typically manually processed, which makes them expensive and frequently small in size. By commonly allocating 1/3 of labeled data for model testing, we reduce the amount of data allocated for training. Moreover, we increase the risks of allocating a non-representative sample of data.

A cross-validation procedure is common in dealing with this problem. In cross-validation, multiple training/testing procedures are performed on different random samples of the labeled data. The averaged performance evaluation measures are then used to understand model performance. Since cross-validation is based on multiple

Cross-validation

The idea of cross-validation is to split dataset into multiple parts ("folds"). In multiple iterations, the model is trained on a combination of several folds and then tested on the remaining fold not used during training.

iterations of training/testing, it is justifiable to allocate a much smaller part of the data for testing during each iteration, commonly only 10%. This particular procedure is called a *10-fold cross-validation*.

In the 10-fold cross-validation, the labeled dataset is separated into ten blocks of equal size. This is followed by ten testing/training iterations, during which one of these blocks is allocated for model testing, and the other nine are used for training. This is a "standard" model evaluation procedure; note, however, that (1) the training/testing computational requirements increase 10-fold, and (2) 10% of the data is still lost for model testing.

One approach to dealing with the first problem is the *leave-one-out* procedure. Leave-one-out is an "extreme" case of cross-validation, which selects only one instance of labeled data for testing while leaving the rest for model training. Essentially, this is an N-fold cross-validation, where N is the size of the labeled dataset. This procedure, however, requires a dramatic (N-times) increase in the computational requirements, even more exacerbating the second problem.

8.4.3 Bootstrapping Procedure

0.632 bootstrap

The idea of bootstrapping is to perform repeated sampling with replacement from the original dataset to create multiple training sets, using non-selected data points for testing. Bootstrapping is the most useful with small datasets.

Bootstrap is a popular performance evaluation approach when the amount of data is limited. Contrasting cross-validation, bootstrapping uses dataset sampling with replacement. In addition, the sampling is done for training rather than testing the dataset. The data point sampled for training is returned to the labeled dataset, meaning that the same data point can be selected for training multiple times. A popular procedure variant called a *0.632 bootstrap* repeats this resampling procedure N times, where N is the number of points in the labeled dataset.

The name for the 0.632 bootstrap reflects the chances for a data point to be selected for the training dataset. We can expect that approximately 63% of labeled data points will be in the training dataset. Some of these points will be selected more than once, bringing the total training data size to N. The remaining labeled data (approximately 37%) will be used for testing. Similar to cross-validation, boot-straping involves multiple iterations of this procedure, with the final evaluation statistics averaged across all iterations.

8.5 Comparing Model Performance at Different Classification Thresholds

8.5.1 Cost-Sensitive Learning

In the ► Chapter 7 lab, we used the Naïve Bayes classifier to group newspaper publications into the relevant ("positive") and irrelevant ("negative") to climate change discussion. The result of the Naïve Bayes classifier is probabilistic, meaning that the final classification depends on the probability threshold that the researcher selects to classify a data point as "positive." For instance, a 0.9 threshold would classify as positive only those data points for which the probability of being positive returned by the classifier is at least 0.9.

Cost-sensitive learning

Cost-sensitive machine learning methods try to incorporate the cost of errors into the learning process.

Selecting a relevant probability threshold is a very important step in model development which should be based on the researcher's expert knowledge. A large threshold would guarantee that the majority of data points in the "positive" group are selected correctly, but many of the "positive" points are placed into the "negative" group. A low threshold would guarantee that most data points in the "negative" group are selected correctly, but many "negative" points are misclassified. A 0.5 threshold value would balance misclassifications in both classes.

Why would a researcher want to set a threshold value other than 0.5? This decision relates to the costs associated with making classification errors. In the climate change articles example, the cost of selecting an irrelevant publication for research is high as it would bias research outcomes. On the other hand, missing some relevant publications is unlikely to affect the analysis of mass media discussion of tourism's impact on climate change. Hence, a high threshold value of 0.8 or even 0.9 is justified.

Conversely, in the mammography example, the cost of missing a cancer positive is high, but the cost of misclassifying a "negative" patient as cancer-positive is low. Hence, a low threshold value of 0.1 or even lower is justified. Finally, a 0.5 threshold value is justified when no additional information regarding misclassification costs exists.

In the previous sections, we discussed approaches to evaluate model performance. Notice that all these approaches rely on the confusion matrix. However, modifying the threshold would, in turn, modify the numbers in the confusion matrix. For example, a high threshold would increase the number of false negatives and true negatives. It will also reduce the number of false positives and true positives. In turn, that would change all model performance measures.

It would be useful to have an overall representation of how selecting the class thresholds affects classification performance. Two popular approaches are the gain/lift charts and the ROC curve.

8.5.2 Gain and Lift Charts

The gain and lift chart approach comes from the marketing and sales applications. It mostly deals with visualization of the model's ability to predict the true positives at different threshold rates. The classical example is from the field of advertisement.

Imagine a company advertising on a search engine. The company decided on the "pay-per-impression" model, paying for every event of their advertisement shown to a user. The company aims to maximize the number of people interested in their advertisement, as measured by the number of clicks on their ad.

One strategy is to show the ads to every person using the search engine, but the cost would be prohibitive. Hence, it makes sense to show the ad only to those users who are the most likely to be interested in the advertised product or service. Hence, search engines use classification models predicting users' interest in specific ads (in reality, in keywords characterizing the ads) based on the users' search history. The businesses can bid on having their ads shown; a higher bidder would have more frequent impressions on the search engine. This advertisement model is used not only on the search engine but on millions of other sites.

Let t represent the classification model probability threshold: $t \in [0, 1]$. The value of t = 1 means that the ad would be shown to all users, and t = 1 means that the ad

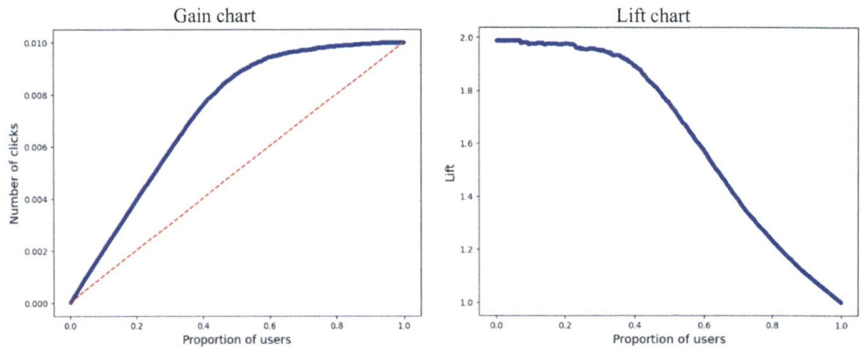

Fig. 8.1 Gain and lift charts

would never be shown. The gain chart is a graph where the y-axis represents the number of true positives (the users who clicked on the ad), and the x-axis represents the proportion of users shown the ad over the total number of users. Since this proportion depends on the model's predictions,

$$x = \frac{tp(t) + fp(t)}{N}; y = tp(t)$$

where tp(t) and fp(t)—the number of true and false positives as a function of the threshold value t; N—the total number of users (Fig. 8.1). The gain chart shows how an increase in the advertisement campaign affects the total number of ad clicks: first, the number of clicks grows rapidly as the ads are shown to the most likely customers only. With an increasing proportion of users involved in the advertisement campaign, the gain from each additional impression slows down and, finally, stalls completely.

The red diagonal line on the gain chart shows the expected total number of clicks from random advertisement impressions. Hence, the distance between the blue and the red lines reflects the model's ability to identify the potential customer. A complementary lift chart shows the ratio of the model gains to the random advertisement impressions' gains. According to the lift chart in Fig. 8.1, a model-based selection of up to 30% of users for the advertisement campaign is twice as effective as selecting a random sample of the same size.

8.5.3 Response Operator Curve and Area Under Curve

Unlike the gain and lift charts developed for business world needs, the Receiver Operating Characteristic (ROC) curve and the associated Area Under Curve (AUC) came from technology. In fact, the ROC curve was initially used in radar development during WWII to describe the tradeoff between the successful target identification rate and false alarms, which explains its name. The y-axis of the ROC chart is the same as in the gain chart, except that it is expressed as a proportion or percentage. The x-axis represents the false positive rate, expressed as a proportion or percentage:

$$x = \frac{fp(t)}{fp(t) + tn(t)};\ y = \frac{tp(t)}{tp(t) + fn(t)}$$

where tp(t) and fp(t)—the number of true and false positives as a function of the threshold value $t \in [0, 1]$; tn(t) and fn(t) are the number of true and false positives (Fig. 8.2).

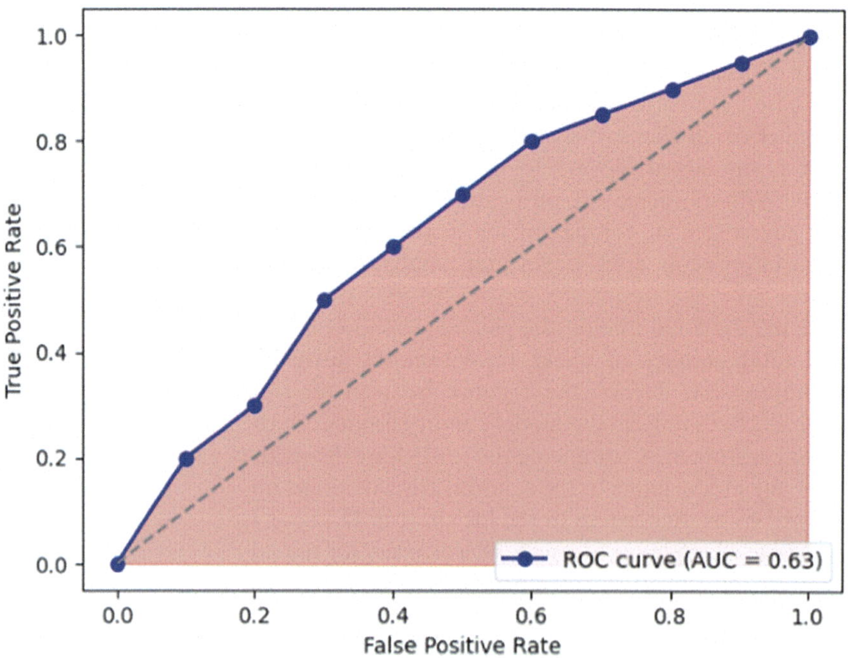

Fig. 8.2 ROC curve. The red-filled area shows the AUC

Note that the true positive rate (the y-axis) is the model's sensitivity, and the false positive rate equals 1—model specificity, meaning that the ROC curve shows the relationship between two commonly used model performance measures that we discussed in ► Sect. 8.2.2. The area under the ROC curve, commonly called the AUC, quantifies the classifier's overall performance, irrespective of the classification threshold. The AUC is the probability that the model will rank a *randomly* chosen true positive higher than a randomly chosen negative instance. An AUC of 1 indicates a perfect classifier, while an AUC of 0.5 suggests that the classifier is no better than a random guess.

AUC values

AUC takes values from 0 to 1 with AUC = 0.5 indicating a random guess. A value below 0.5 indicates a worse than random guess, 0.7–0.8 is acceptable, 0.8–0.9 is excellent, and AUC >0.9 is outstanding. Mandrekar, J. N. (2010). Receiver operating characteristic curve in diagnostic test assessment. *Journal of Thoracic Oncology, 5*(9), 1315–1316.

8.5.4 Comparison of Multiple Models

We have covered multiple models, and every model has multiple parameters affecting its performance. How can we decide which model is better suited for our purpose? For any particular threshold value, models' performance evaluation metrics change, potentially making one model superior to the other with high threshold values and inferior with low thresholds. The ROC and Gain charts can help in making an overall model comparison.

◘ Figure 8.3 shows the ROC curves of three competing classification models. Notice that Model 1 is superior to Model 2 when the classification threshold is 0.3 or smaller. Otherwise, Model 2 is superior to Model 1. Meanwhile, Model 3 is superior to Model 1 and Model 2 and should be retained for further analysis.

8.6 Conclusion

In this section, we mostly dealt with assessing the performance of binary classification models. Slightly modified methods, however, can be modified for multiclass classifiers. For example, ROC and Gain charts can be used to evaluate multiclass classifiers by binarizing the dependent variable. Then, each pair of classes can be compared one to another, or any individual class can be compared to the rest of the classes. Note that there are many more methods of model performance evaluation that we did not cover. These methods and measures are frequently field-specific. For example, to introduce sensitivity and

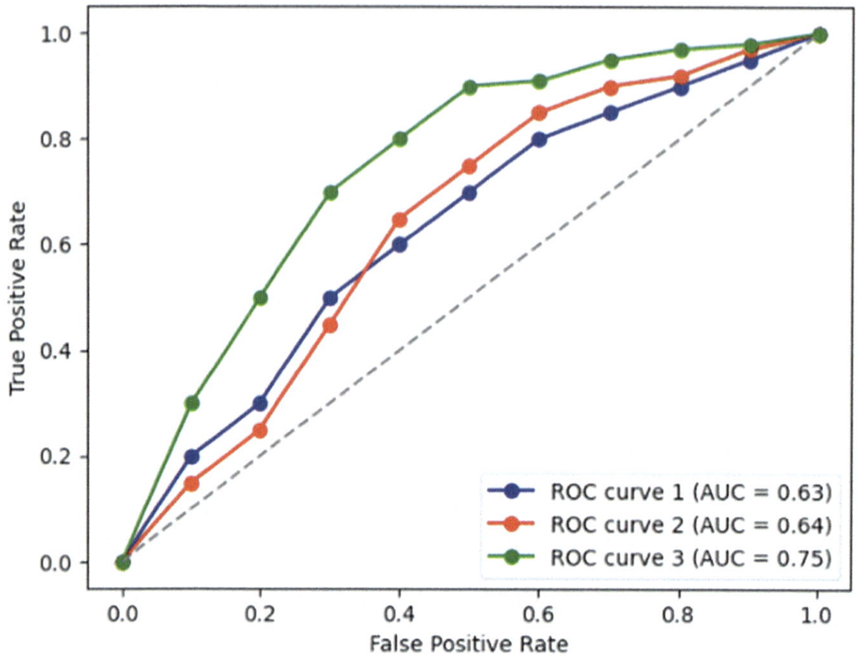

8

Fig. 8.3 ROC curves of three competing models

specificity, we used the mammography case because these two measures are preferred by medical specialists.

We concentrated on the performance measures of the models with categorical labels on purpose, as the measures for the interval/ratio labels are well covered in the research methods and statistics. That includes the mean squared error (MSE), root mean squared error (RMSE), root relative squared error (RRSE), correlation, and many others. The model approach, however, does not change and involves model training and testing on appropriate labeled data samples.

8.7 Review Questions

1. Explain the difference between model validation and model evaluation in model assessment.
2. In a confusion matrix for a binary classification problem, what do the terms True Positive, False Positive, False Negative, and True Negative represent?

3 Explain the meaning of the Type I and Type II errors, and in what scenarios might one be more critical than the other?
4 Why is accuracy alone insufficient to evaluate a classifier's performance, particularly in imbalanced datasets?
5 Define precision, sensitivity, and F1 measures. In which situations would one of these measures be useful?
6 Describe Cohen's kappa. Why could it be useful for model evaluation?
7 Describe the cross-validation method and explain its advantages over the simple holdout method.
8 What is the 0.632 bootstrap method, and how does the data resampling mechanism work?
9 Why might a researcher choose a classification threshold other than 0.5? What factors should influence this decision?
10 Explain the purpose of the Gain and Lift charts in model evaluation in the context of marketing and sales applications.
11 How do the ROC curve and the Area Under Curve (AUC) help compare the performance of different models across varying thresholds?

8.8 Lab

Objectives

1. Practice code markdown skills
2. Practice model evaluation
3. Practice model performance comparison

This lab aims to build, train, and test a model that would predict hotel reservations that are likely to be canceled by hotel guests. The priority is to minimize the incorrect predictions of reservation cancellations. Because of that, to evaluate model performance we will use precision, that is, the ratio of correctly predicted cancellations predictions to the sum of correctly and incorrectly predicted cancellations.

Data

The data for this chapter comes from the Hotel Booking Cancellation Prediction dataset organized by Youssef Aboelwafa (▶ https://www.kaggle.com/datasets/youssefaboelwafa/hotel-booking-cancellation-prediction). The data is stored in a Microsoft Excel file and contains 36,285 records.

The goal is to:

1. Using the labeled dataset, train a classifier to group hotel reservations into two classes: (1) those that are relevant for discussion of climate and tourism interaction and (2) others.
2. Using the trained classifier, classify the unlabeled data.
3. Complete model performance evaluation.

The data is stored as ▶ https://github.com/andykir/DataMiningLectures/raw/main/HotelBookingDataset.xlsx.

Variables

Variable	Explanation	Notes
Booking_ID	Unique identifier for each booking	String
Number of adults	Number of adults included in the booking	Ratio
Number of children	Number of children included in the booking	Ratio
Number of weekend nights	Number of weekend nights included in the booking	Ratio
Number of week nights	Number of week nights included in the booking	Ratio
Type of meal	Type of meal included in the booking	Categorical
Car parking space	Indicates whether a car parking space was requested or included in the booking	0 (No)/1 (Yes)
Room type	Type of room booked	Categorical
Lead time	Number of days between the booking date and the arrival date	Ratio
Market segment type	Type of market segment associated with the booking	Categorical (Online, Offline, etc.)
Repeated	Indicates whether the booking is a repeat booking	Binary (0/1)

Variable	Explanation	Notes
P–C	Number of previous bookings that the customer canceled before the current booking	Ratio
P-not-C	Number of previous bookings not canceled by the customer before the current booking	Ratio
Average price	The average price associated with the booking	Ratio
Special requests	Number of special requests made by the guest	Ratio
Date of reservation	Date of the reservation	Month/day/year
Booking status	Status of the booking (canceled or not canceled)	Binary: Cancelled/ Not_ Cancelled

Part 1. Read, Describe, and Clean the Data

Step 1. Read the Data

Every lab starts with similar steps of reading and describing the data. Because of that, we reuse an older code from ▶ Chapter 7 with minimal modifications. Start with a new Markdown cell. Describe the purpose of the code. Then, describe the purpose of this step.

```
The code for Chapter 8 Lab: Model evaluation
Goal: practice model evaluation
Data: Hotel reservation dataset

Part 1. Read, describe, and clean the data.
Step 1. Read the data
```

```
import pandas as pd

# Read the Excel file from the URL
url = \
"https://github.com/andykir/DataMiningLectures/raw/main/HotelBookingDa
taset.xlsx"
data = pd.read_excel(url, sheet_name="booking")

# Print information about the data DataFrame
print("Data info")
data.info()
print("\n\n Data head")
data.head()
print("\n\n Data description")
data.describe()
```

Notice alternative ways of sending data to the screen or another standard output device:

```
print(data.info())
```

and

```
data.info()
```

The outcomes of these snippets are identical: the value is sent to the screen. The second version, however, is non-standard and specific to the Jupyter. This is why we are using the first version.

Notice how we used the "Line Feed" character '\n' to skip one additional line for better output formatting.

Result:

Data info

```
<class 'pandas.core.frame.DataFrame'>
RangeIndex: 36285 entries, 0 to 36284
Data columns (total 17 columns):
 #   Column                   Non-Null Count  Dtype
---  ------                   --------------  -----
 0   Booking_ID               36285 non-null  object
 1   number of adults         36285 non-null  int64
 2   number of children       36285 non-null  int64
 3   number of weekend nights 36285 non-null  int64
 4   number of week nights    36285 non-null  int64
 5   type of meal             36285 non-null  object
 6   car parking space        36285 non-null  int64
 7   room type                36285 non-null  object
 8   lead time                36285 non-null  int64
 9   market segment type      36285 non-null  object
 10  repeated                 36285 non-null  int64
 11  P-C                      36285 non-null  int64
 12  P-not-C                  36285 non-null  int64
 13  average price            36285 non-null  float64
 14  special requests         36285 non-null  int64
 15  date of reservation      36285 non-null  object
 16  booking status           36285 non-null  object
dtypes: float64(1), int64(10), object(6)
memory usage: 4.7+ MB
```

Notice that there are no missing values, which simplifies data cleaning. In addition, we can exclude two features, *Booking_ID* and *date of reservation*, from the analysis since they don't affect room cancellations.

Data head

	Booking_ID	Number of adults	Number of children	Number of weekend nights	...	Booking status
0	INN00001	1	1	2	...	Not_ Cance- led
1	INN00002	1	0	1	...	Not_ Cance- led
2	INN00003	2	1	1	...	Cance- led
3	INN00004	1	0	0	...	Booking status
4	INN00005	1	0	1	...	Booking status

The label *booking status* is binary, which is suitable for all classification methods. The features, however, are a mix of categorical and ratio types, restricting the classification approaches to those dealing with both types of independent variables.

Data description

```
       number of adults  number of children  number of weekend nights  \
count      36285.000000        36285.000000              36285.000000
mean           1.844839            0.105360                  0.810693
std            0.518813            0.402704                  0.870590
min            0.000000            0.000000                  0.000000
25%            2.000000            0.000000                  0.000000
50%            2.000000            0.000000                  1.000000
75%            2.000000            0.000000                  2.000000
max            4.000000           10.000000                  7.000000

       number of week nights  car parking space     lead time      repeated \
count           36285.000000       36285.000000  36285.000000  36285.000000
mean                2.204602           0.030977     85.239851      0.025630
std                 1.410946           0.173258     85.938796      0.158032
min                 0.000000           0.000000      0.000000      0.000000
25%                 1.000000           0.000000     17.000000      0.000000
50%                 2.000000           0.000000     57.000000      0.000000
75%                 3.000000           0.000000    126.000000      0.000000
max                17.000000           1.000000    443.000000      1.000000

                P-C       P-not-C  average price  special requests
count  36285.000000  36285.000000   36285.000000      36285.000000
mean       0.023343      0.153369     103.421636          0.619733
std        0.368281      1.753931      35.086469          0.786262
min        0.000000      0.000000       0.000000          0.000000
25%        0.000000      0.000000      80.300000          0.000000
50%        0.000000      0.000000      99.450000          0.000000
75%        0.000000      0.000000     120.000000          1.000000
max       13.000000     58.000000     540.000000          5.000000
```

There are potential outliers and features with highly skewed distributions. We will deal with these problems in the next step.

Step 2. Describe, Clean, and Transform Data

We will start with visualizing data distribution to understand the need to remove outliers and transform variables.

USER:

Draw histograms for the distribution of the following columns of DataFrame data: "number of adults", "number of children", "number of weekend nights", "number of week nights", "car parking space", "lead time", "repeated", "P–C", "P-not-C", "average price", and "special requests".

Assistant:


```
import matplotlib.pyplot as plt
import pandas as pd
import math

# Assuming 'data' is the DataFrame containing the mentioned columns

columns_to_plot = ["number of adults", "number of children", \
"number of weekend nights", "number of week nights", \
"car parking space", "lead time", "repeated", "P-C", \
"P-not-C", "average price","special requests"]

num_plots = len(columns_to_plot)
num_cols = 4
num_rows = math.ceil(num_plots / num_cols)

fig, axes = plt.subplots(num_rows, num_cols, figsize=(15, 15))

for i, column in enumerate(columns_to_plot):
    row = i // num_cols
    col = i % num_cols
    ax = axes[row, col]
    ax.hist(data[column], bins=20, color='skyblue')
    ax.set_title(f'Distribution of {column}')
    ax.set_xlabel(column)
    ax.set_ylabel('Frequency')

# Remove any unused subplots
for i in range(num_plots, num_rows * num_cols):
    fig.delaxes(axes.flatten()[i])

plt.tight_layout()
plt.show()
```

Explanation:

In this code, we calculate the number of rows needed based on the number of columns to plot and then create subplots using *plt.subplots*(). We then loop through each column in the list *columns_to_plot* and create a histogram for each column using the appropriate subplot.

Result:

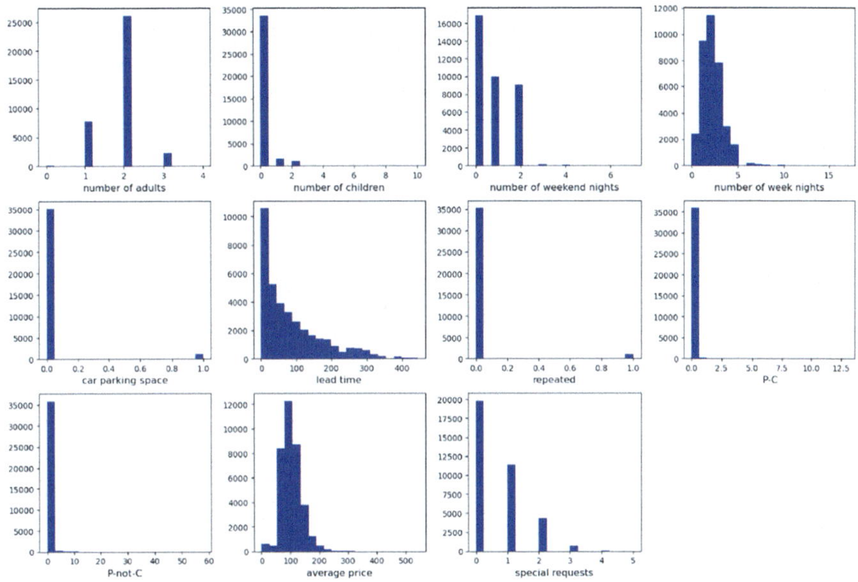

The Naïve Bayes classifier allows both categorical and ratio variables but also assumes that ratio variables are normally distributed. This makes the feature "lead time" a clear candidate for logarithmic transformation. The variables "number of children", "car parking space", "repeated", "P–C", "P-not-C", and "special requests" can be transformed into binary (Yes/No) features. This will remove the problem with normal distribution assumptions and with outliers. Finally, the features "number of weekend nights" and "number of week nights" can be transformed into "reservation length" (ratio) and "weekend reservation" (Yes/No). The outlier values of the "reservation length" feature will be defined as those exceeding seven days and replaced with the maximum value (7).

USER:

Apply logarithmic transformation to column "lead time" + 1. Transform the columns "number of children", "car parking space", "repeated", "P–C", "P-not-C", and "special requests" as follows: all values exceeding zero should be transformed to 1. Summarize columns "number of weekend nights" and "number of week nights", replace all values exceeding 7 with 7, and save as "reservation length". Replace all values of the column "number of week nights" equal to 1 or 2 with 1 and save as column "weekend reservation".

Assistant:

```
import numpy as np

# Apply logarithmic transformation to column "lead time" + 1
data['lead time'] = np.log1p(data['lead time'])

# Transform the columns as specified
columns_to_transform = ["number of children", "car parking space",
"repeated", "P-C", "P-not-C", "special requests"]

for column in columns_to_transform:
    data[column] = np.where(data[column] > 0, 1, 0)

# Summarize columns "number of weekend nights" and "number of week
nights"
data['reservation length'] = data['number of weekend nights'] +
data['number of week nights']
data['reservation length'] = np.where(data['reservation length'] > 7,
7, data['reservation length'])

# Replace values in "number of week nights" equal to 1 or 2 with 1 and
save as "weekend reservation"
data['weekend reservation'] = np.where(data['number of week
nights'].isin([1, 2]), 1, 0)
```


Explanation:

In this code, we use the *np.log1p*() function from the *numpy* package to apply a logarithmic transformation to the "lead time" column. We then use the *np.where*() function to transform the specified columns to 1 if the value exceeds 0, and 0 otherwise. We also create a new column "reservation length" based on the summarized values of "number of weekend nights" and "number of week nights", and replace values exceeding 7 with 7. Finally, we create a new column "weekend reservation" based on the values in the "number of week nights" column.

We can reuse the histogram code to see the distribution of the modified features. **Do not forget to change variable names!**

```
columns_to_plot = ["number of adults", "number of children",\
'reservation length', 'weekend reservation',\
"car parking space", "lead time", "repeated", "P-C", "P-not-C",\
"average price","special requests"]

num_plots = len(columns_to_plot)
num_cols = 4
num_rows = math.ceil(num_plots / num_cols)

fig, axes = plt.subplots(num_rows, num_cols, figsize=(15, 10))

for i, column in enumerate(columns_to_plot):
    row = i // num_cols
    col = i % num_cols
    ax = axes[row, col]
    ax.hist(data[column], bins=20, color='blue')
   # ax.set_title(f'{column}')
    ax.set_xlabel(column, fontsize=12)
    #ax.set_ylabel('Frequency')

# Remove any unused subplots
for i in range(num_plots, num_rows * num_cols):
    fig.delaxes(axes.flatten()[i])

plt.tight_layout()
plt.show()
```

Result:

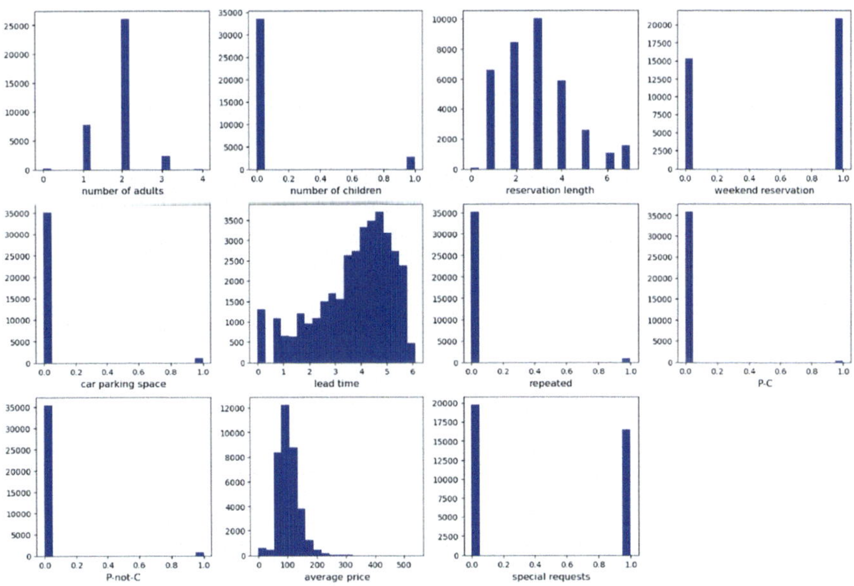

Data transformation was not entirely successful as the feature "lead time" is still heavily skewed. We can replace the logarithmic transformation with the Box-Cox power transformation. Since Box-Cox, similar to the logarithmic function, requires a positive argument, we add one to the lead time.
Replace the code

```
data['lead time'] = np.log1p(data['lead time'])
```

with

```
from scipy.stats import boxcox

data['lead time'] = boxcox(data['lead time']+1)[0]
```


Now, run all cells from the very beginning of the session. You need to do that because the features have been transformed.

Result:

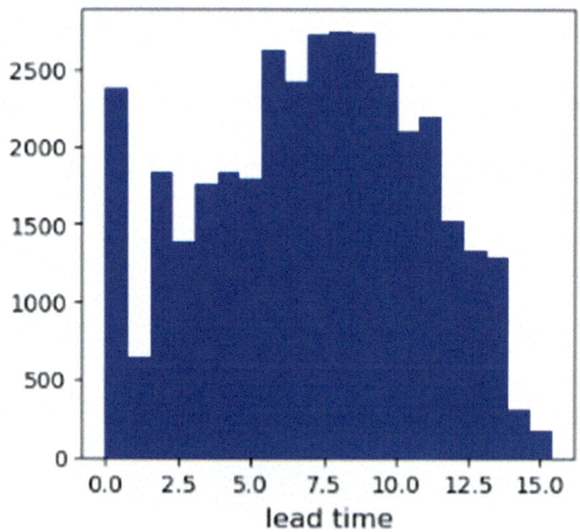

Part 2. Use Naïve Bayes Method to Train a Classifier to Identify Potential Cancellations

Step 1. Write a Draft Code

USER:

Use the DataFrame "data". Train and test a Naive Bayes classifier. The features are "number of adults", "number of children", 'reservation length', 'weekend reservation', "car parking space", "lead time", "repeated", "P–C", "P-not-C", "average price", "special requests". The labels are "booking status". Print accuracy, precision, sensitivity, specificity, and f1 measures.

Assistant:

```
from sklearn.model_selection import train_test_split
from sklearn.naive_bayes import GaussianNB
from sklearn.metrics import accuracy_score, precision_score,
recall_score, f1_score, confusion_matrix
import pandas as pd

# Assuming 'data' is the DataFrame containing the specified features
and labels

# Selecting features and labels
features = ["number of adults", "number of children", 'reservation
length', 'weekend reservation',
            "car parking space", "lead time", "repeated", "P-C", "P-
not-C", "average price", "special requests"]
labels = "booking status"

X = data[features]
y = data[labels]

# Splitting the data into training and testing sets
X_train, X_test, y_train, y_test = train_test_split(X, y,
test_size=0.2, random_state=42)

# Training the Naive Bayes classifier
nb_classifier = GaussianNB()
nb_classifier.fit(X_train, y_train)

# Testing the classifier
y_pred = nb_classifier.predict(X_test)

# Calculating and printing accuracy, precision, sensitivity,
specificity, and f1 measures
accuracy = accuracy_score(y_test, y_pred)
precision = precision_score(y_test, y_pred, pos_label='Canceled')
recall = recall_score(y_test, y_pred, pos_label='Canceled')
f1 = f1_score(y_test, y_pred, pos_label='Canceled')
conf_matrix = confusion_matrix(y_test, y_pred)
tn, fp, fn, tp = conf_matrix.ravel()
specificity = tn / (tn + fp)

print(f'Accuracy: {accuracy}')
print(f'Precision: {precision}')
print(f'Sensitivity/Recall: {recall}')
print(f'Specificity: {specificity}')
print(f'F1 Measure: {f1}')
```

Explanation:

In this code, we first select the specified features and labels from the DataFrame. We then split the data into training and testing sets using *train_test_split*. Next, we train the Naive Bayes classifier using the training data and test it using the testing data. Finally, we calculate and print the accuracy, precision, sensitivity, specificity, and f1 measures using the predicted and actual labels.

Result:

```
Accuracy: 0.4296541270497451
Precision: 0.364656381486676
Sensitivity/Recall: 0.974188176519567
Specificity: 0.974188176519567
F1 Measure: 0.5306724118380769
```

USER:

Print the confusion matrix. Use easily understandable formatting.

Assistant:

```
print('Confusion Matrix:')
print(pd.DataFrame(conf_matrix, columns=['Predicted Not Canceled',
'Predicted Canceled'], index=['Actual Not Canceled', 'Actual
Canceled']))
```

Result:

```
Confusion Matrix:
                     Predicted Not Canceled  Predicted Canceled
Actual Not Canceled                    2340                  62
Actual Canceled                        4077                 778
```

There are several problems with this default code returned by the Assistant.

First, the accuracy and precision are low. Observe the confusion matrix. The majority of reservations predicted to be not canceled in reality were canceled. The reason is that the Gaussian Naïve Bayes the Assistant uses by default is unsuitable for categorical data. We will replace it with the Mixed Naïve Bayes, which can accept both interval/ratio and categorical predictors.

Second, the labels are strings. While the GaussianNB accepts string-type labels, most classifiers, including the Mixed Naïve Bayes, require numerical labels. In general, replacing all string-type variables with numerical codes is a good practice.

Third, the Assistant suggested the "hold out" testing method for model evaluation, which is suboptimal. Adding to that, only 20% of the labeled data is used for testing:

```
# Splitting the data into training and testing sets
X_train, X_test, y_train, y_test = train_test_split(X, y, \
test_size=0.2, random_state=42)
```

We will start by increasing the percentage of data allocated for testing to 30%. After selecting the final model, we replace the holdout validation with 10-fold cross-validation. This decision is made because cross-validation is significantly slower than the holdout method; we will use it only at the end.

In the following steps, these deficiencies will be removed.

Step 2. Replace Gaussian Naïve Bayes with Mixed Naïve Bayes Classifier

Start with installing the Mixed Naïve Bayes module. Run the installation code from a new cell.

```
!pip install mixed-naive-bayes
```


After a successful library install, you need to restart the kernel; otherwise, the new installation will not be recognized. Since restarting the kernel cleans the memory from all preceding calculations, you need to repeat running all cells.

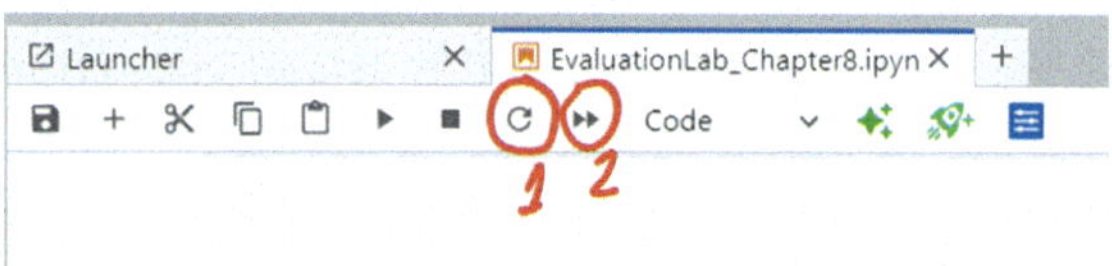

Make sure that no errors are raised.

Edit the cell with the classifier code.

1. **Recode the label**

```
# Selecting features and labels
features = ["number of adults", "number of children", \
'reservation length', 'weekend reservation', \
"car parking space", "lead time", "repeated", "P-C", "P-not-C", \
"average price", "special requests"]
labels = "booking status"

X = data[features]
y = data[labels]

y = [ 1 if n=="Canceled" else 0 for n in y]
print("0: reservation not canceled; 1: reservation canceled")
```

Notice how we used list comprehension for recoding. The new code (in red font) is equivalent to the following loop:

```
y_recoded=[]
for n in y:
  if n=="Canceled":
    recoded.append(1)
  else:
    recoded.append(0)
y= recoded
print("0: reservation not canceled; 1: reservation canceled")
```

List comprehension is a great way of creating a new list based on the values of an existing list.

2. **Increase test data allocation from 20 to 30%:**

```
# Splitting the data into training and testing sets
X_train, X_test, y_train, y_test = \
train_test_split(X, y, test_size=0.3, random_state=42)
```

3. **Replace the Gaussian Bayes with mixed Bayes.**

You can find details on the mixed Bayes method at ► https://pypi.org/project/mixed-naive-bayes/.

```
# Training the Naive Bayes classifier
#nb_classifier = GaussianNB()
#nb_classifier.fit(X_train, y_train)

# Testing the classifier
#y_pred = nb_classifier.predict(X_test)

from mixed_naive_bayes import MixedNB
classifier = MixedNB(categorical_features=[1,3,4,6,7,8,10])
classifier.fit(X_train, y_train)

# Testing the classifier
y_pred =classifier.predict(X_test)
```

The categorical_features parameter in the MixedNB method defines which of the features are categorical with the numeration starting from zero. We selected the following features: "number of children", "weekend reservation", "car parking space", "repeated", "P–C", "P-not-C", and "special requests". The final code is as follows:

```
# Selecting features and labels
features = ["number of adults", "number of children", \
'reservation length', 'weekend reservation', \
"car parking space", "lead time", "repeated", "P-C", "P-not-C", \
"average price", "special requests"]
labels = "booking status"

X = data[features]
y = data[labels]

y = [ 1 if n=="Canceled" else 0 for n in y]
print("0: reservation not canceled; 1: reservation canceled")

# Splitting the data into training and testing sets
X_train, X_test, y_train, y_test = \
train_test_split(X, y, test_size=0.3, random_state=42)

# Training the Naive Bayes classifier
from mixed_naive_bayes import MixedNB
classifier = MixedNB(categorical_features=[1,3,4,6,7,8,10])
classifier.fit(X_train, y_train)

# Testing the classifier
y_pred =classifier.predict(X_test)

# Calculating and printing accuracy, precision, sensitivity,
specificity, and f1 measures
accuracy = accuracy_score(y_test, y_pred)
precision = precision_score(y_test, y_pred, pos_label=1)
recall = recall_score(y_test, y_pred, pos_label=1)
f1 = f1_score(y_test, y_pred, pos_label=1)
conf_matrix = confusion_matrix(y_test, y_pred)
tn, fp, fn, tp = conf_matrix.ravel()
specificity = tn / (tn + fp)

print(f'Accuracy: {accuracy}')
print(f'Precision: {precision}')
print(f'Sensitivity/Recall: {recall}')
print(f'Specificity: {specificity}')
print(f'F1 Measure: {f1}')

print('Confusion Matrix:')
print(pd.DataFrame(conf_matrix, columns=['Predicted Not Canceled',
'Predicted Canceled'], index=['Actual Not Canceled', 'Actual
Canceled']))
```

8

Run the code. Make sure there are no errors.

Results:

```
0: reservation not canceled; 1: reservation canceled
Accuracy: 0.7637332353481536
Precision: 0.6622222222222223
Sensitivity/Recall: 0.5804117974401781
Specificity: 0.8540866703236424
F1 Measure: 0.6186239620403321
```

Confusion Matrix:

```
                     Predicted Not Canceled  Predicted Canceled
Actual Not Canceled                    6228                1064
Actual Canceled                        1508                2086
```

Significant improvements in the model compared with the Gaussian Bayes are evident. The accuracy has increased from 0.43 to 0.76, and the precision (the ratio of correct cancellation predictions) has increased from 0.36 to 0.66. Model sensitivity (the ratio of actual cancellation that the model could predict) has decreased from 0.97 to 0.58, but this measure is of secondary importance. The "balanced" F1 measure has increased from 0.53 to 0.62.

Step 3. Replace the Hold-Out Validation with a 10-Fold Validation Method

Previously, we used Anaconda Assistant to help write and modify the code. In this step, we will demonstrate how to do the same using ChatGPT (or another AI assistant).

Login to ChatGPT: ▸ https://chatgpt.com/.

In the ChatGPT prompt, follow the dialog:

USER:

Act as a computer programmer. Learn the context from the following Python code.

ChatGPT:

Please provide the Python code you'd like me to learn the context from, and I'll be happy to assist!

USER:

<Copy and paste the Mixed Naïve Bayes classifier code from Part 2 Step 2 to the ChatGPT prompt>

ChatGPT:

<Provides analysis of the code>

USER:

Modify the code to use 10-fold cross validation instead of the hold out validation.

ChatGPT:

Here's the modified code to use 10-fold cross-validation instead of the hold-out testing approach:

```
from sklearn.model_selection import cross_val_score, cross_val_predict
from sklearn.metrics import accuracy_score, precision_score,
recall_score, f1_score, confusion_matrix, make_scorer
import pandas as pd
from mixed_naive_bayes import MixedNB

# Assuming 'data' is the DataFrame containing the specified features
and labels

# Selecting features and labels
features = ["number of adults", "number of children", 'reservation
length', 'weekend reservation',
            "car parking space", "lead time", "repeated", "P-C", "P-
not-C", "average price", "special requests"]
labels = "booking status"

X = data[features]
y = data[labels]

y = [1 if n == "Canceled" else 0 for n in y]
print("0: reservation not canceled; 1: reservation canceled")

# Creating the Naive Bayes classifier
classifier = MixedNB(categorical_features=[1, 3, 4, 6, 7, 8, 10])

# Using 10-fold cross-validation
y_pred = cross_val_predict(classifier, X, y, cv=10)

# Calculating and printing accuracy, precision, sensitivity,
specificity, and f1 measures
accuracy = accuracy_score(y, y_pred)
precision = precision_score(y, y_pred, pos_label=1)
recall = recall_score(y, y_pred, pos_label=1)
f1 = f1_score(y, y_pred, pos_label=1)
conf_matrix = confusion_matrix(y, y_pred)
tn, fp, fn, tp = conf_matrix.ravel()
specificity = tn / (tn + fp)

print(f'Accuracy: {accuracy}')
print(f'Precision: {precision}')
print(f'Sensitivity/Recall: {recall}')
print(f'Specificity: {specificity}')
print(f'F1 Measure: {f1}')

print('Confusion Matrix:')
print(pd.DataFrame(conf_matrix, columns=['Predicted Not Canceled',
'Predicted Canceled'], index=['Actual Not Canceled', 'Actual
Canceled']))
```

Changes Made:

1. Removed Hold-Out Splitting: The *train_test_split* function is no longer used since we're switching to cross-validation.
2. 10-Fold Cross-Validation: The *cross_val_predict* function is used with $cv = 10$ to generate predicted labels using 10-fold cross-validation.
3. Evaluation Metrics: The performance metrics (accuracy, precision, recall, specificity, and F1 score) are calculated based on the predictions obtained from the cross-validation.

This version of the code uses the entire dataset for training and testing in a cross-validated manner, which helps in getting a more robust estimate of the model's performance.

Copy the ChatGPT code to a new cell and run it.

Result:

```
0: reservation not canceled; 1: reservation canceled
Accuracy: 0.7629323411878186
Precision: 0.6523028449633954
Sensitivity/Recall: 0.592059887290773
Specificity: 0.8462042957861945
F1 Measure: 0.6207231040564374
Confusion Matrix:
                     Predicted Not Canceled  Predicted Canceled
Actual Not Canceled                   20644                3752
Actual Canceled                        4850                7039
```

The changes in the model performance measures are very small because the amount of labeled data is very large (36,285 data points).

Step 4. Visual Analysis. Draw ROC

We will continue to use ChatGPT for this step.

USER:

Draw a ROC curve. Use probabilities for class “1”.

ChatGPT:

<Provides a code and an explanation>

USER:

<Copy and paste the code into the Anaconda environment. Run the code. The code returns an error.>

USER:

The code returns an error "AttributeError: 'MixedNB' object has no attribute 'classes_'". Correct the error.

ChatGPT:

The error you are encountering may be related to the use of *cross_val_predict* with *method='predict_proba'* for a custom classifier like *MixedNB*. Some custom classifiers might not support the *predict_proba* method in the context of cross-validation, or there could be an issue with how the *MixedNB* class handles the *predict_proba* method.

To address this, we can try the following alternative approach: manually perform cross-validation to calculate the predicted probabilities. Then, plot the ROC curve based on the collected probabilities.

Modified code:

```
import matplotlib.pyplot as plt
from sklearn.metrics import roc_curve, roc_auc_score
from sklearn.model_selection import StratifiedKFold

# Initialize StratifiedKFold for 10-fold cross-validation
skf = StratifiedKFold(n_splits=10)

# List to store predicted probabilities and true labels
y_pred_class_1 = []
y_true = []

# Perform 10-fold cross-validation manually
for train_index, test_index in skf.split(X, y):
    X_train, X_test = X.iloc[train_index], X.iloc[test_index]
    y_train, y_test = [y[i] for i in train_index], [y[i] for i in
test_index]

    # Train the classifier
    classifier.fit(X_train, y_train)

    # Predict probabilities for class 1
    y_pred_proba = classifier.predict_proba(X_test)[:, 1]

    # Append results to lists
    y_pred_class_1.extend(y_pred_proba)
    y_true.extend(y_test)

# Calculate the ROC curve
fpr, tpr, thresholds = roc_curve(y_true, y_pred_class_1)

# Calculate the AUC (Area Under the Curve)
roc_auc = roc_auc_score(y_true, y_pred_class_1)

# Plotting the ROC curve
plt.figure(figsize=(8, 6))
plt.plot(fpr, tpr, color='blue', lw=2, label=f'ROC curve (AUC =
{roc_auc:.2f})')
plt.plot([0, 1], [0, 1], color='gray', lw=2, linestyle='--',
label='Chance')
plt.xlim([0.0, 1.0])
plt.ylim([0.0, 1.05])
plt.xlabel('False Positive Rate')
plt.ylabel('True Positive Rate')
plt.title('Receiver Operating Characteristic (ROC) Curve')
plt.legend(loc='lower right')
plt.grid(True)
plt.show()
```

Explanation:

1. Manual Cross-Validation: Instead of using *cross_val_predict*, we manually perform cross-validation using *StratifiedKFold*. This ensures that the predicted probabilities for each fold are collected and used to plot the ROC curve.
2. Model Training and Prediction: The model is trained on each training fold and tested on the corresponding

test fold. The predicted probabilities for class 1 ("Canceled") are collected from each fold.

3. ROC Curve and AUC Calculation: After collecting the predicted probabilities across all folds, the ROC curve and AUC are calculated. The ROC curve is then plotted using *matplotlib*.

This method should work without errors.

Result:

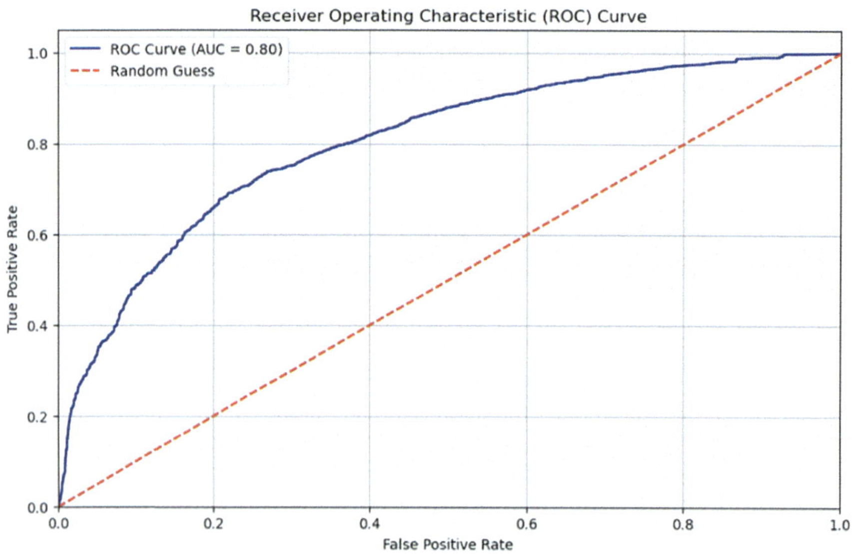

Notice that AUC=0.8, which is considered exceptional: our model *generally* performs much better than a random guess. The ROC curve is significantly higher than the red line, indicating that the model performs better than a random guess not only in general but also at all probability threshold levels.

Part 3. On Your Own

It is unsurprising that model performance evaluations are very similar using the holdout and cross-validation methods since we use over 30,000 data points for model training and testing. However, outcomes may differ for smaller datasets.

1. Modify the code as follows: Select a 5% random sample of DataFrame 'data' (approximately 1800 data

points) and run model training and testing on this moderate-size dataset. Compare the outcomes of the holdout and tenfold validation.

2. Select an even smaller 1% random sample of DataFrame 'data' (approximately 360 data points) and run model training and testing on this moderate-size dataset. Compare outcomes of the hold out, tenfold, **AND bootstrap** validation.

Build Your Python Skills

Chapter 16 of the free book Learn Python with Jupyter by Serena Bonaretti. ▶ https://www.learnpythonwithjupyter.com/.

Additional Reading

Mandrekar, J. N. (2010). Receiver operating characteristic curve in diagnostic test assessment. *Journal of Thoracic Oncology, 5*(9), 1315–1316.

Sokolova, M., & Lapalme, G. (2009). A systematic analysis of performance measures for classification tasks. *Information processing & management, 45*(4), 427–437.

Web Data Scraping

A. P. Kirilenko, *Practical Data Mining with AI for Social Scientists*,
Springer Texts in Social Sciences, https://doi.org/10.1007/978-3-031-89689-7_9

9.1 Introduction

The next few chapters deal with text analysis. In digital social sciences, this frequently means analyzing textual data coming from the Web: social media, customer reviews, websites, newspaper publications, and so on. This chapter introduces methods for extracting large volumes of online data for research purposes.

Web data scraping can be broadly categorized into content mining, structure mining, and usage mining. Content mining, as the name suggests, involves extracting the content of web pages: the texts, user profiles, images, videos, or any other form of data that is part of the website's content. For instance, a university extension office scientist might use Reddit content mining to analyze social media data related to urban gardening in order to provide professional advice to grassroots initiatives. Political scientists might be interested in public acceptance of solar energy; for this purpose, they would extract solar energy-related discussions on Twitter (X). Public health specialists interested in vaping epidemics in youth would use TikTok data to investigate vaping product advertisements on younger-generation social media platforms. All of those projects would involve content web mining.

While web content mining extracts textual and visual data, structure mining focuses on organizing the data on and between the websites. It involves analyzing the structure of web pages and the hyperlinks between them. For instance, a tourism researcher may be interested in how an official tourism organization integrates and promotes local businesses by researching its connectivity with other governmental, commercial, and not-for-profit entities.

Web scraping types

Content mining: scraping the content of web pages.

Structure mining: scraping the hyperlink connectivity of the website and between websites.

Usage mining: analysis of the pattern of users' navigation through the website.

Finally, usage mining investigates the behavior of website visitors. This could involve tracking visitors' navigation patterns through multiple pages of a website or between different websites, analyzing the time a visitor spends on different web pages, links, and buttons clicked, and so on. Together, this type of information is frequently called *clickstream data*. The usage mining could be used to search for popular and unpopular pages, underutilized information, website design faults, drop-off points where users lose interest or encounter issues, leading them to leave the site, and conversion funnels where the users tend to move toward completing a desired action, such as making a purchase.

To accommodate these diverse goals, web data scraping employs diverse approaches and tools. In this book, we target content analysis. Therefore, the rest of this chapter will deal with content mining as the most typical interest of a social scientist.

9.2 API Scraping

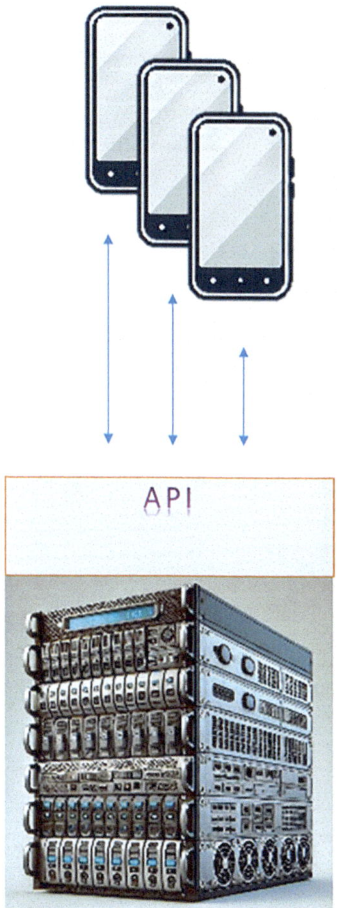

The easiest way of collecting data from the Internet is by using APIs. APIs, or Application Programming Interfaces, are sets of protocols and tools that allow software applications to communicate with each other. In web mining, APIs serve as a bridge between the data source and the end-user, enabling the automated retrieval of web data without manual intervention. This automation is particularly useful for handling large datasets, such as a multi-day collection of social media discussions on a particular topic or hotel reviews for an entire city. In many of these cases, manual data collection would be impractical or overall impossible.

9.2.1 API Example

Consider a hotel booking website. As we enter the parameters of a desirable reservation and press the "Search" button, a *request* is sent to an online server of the booking company. After receiving a valid request, the server executes a relevant program to process it. In our example, the program would search a reservations database for matching room availability. Possibly, it would be using another API request to a third-party data server to retrieve this information. Finally, it will respond with data reflecting search results. Notably, all this data exchange will be hidden from the end user, who would just see the web page updated with hotel availability information (◘ Fig. 9.1).

The APIs provide a structured and standardized method for accessing data, making data retrieval much easier as compared to parsing through a web site looking for relevant information. In fact, many websites allow free (usually limited) API access to anyone interested in the data they provide. For instance, consider the site ▸ dataUSA.io. This site provides access to the US government data about the country, available through

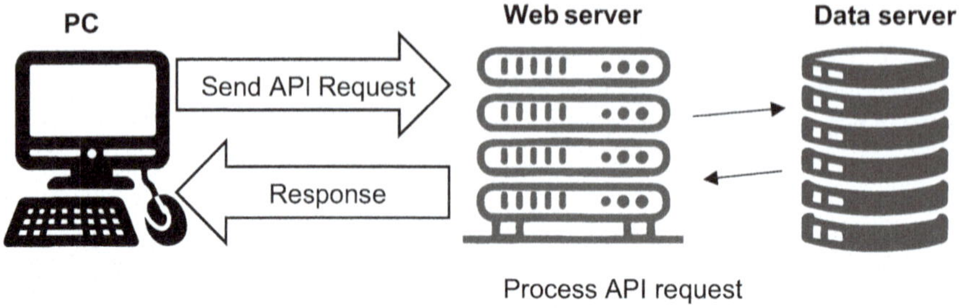

Fig. 9.1 API request processing

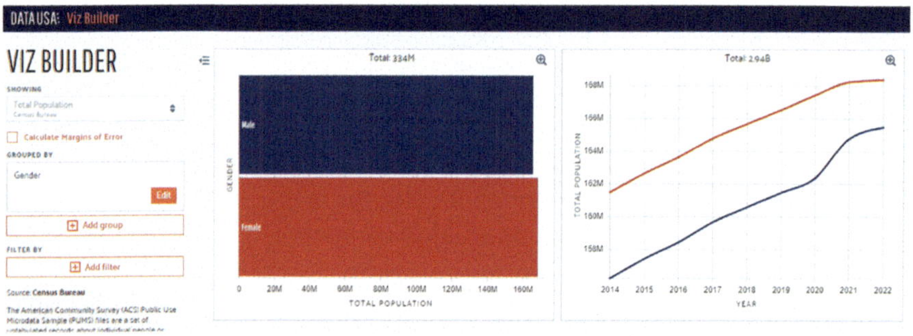

Fig. 9.2 VIZ interface of the DATAUSA.io website

a graphic user interface. For instance, we can use the site's Viz Builder interface (▶ https://datausa.io/visualize) to find the USA population data for the past year (Fig. 9.2). Then, we can write down the numbers we need. This is not the most effective way of retrieving the data.

Another way of accessing the data is through API:

▶ https://datausa.io/api/data?drilldowns=Nation&measures=Population&year=latest.

This call has the following elements:

- The **base URL**: ▶ https://datausa.io/api/data. This is the *base endpoint* of the API, which indicates that the request is being made to the data resource of the datausa.io API. The base URL points to the main service that handles API requests and serves data from the website.
- Three **query parameters**. Everything after the ? in the URL is a series of query parameters that modify the API request and define what data should be returned. The query parameters are key-value pairs separated by the & symbol.

- The parameter **drilldowns** specifies the level at which the data is requested. We request the data for the entire nation, but other values such as "State" or "County" are also available.
- The parameter **measures** specifies which measure is requested. Measures could include other metrics like GDP, employment, or educational attainment, but this query requests the population.
- The parameter **year** specifies the data collection period. The value "latest" means the API should return data for the most recent year available in the dataset, but other possible values could be specific years like "year=2020".

There are many more query parameters described on the DATAUSA website. Following this link returns the following data in *JSON* (JavaScript Object Notation) format, which follows the *attribute:value* specification:


```
response=
{
  "data": [
    {
      "ID Nation": "01000US",
      "Nation": "United States",
      "ID Year": 2022,
      "Year": "2022",
      "Population": 331097593,
      "Slug Nation": "united-states"
    }
  ],
  "source": [
    {
      "measures": [
        "Population"
      ],
      "annotations": {
        "source_name": "Census Bureau",
        "source_description": "The American Community
Survey (ACS) is conducted by the US Census and sent to a
portion of the population every year.",
        "dataset_name": "ACS 5-year Estimate",
        "dataset_link": "http://www.census.gov/programs-
surveys/acs/",
        "table_id": "B01003",
        "topic": "Diversity",
        "subtopic": "Demographics"
      },
      "name": "acs_yg_total_population_5",
      "substitutions": []
    }
  ]
}
```

Comparing this response to the VIS interface (◘ Fig. 9.2), the data returned by API is very specific, free of unneeded information, highly structured, and has details missing from VIS, such as the data source and details on the tables from which the data was extracted. Using the API interface, one can easily automate querying the DATAUSA database, retrieving large amounts of structured data. Further, one can utilize the DATAUSA API to create new web applications and services, perhaps repackaging DATAUSA data for their local communities.

The DATAUSA website provides access to its entire database, which one can download, filter out, and use for research. However, this capability is uncommon and mostly restricted to public data provided by governments, universities, and similar entities. Meanwhile, many organizations provide open APIs allowing access to their data. These services are often free for light users or limited data access.

JSON format

JSON (JavaScript Object Notation) is a very popular format transmitting data between a server and web applications, especially in APIs. Two main building blocks of JSON are objects (key:value pairs) and arrays. Each object is enclosed in {} braces and each array is enclosed in [] brackets.

Notably, the key values can also be objects or arrays, creating a nested structure. For example, the key "Population" could contain the objects "Males" and "Females".

9.2.2 JSON Format

DATAUSA API request returned data in JSON format. JSON is a lightweight, text-based data interchange format that is widely used in web applications to transmit data between a server (in our example, DATAUSA) and a client (your PC). It is designed to be transparent for humans to understand while providing a compact and structured way to represent data. Because of this efficiency, JSON has become a de facto standard in web communication. For instance, many public APIs return responses in JSON format.

JSON is composed of two main structures: *objects* and *arrays*. A JSON array is an ordered list of values of any type enclosed in square brackets [ ]. A JSON object is a collection of key-value pairs enclosed in curly braces {}. Keys are **always strings** and must be enclosed in double-quotes. Values can be any valid JSON data type (string, number, boolean, array, object, or null). Key-value pairs are separated by commas. The following is an example of an object taken from the DATAUSA response:

```
{
  "ID Nation": "01000US",
  "Nation": "United States",
  "ID Year": 2022,
  "Year": "2022",
  "Population": 331097593,
  "Slug Nation": "united-states"
}
```

In this example, "IDNation", "Nation", "ID-Year", "Year", "Population", and "SlugNation" are the keys, while "01000US", "UnitedStates", 2022, "2022", 331097593, and "united-states" are the corresponding key values. Notice that the value of the key "ID Year" is a number, while the value of the key "Year" is a string.

The DATAUSA example shows how JSON excels in capturing the diverse hierarchical structure of web data. The response was an object *response*, which contained two keys: "data" and "source". The key "data" value was an array containing one object described above. The value of the key "source" was also an array containing one object with four keys, "measures", "annotations", "name", and "substitutions". In turn, the values of these keys are other arrays, objects, and strings.

This structure can be easily parsed following the bracket notation for the key values and the index notation for the array elements. For instance, to get the USA population value from the object "response", one needs first to look up the value of the key "data" (which is an array), then take the first element of this array (which is an object), and then look up the value of the key "Population" of this object. The Python notation for this operation is as follows:

population = response["data"][0]["Population"]

Here, response["data"] retrieves the entire array; response["data"][0] retrieves the first element of this array (recall that the indexing starts from 0); and response["data"][0]["Population"] retrieves the value of the key "Population" from this element.

9.2.3 Authentication and Authorization

The API used as an example in the previous chapter did not require any form of user authentication. Meanwhile, many service providers only allow API access to a regis-

Authentication and authorization

Authentication identifies a user, device, or program before granting access to a resource. It ensures that the entity requesting access is who it claims to be.

Authorization determines what an authenticated entity is allowed to do. It ensures that the authenticated user, device, or program only has access to the permitted resources or actions.

Fig. 9.3 An example of a client ID and a client secret

tered user. The API authorization adds a layer of security by determining who or what can interact with the data, under what conditions, and the access level and limits. For example, an API may allow any authenticated user to view public data, limiting the number of API requests to 10 per day. Meanwhile, a paying user may be authorized to 10,000 requests per day. Finally, a user with administrative privilege would have an additional ability to edit or delete the data.

There are various ways of handling API authorization. For instance, you are developing an application that accesses hotel data from the global travel reservation system Amadeus. You start by registering for API access at the page ▶ https://developers.amadeus.com/register. Upon registration completion, you get access to the Amadeus self-service API, which grants free access to a limited version of their reservation system.

The first step is to create a new app. Select the name for your data access project. This will generate an API key and an API secret for your app, which are required for the authorization flow (Fig. 9.3). You can think of these two values as your application ID and password; because of that, the API secret should be kept secure.

To start an authorization process, the application has to send the API key and secret to the Amadeus API authorization server. The API server first checks the API key included in the request. The API key is compared against the API provider's database to identify the client making the request. That ensures that the key is valid and not revoked. Then, the API secret is compared with the one stored in the API provider's database for the matching API key. After the API server authenticates the re-

9

API key and API secret

API Key identifies the client or application making the API request. The API Key is similar to a username.

API Secret authenticates the app that makes an API request. The API Secret is similar to a password.

quest, it checks whether the client has the appropriate permissions. Finally, the API server processes the request and returns the requested data. An error message is returned if any step of this process fails. Hence, the entire process works as follows:

- Create an account in the Amadeus for Developers portal;
- Register a new app, obtain an API Key and an API secret;
- The app sends the API Key and API secret along with a request for data;
- The server completes the app authorization process using the provided API Key and API secret and responds with the requested data.

This authorization process is simple, straightforward, and fast. For instance, the following few lines of Python code can be used to obtain all 487 flight destinations of British Airlines (code BA):

```
from amadeus import Client
amadeus = Client(client_id='YOUR_ KEY',
      client_secret='YOUR_ SECRET')
response =
amadeus.airline.destinations.get(airlineCode='BA')
print(response.data)
```

In this example, 'YOUR_ KEY' and 'YOUR_ SECRET' should be replaced with the API Key and the API secret received during your app registration with Amadeus. Running this code will return a list of airports in JSON format, along with their latitude and longitude, IATA code, address, time zone, and local time. In this example, we used a Python module *amadeus* specifically developed for the API access to Amadeus data. In the lab, we will use more flexible module *requests* to send HTTP requests to interact with websites or APIs.

API access frequently requires user authorization. Probably the most common user authorization mechanism is OAuth (Open Authorization). The OAuth 2.0 standard enables third-party applications (such as your Python code) to obtain limited access to provider resources without exposing the credentials. Instead, OAuth 2.0 relies on access tokens as the primary mechanism for authorizing requests.

OAuth

OAuth is an authorization protocol. The current version OAuth 2.0 is a de-facto standard for online authorization.

Similar to what has been described, the user starts with registration, obtaining a combination of credentials like client ID and client secret. The client application sends the client ID and client secret to the authorization server. The authorization server will respond with an access token if the credentials are valid. The access token is a short-lived token that authorizes your application to access the API. This token needs to be included in your further API requests. The access token has an expiration time. After it expires, your application will no longer be authorized to access the API. You will need to request a new access token by repeating the token request process and sending your API key and secret to get a fresh token. Hence, the entire process works as follows:

- Create an account in the Amadeus for Developers portal;
- Register a new app, obtain a client ID and client secret;
- The app sends the client ID and client secret to the authorization server;
- The server completes the app authorization process using the provided client ID and client secret and responds with an access token;
- The app sends requests for data or service along with the access token;
- When the access token expires, the app resends the client ID and client secret to the authorization server in exchange for a new access token.

HTTP request

HTTP request methods are the different ways a client (a web browser or API client) can communicate with a web server. Each method defines a specific action for the server to perform. The most common HTTP request methods are GET, used to retrieve data from the server, and POST, used to send data to the server.

9.2.4 GET and POST HTTP Request Methods

Open a list of Amadeus APIs, for instance ▶ https://developers.amadeus.com/pricing.

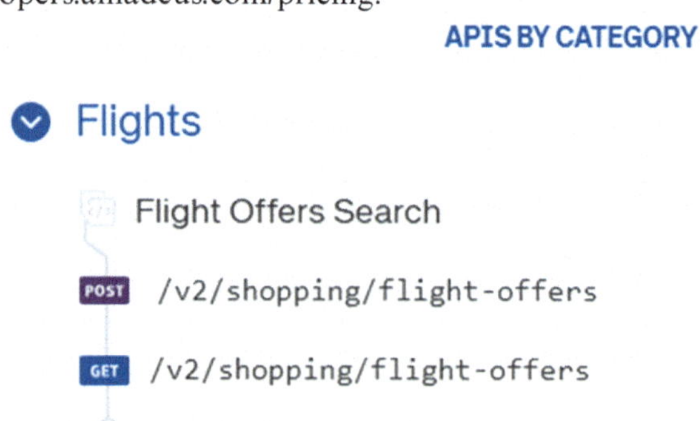

Notice that many APIs are marked as POST or GET.

These are the methods used to communicate between the client (your computer) and the web server. There are five methods in total: GET, POST, PUT, PATCH, and DELETE. GET and POST are the most commonly used methods.

GET is used to retrieve data from the server. The parameters of GET requests are passed in the URL string following the "?" sign. For instance, the API call discussed in ▶ Sect. 9.2.1.

▶ https://datausa.io/api/data?drilldowns=Nation&measures=Population&year=latest

passed the parameters *drilldown*, *measures*, and *year* separated by the "&" sign. This makes it easy to understand, bookmark, or share GET requests. However, this also makes this method not very secure; because of that, GET requests are typically used with publicly available information.

As opposed to GET requests, the POST method is used to submit data to the server. The POST method sends data in the body of the request, separate from the URL. This allows for radically more data to be sent to the server in a secure way. Since request parameters are not exposed in the URL, POST requests are used when transmitting sensitive information, such as user login credentials.

9.2.5 API Limitations

API provides an easy and straightforward way of collecting and storing web-based data in a structured format. Unfortunately, the majority of websites don't offer APIs, and even when they do, the data accessible through these APIs may be restricted to a limited group of users and/or severely curated. For instance, social media platforms often provide APIs, but they typically restrict access to certain types of data and impose rate limits on data requests. As a result, the data obtained through APIs may differ from the content delivered through a web browser. In addition, web services frequently update or discontinue their APIs without notice, which can disrupt ongoing data collection efforts. The changes to the API can result in changes to the structure or format of the data. Consequently, researchers relying solely on APIs may encounter a significant barrier in obtaining a comprehensive dataset.

In comparison, web scraping bypasses the need for an API by directly extracting data from a web page. This approach offers much higher control over the data collection process, yet at the cost of increased complexity and possible ethical and legal implications. The following section provides an introduction to direct data scraping from web pages.

9.3 Web Scraping

9.3.1 Web Page Structure

HTML

HTML is the language that defines the content and structure of a web page. It uses a series of **tags** to indicate how different parts of the page (text, images, links, etc.) should be displayed in a browser. HTML is often used in combination with **CSS** for styling and **JavaScript** for web page interactivity.

9

Open any web page in the Google Chrome browser, for instance, ▶ https://books.toscrape.com/. This webpage, specifically designed for those learning scraping, emulates an online bookstore. Right-click on any place on the page. From the opening menu, select the option “Inspect”. This opens a set of web development tools built into the Google Chrome browser, which allows developers to inspect, debug, and analyze a webpage. The Elements tab of the newly opened window shows the HTML structure of the page. It can look as follows:

```
<!DOCTYPE html>
▶<head>…</head>
▶<body>…</body>
…</html>
```

Clicking the ▶ symbol expands the elements to reveal the nested elements. Conversely, clicking the ▼ symbol collapses the structure under the element.

Notice two sections of the HTML document: <head> and <body>. Click the ▶ symbol to expand the <head>. This element provides information about the web page to the web browser and web services. For instance, the

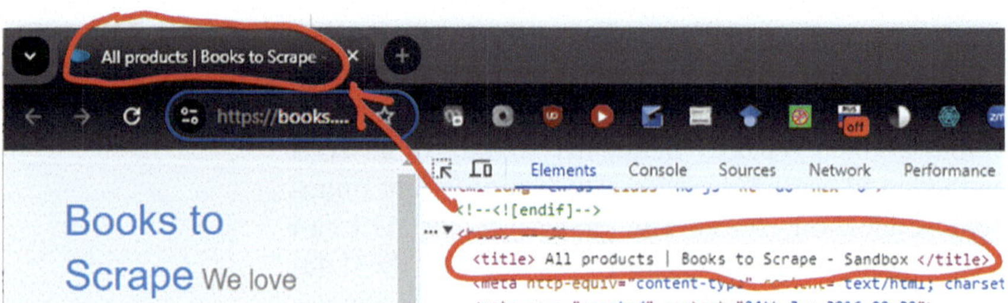

Fig. 9.4 Illustration of the functionality of the <title> element

<title> element defines the title of the page, which is displayed in the browser's title bar or tab (Fig. 9.4).

Collapse the <head> and expand the <body>. This element contains the content visible to the user in the browser window: the text, images, videos, links, forms, buttons, and so on. This is the part of the web page targeted by the web scraping software.

Notice a hierarchical structure of the nested elements that employs a system of tags and attributes to define the layout of a web page. The specific types of content are delineated by the tags, enclosed in angle brackets. For instance, the <p> … </p> pair of tags delineate a paragraph, the <a> … </a> tags define a hyperlink, the <li> … </li> tags encompass a list element, the <img> … </img> tags are set around a link to an image, and <h1>, <h2> and similar tags format the enclosed text as a header. Other tags, such as <div>, group the content together.

There could be additional key/value pairs enclosed in the angle brackets, called the element attributes. The attributes provide additional information essential for a proper web page display by a browser. For instance, the *id* attribute provides a unique identifier for an element. The *class* attribute is frequently used to define specific formatting for a group of elements. Other commonly used attributes are *style*, *href*, *src*, *alt*, *title*, and so on.

In addition to the static elements, a web page frequently includes dynamic features that change page content on your computer without the need to request a new full page sent from a web server. The code providing this functionality is frequently included at the end of the web page. For instance, the code written in the JavaScript programming language can be added to the HTML document enclosed in the <script> … </script> tags.

The next section considers how this information on the web page structure can be used to scrape its data.

HTML tag examples

<html>: document root
<body>: content; visible part of webpage
<h1> to <h6>: headings
<p>: a paragraph

: line break
<b>: bold font
<i>: italic font
<a>: a hyperlink
<img>: embeds image
<video>: Embeds a video.
<ul>: bullet point list
<ol>: numbered list
<li>: defines a list item
<button>: a button
<div>, <span>, <article>, <section>: a group of elements
<style>: Embeds internal CSS
<script>: Embeds or links JavaScript

9.3.2 Scraping Basics

Scraping websites is essentially reverse engineering: deconstructing a product, software, or web page to understand how it works without access to the original design documentation. Let us outline how we could scrape book names from the site ▶ https://books.toscrape.com/. Right-click the name of a book on the web page. In the opening menu, select the option "inspect". The page code

Fig. 9.5 Book title (left) and the respective HTML code (right)

shown in the Elements tab of the Inspector window will scroll to the element responsible for book name display (Fig. 9.5):

<a href="catalogue/a-light-in-the-attic_1000/index.html" title="A Light in the Attic"> A Light in the …</a>

The tag <a> in HTML defines hyperlinks: clicking on the element directs users to the page "catalogue/a-light-in-the-attic_1000/index.html". The link text between the opening and closing <a> tags is what the user clicks on. Notice the attribute "title"; the value of this attribute is the target of our scraping.

Inspecting a few other book titles confirms this initial guess. Hence, our goal of scraping book titles can be achieved by parsing the web page for all <a> elements with the attribute "title" and collecting the attributes' values. This approach, however, collects only the names of the first 20 books from the first page of the website. We need to repeat the scraping process for all pages to scrape all books.

Click the "next" button at the web page's bottom right. Notice that the name of the new web page has changed to ▶ https://books.toscrape.com/catalogue/page-2.html. Similarly, the following web page name is ▶ https://books.toscrape.com/catalogue/page-3.html, and so on. Now, we can formulate the scraping strategy:

1. Set page tracker value $i = 1$;
2. Go to the page ▶ https://books.toscrape.com/;
3. Find all elements with the tag <a> containing attribute "title";
4. Collect the values of these attributes;
5. Increment the page tracker value $i = i + 1$;
6. Go to the page ▶ https://books.toscrape.com/catalogue/page-i.html
7. If the page does not exist, stop data collection; otherwise, return to step 3.

9.3.3 Scraping Multiple Related Elements

In the previous section, we learned how to scrape one attribute of a product (book titles), but we are usually interested in multiple attributes, such as the title, price, and star rating. One strategy would be to run the outlined scraping process three times for different attributes. A better solution, however, would be to capture the element enclosing all information related to a single product first. Then, you can extract the nested elements containing title, price, and star rating information.

Right-click on the title of the first book and open the Elements tab of the Inspector window. Now, move the cursor up in the Inspector window and observe the change in highlighted parts of the book. Stop when the entire product is highlighted. This is the element that contains all data related to the book.

Note

The related items from the same group should be scraped together rather than sequentially to preserve their relations.

This gives us an idea that the tag <article class="product_pod"> can be used to search for book titles. Experiment with other books to confirm that information related to a single book is stored in the element <article class="product_pod">. Notice the information about the book star rating:

<p class="star-rating *">

and about the price:

<div class="product_price">

p class="price_color"> £51.77 </p>

Now, we can modify the scraping strategy outline in the previous section:

1. Set page tracker value $i=1$;
2. Go to the page ▶ https://books.toscrape.com/;
3. Find all elements <article class="product_pod">;
4. For each element:
   a. Find the nested element with the tag <a> containing attribute "title" and store the value of the attribute "title";

Hint

Websites frequently update their HTML structure, which can break the scraper's logic. Always check if your code is up to date.

   b. Find the nested element with the tag <p class="star-rating *">, where * is any sequence of letters, and store the value of the attribute "class";
   c. Find the nested element with the tag <p class="price_color"> and store the text;
5. Increment the page tracker value $i = i + 1$;
6. Go to the page ▶ https://books.toscrape.com/catalogue/page-i.html
7. If the page does not exist, stop data collection; otherwise, return to step 3.

9.3.4 Navigating the Web Page Structure

Identifying the elements containing data to be scraped requires a set of tools that can parse the tree-like structure of a web page, search for specific tags, access the nested elements, and extract the attributes. Two frequently used tools are XPath and CSS selector.

The XPath (XML Path) is a language designed to navigate the elements and attributes in an XML or HTML document. XPath views a document as a tree of nodes (elements), which it can search, and navigate up and down the structure. For instance, the following XPath expression selects <article class="product_pod"> elements on a web page:

//article[@class='product_pod'].

The nested elements can be then selected with the following expressions:

.//a[@title]
.//p[contains(@class, "star-rating")]
.//p[@class="price_color"]

Another popular way of searching for and selecting a particular element is the CSS (Cascading Style Sheets) query. CSS selectors are often more transparent and lesser error-prone when used for web scraping. Return-

Exception handling

When scraping, you may encounter various errors—for instance, "404 Page not found," "Request timed out," etc. Normally, the scraper's code crashes when an error is encountered. Exception handling ensures that a web scraper can handle errors predictably. In this chapter's lab, we consider a practical example of exception handling.

ing to the book scraping example, the following CSS selectors can be used:

- To select <article class="product_pod"> element: article.product_pod
- To select <a> element with the title attribute: a[title]
- To select <p> element with a class that contains star-rating: p.star-rating
- To select <p class="price_color"> element: p.price_color

We can observe how the CSS selectors work using the Inspector. Click the Console tab. Then type:

document.querySelectorAll('article.product_pod').

You should see the element containing the data describing the first book:

```
> document.querySelector('article.product_pod')
<· ▼<article class="product_pod">
     ▶<div class="image_container">…</div>
     ▶<p class="star-rating Three">…</p>
     ▶<h3>…</h3>
     ▶<div class="product_price">…</div>
    </article>
```

The following code captures all 20 <article> elements and then extracts the title, rating, and price of the first book in the list:

```
> x=document.querySelectorAll('article.product_pod')
<·  NodeList(20) [article.product_pod, article.product_pod, article.product_pod, article.product_pod,
    article.product_pod, article.product_pod, article.product_pod, article.product_pod, article.product_pod
  ▶ , article.product_pod, article.product_pod, article.product_pod, article.product_pod,
    article.product_pod, article.product_pod, article.product_pod, article.product_pod, article.product_pod
    , article.product_pod, article.product_pod]
> title=x[0].querySelector('a[title]').getAttribute('title')
<· 'A Light in the Attic'
> rating=x[0].querySelector('p.star-rating').getAttribute('class')
<· 'star-rating Three'
> price=x[0].querySelector('p.price_color').textContent
<· '£51.77'
>
```

9.4 Scraping Tools

9.4.1 No-Code Web Scraping Tools

Sample Octoparse templates

Social media scraper
Google maps scraper
Amazon product scraper
YouTube video scraper
Reddit scraper
Etsy scraper
TikTok scraper
Airbnb scraper
Booking.com scraper
TripAdvisor scraper

Multiple web scraping tools require little or no coding. Typically, these tools come with user-friendly interfaces and sometimes even data cleaning features. They can vary in complexity, from simple point-and-click applications to advanced frameworks for experienced developers. Here is a list of a few popular applications.

Octoparse is a no-code web scraping tool that provides an easy-to-use visual interface to select the relevant web page elements you want to scrape. The tool automatically searches and highlights page nodes similar to the user-selected ones, simplifying the element selection process. It also has advanced features such as pagination handling, infinite scrolling, CAPTCHA, and clicking the buttons. The tool offers automatic data extraction and scraping scheduling. One important feature of Octoparse is a set of templates pre-built for popular sites such as YouTube, Booking.com, and Airbnb.

ParseHub is a cloud-based web scraping tool with features similar to those of the desktop and cloud-based Octoparse. It has better functionality and can scrape sites Octoparse struggles with, yet it is considerably more pricy.

Instant Data Scraper is a simpler tool with limited functionality that operates as a browser extension. Using a simple point-and-click interface, it is designed to quickly extract information like tables, lists, and other structured data directly from web pages. This makes it ideal for quickly grabbing data from a web page without the need to analyze its structure, as Octoparse and ParseHub require.

Scraper options

Scraper options

Static page: Requests for loading web pages + BeautifulSoup for page parsing.

Dynamic page: Selenium for loading web pages + BeautifulSoup for page parsing.

9.4.2 Building Your Own Scraper

The no-code web scraping tools are relatively easy to use, yet are rarely suitable for scraping data from websites with a complex structure. When the no-code tool of choice fails, a custom code could be a better solution. **BeautifulSoup** is a Python library used for web scraping. This library provides a comprehensive list of methods for parsing HTML documents and searching for specific tags, nodes, attributes, and texts. It also provides an easy way to traverse the web page hierarchy from parent nodes to children and from children to parents. BeautifulSoup methods support the XPath and CSS selectors, adding to the library's flexibility.

BeautifulSoup, in combination with the **Requests** library used for loading web pages for parsing, is a powerful tool for customized web scraping. However, it fails when web content is dynamic. For instance, suppose the page includes a Java script that updates page content based on user interaction. Unlike static web pages, which can be received from the server in full and processed with BeautifulSoup, a dynamic page simply has no data that should be generated in interaction with the user.

Selenium is a powerful web testing library that can handle static and dynamic web pages. In essence, Selenium emulates a web browser, allowing the analyst to program "user's" actions: typing on the keyboard, mouse hovering, scrolling, clicking buttons, and navigating between pages. Because of that, Selenium is particularly useful when dealing with JavaScript-heavy websites where the data is loaded dynamically. It can interact with all the elements in a webpage, simulating a human-like interaction with the website.

In web scraping, Selenium is often used together with BeautifulSoup or another HTML parsing library. Selenium interacts with the webpage to access data that is not readily available, and BeautifulSoup is used for parsing the HTML. For example, Selenium can be used to click on a button that loads more data, and then BeautifulSoup can be used to parse the newly loaded data.

Even though Selenium is a powerful and flexible tool that can be used to scrape most websites, it is large and slow. It also opens ways for abusing web scraping. With Selenium, it is easy to make a scraping program mimic human behavior, such as hovering the mouse across the site, clicking, and taking random waiting intervals between actions. This allows it to bypass many (but not all) anti-scraping mechanisms, especially when combined with proxy rotation and user-agent spoofing.

9.5 Ethical and Legal Considerations

Web scraping can return a treasure of data for a researcher, but it also has ethical and legal implications. Extracting private data, even if publicly available, can lead to privacy violations if used without consent or for malicious purposes. Complicating the issue, laws and regulations are complex, patchy, and varied across jurisdictions.

One source to consult is the site's **robots.txt** file. Websites use this file's content to communicate with web crawlers and scrapers, such as Googlebot and GPTbot, about which parts of the site they can or cannot access. It acts as a guide for responsible web scraping by providing rules that specify which pages, directories, or files should be restricted from automated crawling.
For instance, look up the following URL: ▶ https://www.airbnb.com/robots.txt. The following generic information is on the top of the page: "If you're a bot who likes crawling webpages, please mind the house rules and avoid accessing any disallowed subfolders to earn a 5-star review from us." Following that, there is a long list of sub-folders that are allowed and disallowed to crawl for various bots (User-agents).

Another source of information on scraping policy is the Terms of Services (TOS) and Privacy Policy of the website. For instance, Airbnb TOS states: "Do not use bots, crawlers, scrapers, or other automated means to access or collect data or other content from or otherwise interact with the Airbnb Platform," effectively disallowing scraping activity (▶ https://www.airbnb.com/help/article/2908). In comparison, Booking.com TOS forbids commercial use: "You're not allowed to monitor, copy, scrape/crawl, download, reproduce or otherwise use anything on our Platform for any commercial purpose without written permission of Booking.com or its licensors" (▶ https://www.booking.com/content/terms.en-gb.html).

Even when data scraping is allowed, it should be done responsibly. Scraping large amounts of data may interfere with main activities and is definitely discouraged. Similarly, requests for new web pages should be paced to avoid overloading the server. This creates a denial-of-service (DoS)-like effect, exposing your IP address to the risk of being blacklisted. Whenever possible, APIs should be used over scraping. Not only is programming an API access easier and provides better-quality, structured data, but it is also specifically designed for automated data extraction.

Web page scraping steps

Choose the webpage to scrape.

Check robots.txt file and ToS (terms of service) to ensure you have permission to scrape.

Choose a scraping tool: a no-code tool (Octoparse, Parsehub) or a coding library (BeautifulSoup, Selenium).

Understand web page structure and locate the desired content.

Develop a scrapping scheme.

9.6 Conclusion

Web scraping is a complex topic with many different approaches and tools in existence. We have covered only very few basic techniques. With web technologies developing fast, new web scraping methods appear every year. In addition, the websites are also fast updating; a web scraping application that works today may stop functioning in a year or even a month's time. Finally, scraping is not necessarily legal, especially when conducted for commercial purposes. Meanwhile, while many websites disallow data scraping, they may still make the data available through their API or through partnering organizations for a fee.

9.7 Review Questions

1. Describe the key differences between content mining, structure mining, and usage mining in web scraping.
2. What is an API, and how does it facilitate data collection from online resources?
3. Explain the main features of JSON format.
4. Outline the steps involved in making an API request to retrieve web data.
5. How do authentication and authorization work in API scraping?
6. Compare the GET and POST methods.
7. Discuss the limitations of using APIs for data scraping. Why might a researcher choose to scrape data directly from a website instead?
8. Explain how CSS selectors and XPath are used to navigate web page structures in web scraping.
9. Explain a step-by-step strategy for scraping multiple attributes (e.g., title, price, and rating for each book from an e-commerce website).
10. What are some common no-code web scraping tools, and what are their advantages and limitations?
11. Describe the problems arising when scraping the websites with dynamic content.
12. What ethical and legal considerations must be considered when conducting web scraping?

9.8 Lab

Objectives

1. Practice code markdown skills
2. Practice API data access
3. Practice scraping with Request and BeautifulSoup
4. Practice exception handling

The overall goal of this assignment is to make initial steps in scraping data online.

Data

The data for the first part of this lab comes from Amadeus, the largest GDS service provider for the travel industry. GDS (Global Distribution System) enables connections between the inventory of hotels, airlines, and similar services and travel agencies. We will use Amadeus API data access to retrieve a list of hotels in Tokyo.

Amadeus API registration: ► https://developers.amadeus.com/register.

API documentation: ► https://developers.amadeus.com/self-service/apis-docs.

The second part of the lab will cover data scraping. To prevent a possible impact on a real business, we will use web site ► https://books.toscrape.com/, which is specifically developed for web scraping training.

Part 1. Retrieve Hotel Data from Amadeus

Step 1. Register for API Data Access

Start with a new Markdown cell. Describe the purpose of the code.

```
The code for Chapter 9 Lab: Web scraping
Part 1. Obtain data from Amadeus using API access.
Step 1. Register for API access and get the API key and API secret
```

Now, consult with Amadeus API Get Started page: ▶ https://developers.amadeus.com/get-started/get-started-with-self-service-apis-335.

Notice two types of APIs: Self-Service and Enterprise level. **You will be using the self-service API**.

There are three main steps in the process:

1. Register and create an account;
2. Get the API key and secret;
3. Write the code that uses the API key and secret to request data from Amadeus.

1. Start with creating a developer account at Amadeus: ▶ https://developers.amadeus.com/register. activate the account. sign in to your account.

2. Sign in to your account. You will create a self-service app. Go to ▶ https://developers.amadeus.com/my-apps. Click "Create new app." Give your application a name and a description. Select the application target as "Business to consumer."

In the new screen, you will be issued the API key and API secret. Notice the test keys and production keys. The test keys give free access to a small part of Amadeus data you can use to train and test your app. Your free account is limited to a certain number of monthly requests—for instance, no more than 3000 searches for the flight's cheapest data and 2000 searches for the hotel searches.

Once you are confident in your app, you can move it to production and obtain access to all data. In the production environment, you need to pay for data access. The typical fees are a few cents per 1,000 transactions. Since our purpose is limited to training, we will limit our data access to the testing environment.

Copy and save the API key and secret—you will need them to access the data. Also, notice the API endpoint, which your app will use to access Amadeus.

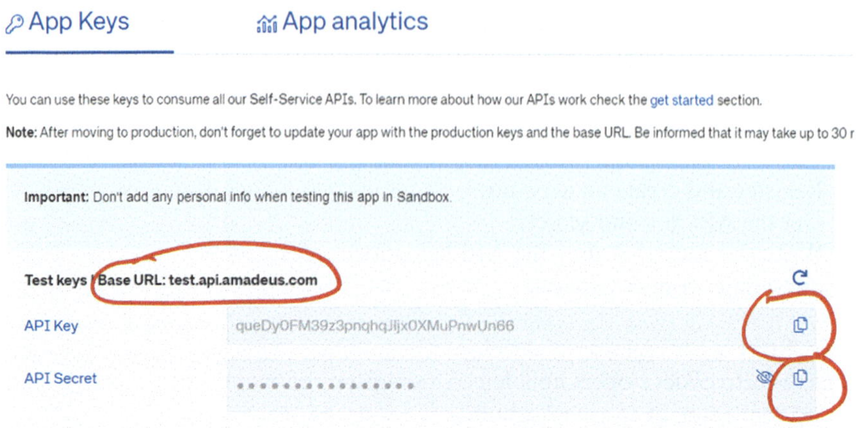

Step 2. Request Hotel Rating Data for Tokyo: First Attempt

We will use Anaconda AI Assistant.

USER:

Access Amadeus API to retrieve a list of Tokyo hotels.

Assistant:

```
import requests

# Replace 'YOUR_API_KEY' and 'YOUR_CLIENT_SECRET' with your actual Amadeus API
credentials
api_key = 'YOUR_API_KEY'
client_secret = 'YOUR_CLIENT_SECRET'
url = 'https://test.api.amadeus.com/v2/shopping/hotel-offers'

# Set the parameters for the request
params = {
    'cityCode': 'TOKYO',
    'checkInDate': '2023-10-01',  # Example check-in date
    'checkOutDate': '2023-10-05',  # Example check-out date
    'adults': 1
}

# Get the access token
auth_response = requests.post('https://test.api.amadeus.com/v1/security/oauth2/token',
            data={'grant_type': 'client_credentials',
            'client_id': api_key,
            'client_secret': client_secret})
auth_response_data = auth_response.json()
access_token = auth_response_data['access_token']

# Make the request to get hotel offers
headers = {'Authorization': f'Bearer {access_token}'}
response = requests.get(url, headers=headers, params=params)

# Print the response
print(response.json())
```

Prior to running this code, do not forget to install the module “requests”:

```
!pip install requests
```

Explanation:

First, you need to replace the 'YOUR_API_KEY' and 'YOUR_CLIENT_SECRET' placeholders with the API key and API secret you obtained during the app registration process.

```
# Replace 'YOUR_API_KEY' and 'YOUR_CLIENT_SECRET' with your actual Amadeus API
credentials
api_key = 'YOUR_API_KEY'
client_secret = 'YOUR_CLIENT_SECRET'
```

Then, you define the API endpoint:

```
url = 'https://test.api.amadeus.com/v2/shopping/hotel-offers'
```

You also set the parameters used to request hotel data:

```
# Set the parameters for the request
params = {
    'cityCode': 'TOKYO',
    'checkInDate': '2023-10-01',  # Example check-in date
    'checkOutDate': '2023-10-05',  # Example check-out date
    'adults': 1
}
```

You establish a connection with the server by calling a function that takes the API key and API secret as parameters to request an access token. The API key and API secret are passed as arguments to this function. This access token is then used as an attribute or a value stored in the variable access_token during the session to request data from the server until it expires. Note that the secure HTTP method POST is used to transmit the API key and API secret to the server.


```
# Get the access token
auth_response = requests.post('https://test.api.amadeus.com/v1/security/oauth2/token',
            data={'grant_type': 'client_credentials',
            'client_id': api_key,
            'client_secret': client_secret})
auth_response_data = auth_response.json()
access_token = auth_response_data['access_token']
```

Finally, you call the function to send the access token (as an argument) and other request parameters to the API endpoint, then print the response:

```
# Make the request to get hotel offers
headers = {'Authorization': f'Bearer {access_token}'}
response = requests.get(url, headers=headers, params=params)

# Print the response
print(response.json())
```

At this time, the server returns an error: "The targeted resource doesn't exist".

Result:

```
{'errors': [{'code': 38196, 'title': 'Resource not found', 'detail':
"The targeted resource doesn't exist", 'status': 404}]}
```

To understand this error, we need to consult with the API documentation.

Step 3. Understanding the Error

Open the API documentation and search for the section relevant to the hotel list search. Go to the "Hotel APIs" section:

▶ https://developers.amadeus.com/self-service/category/hotels.

Then go to the section "Hotel Search":

▶ https://developers.amadeus.com/self-service/category/hotels/api-doc/hotel-search.

Then, click on "Hotel List API":

▶ https://developers.amadeus.com/self-service/category/hotels/api-doc/hotel-list.

Finally, go the the "API Reference":

▶ https://developers.amadeus.com/self-service/category/hotels/api-doc/hotel-list/api-reference.

The API reference page shows documentation on the relevant APIs. Notice the method (GET) and the API endpoints:

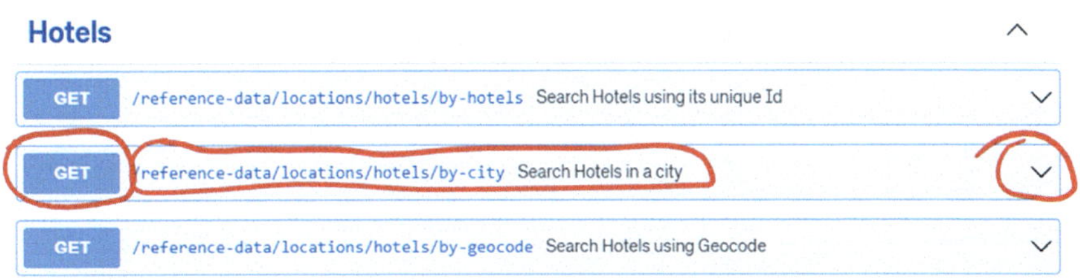

Notice the differences between the API documentation and the Assistant-provided code. First, the URL for requesting hotel data is incorrect.

Assistant: url = '▶ https://test.api.amadeus.com/v2/shopping/hotel-offers'

Documentation: url = '▶ https://test.api.amadeus.com/v1/reference-data/locations/hotels/by-city'

Expand the "Search Hotels in the City" section by clicking the ⌄ symbol. A list of allowed parameters will open. Notice the required parameter cityCode:

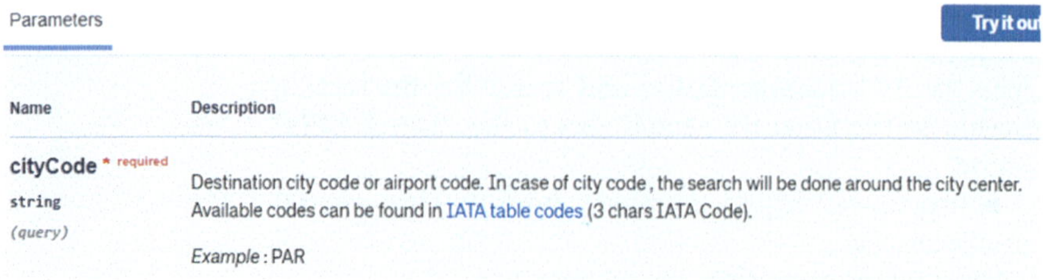

This parameter differs from the one provided by the Assistant. The Assistant used the city name. However, the documentation states that the cityCode parameter should be an IATA (International Air Transport Association) city code. To learn the code for Tokyo city (TYO), we can search IATA website:

- https://www.iata.org/en/publications/directories/code-search/

Finally, the Assistant used the parameters not covered in the documentation. These parameters have to be removed.

Step 4. Correct the Code

After modification, the code is as follows:

```
import requests

# Replace 'YOUR_API_KEY' and 'YOUR_CLIENT_SECRET' with your actual Amadeus API
credentials
api_key = 'queDy0FM39z3pnqhqJljx0XMuPnwUn66'
client_secret = 'OnWUyzZpRvbYEzu9'
url = 'https://test.api.amadeus.com/v1/reference-data/locations/hotels/by-city'

# Set the parameters for the request
params = {
    'cityCode': 'TYO',
}

# Get the access token
auth_response = requests.post('https://test.api.amadeus.com/v1/security/oauth2/token',
            data={'grant_type': 'client_credentials',
            'client_id': api_key,
            'client_secret': client_secret})
auth_response_data = auth_response.json()
access_token = auth_response_data['access_token']

# Make the request to get hotel offers
headers = {'Authorization': f'Bearer {access_token}'}
response = requests.get(url, headers=headers, params=params)

# Print the response
print(response.json())
```

Result (reformatted using ▶ https://jsonformatter.curiousconcept.com/# and abridged):


```
{
      "data":[
            {
                  "chainCode":"AZ",
                  "iataCode":"TYO",
                  "dupeId":700085152,
                  "name":"SOMERSET ROPPONGI",
                  "hotelId":"AZTYO153",
                  "geoCode":{
                        "latitude":35.66478,
                        "longitude":139.73583
                  },
                  "address":{
                        "countryCode":"JP"
                  },
                  "lastUpdate":"2023-06-15T10:09:54"
            },
            ...
```


We obtained a list of records for all Tokyo hotels registered with Amadeus in JSON format.

Step 5. Save the Hotel List

USER:

From response.json(), extract and save in Pandas DataFrame "hotel_df" the following features: "iataCode", "name", "hotelId", "geoCode". Save "hotel_df" in an Excel file "TokyoHotels.xlsx".

Assistant:

```
import pandas as pd

data = response.json()

# Extract the relevant features from the JSON response
hotels = data.get('data', []) # Get the list of hotels from the response
extracted_data = []

for hotel in hotels:
    extracted_data.append({
        'iataCode': hotel.get('iataCode'),
        'name': hotel.get('name'),
        'hotelId': hotel.get('hotelId'),
        'geoCode': hotel.get('geoCode')
    })

# Create a DataFrame from the extracted data
hotel_df = pd.DataFrame(extracted_data)

# Save the DataFrame to an Excel file
hotel_df.to_excel("TokyoHotels.xlsx", index=False)
```


Run the code. You should see a new Excel file created in the list of the folders. Download the file and open in Excel on your computer.

Result:

iataCode	name	hotelId	geoCode
TYO	SOMER-SET ROP-PONGI	AZTYO153	{'latitude': 35.66478, 'longitude': 139.73583}
TYO	NEW OTANI MAKU-HARI	BCTYO-MAK	{'latitude': 35.64551, 'longitude': 140.03821}
TYO	NEW OTANI TO-KYO THE MAIN	BC-TYONOH	{'latitude': 35.68087, 'longitude': 139.73401}
...			

Part 2. Scrape Book Data from ▶ https://books.toscrape.com/

Step 1. The Initial Code

USER:

Analyze the site ▶ https://books.toscrape.com/. Scrape the following data: book star rating, title, and price.

Assistant:

```
import requests
from bs4 import BeautifulSoup
import pandas as pd

# Step 1: Fetch the webpage
url = 'https://books.toscrape.com/'
response = requests.get(url)

# Step 2: Parse the HTML content
soup = BeautifulSoup(response.text, 'html.parser')

# Step 3: Extract data
books = []
for book in soup.find_all('article', class_='product_pod'):
    book_title = book.h3.a['title']  # Book title
    book_price = book.find('p', class_='price_color').text  # Book price
    book_star_rating = book.find('p', class_='star-rating')['class'][1]  # Book star rating class

    # Convert star rating class to a numerical value
    star_rating_map = {
        'One': 1,
        'Two': 2,
        'Three': 3,
        'Four': 4,
        'Five': 5
    }
    book_star_rating = star_rating_map.get(book_star_rating, 0)  # Default to 0 if not found

    books.append({
        'book_title': book_title,
        'book_price': book_price,
        'book_star_rating': book_star_rating
    })

# Step 4: Create a DataFrame
book_df = pd.DataFrame(books)

# Display the DataFrame
print(book_df)

# Optionally, save the DataFrame to an Excel file
book_df.to_excel("BooksData.xlsx", index=False)
```

Prior to running the code, do not forget to install missing libraries, for example:

```
!pip install requests beautifulsoup4
```

Explanation:

Import Libraries: The script imports the necessary libraries: *requests* (a library providing functions to make HTTP requests); *BeautifulSoup* from *bs4* (a class used for parsing and navigating HTML content); and *pandas*.

Fetch the Webpage: The *requests.get*() function is called with the argument *url* (a string representing the webpage URL). It returns a response object that contains the server's response to the request. The text attribute of the response object contains the raw HTML content.

Parse the HTML: *The BeautifulSoup* class is called with the HTML content (*response.text*) as an argument and a parsing method ('*html.parser*') as another argument. This creates a *BeautifulSoup* object, which provides methods and attributes to navigate and extract data from the HTML.

Extract Data: A **for** loop iterates over each book element found in the HTML. Using the methods and attributes of BeautifulSoup:

Book Title: Extracted from the "a" tag nested inside the "h3" tag using the *get*() method to access the title attribute.

Book Price: Extracted from the "p" tag with the class *price_color* using the *text* attribute.

Book Star Rating: Extracted from the "p" tag with the class *star-rating*. The star rating is stored as a class name (e.g., "Three"), and a mapping dictionary is used to convert these class names into numerical values.

Create a DataFrame: A list of dictionaries (each representing a book's data) is passed as an argument to the *pandas.DataFrame( )* function. This function returns a DataFrame *book_df*.

Display and Save Data: The DataFrame *book_df* is printed to the console using the *print*() function. It is also saved to an Excel file named "BooksData.xlsx" using the *to_excel*() method of the DataFrame.

Result (the Excel file):

book_title	book_price	book_star_rating
A Light in the Attic	Â£51.77	3
Tipping the Velvet	Â£53.74	1
Soumission	Â£50.10	1
Sharp Objects	Â£47.82	4
Sapiens: A Brief History of Humankind	Â£54.23	5
… 20 books …		

The code scraped the book's star rating, title, and price, but only from the website's first page.

Step 2. Modify the Code

There are two problems with this code. First, the book price contains an incorrect character, likely because of an incorrect code page. We will change the encoding to UTF-8 to prevent this type of error.

Second, only the first page is scraped. We will change the code to scrape the first 20 pages.

USER:

Analyze the site ▶ https://books.toscrape.com/. Scrape the following data: book star rating, book title, and book price. Use UTF-8 encoding for scraped data. Scrape the first 20 pages of the site.

Assistant:

```
import requests
from bs4 import BeautifulSoup
import pandas as pd

# Base URL for the website
base_url = 'https://books.toscrape.com/catalogue/page-{}.html'

# List to store book data
books = []

# Loop through the first 20 pages
for page in range(1, 21):
    # Fetch the webpage
    response = requests.get(base_url.format(page))
    response.encoding = 'utf-8'  # Ensure UTF-8 encoding

    # Parse the HTML content
    soup = BeautifulSoup(response.text, 'html.parser')

    # Extract data from each book on the page
    for book in soup.find_all('article', class_='product_pod'):
        book_title = book.h3.a['title']  # Book title
        book_price = book.find('p', class_='price_color').text # price
        book_star_rating =
                      book.find('p', class_='star-rating')['class'][1] # rating

        # Convert star rating class to a numerical value
        star_rating_map = { 'One': 1, 'Two': 2, 'Three': 3, 'Four': 4, 'Five': 5}
        book_star_rating = star_rating_map.get(book_star_rating, 0)
                      # Default to 0 if not found

        # Append the data to the list
        books.append({
            'book_title': book_title,
            'book_price': book_price,
            'book_star_rating': book_star_rating})

# Create a DataFrame from the collected data
book_df = pd.DataFrame(books)

# Display the DataFrame
print(book_df)

# Save the DataFrame to an Excel file with UTF-8 encoding
book_df.to_excel("BooksData.xlsx", index=False, encoding='utf-8')
```

Notice that the Assistant code has an error in the last string of code: the parameter *encoding* is incorrect. Remove it.

Explanation:

```
base_url = 'https://books.toscrape.com/catalogue/page-{}.html'
```

This line defines a string named base_url that serves as a template URL for web scraping. The curly braces {} act as a placeholder, which is dynamically replaced with a page number during execution using Python's *format*() method.

```
for page in range(1, 21):
    # Fetch the webpage
    response = requests.get(base_url.format(page))
```

This code snippet performs the following steps:

Loop Initialization: The *for* loop iterates over the sequence of numbers generated by the *range*() function, which here produces integers from 1 to 20 (inclusive of 1 and exclusive of 21).

Dynamic URL Generation: Within each iteration, the page variable takes on the next value in the range. The *format*() method of the base_url string is called with the current value of page as its argument. This method replaces the {} placeholder in base_url with the value of page. For example:

When page = 1, the resulting URL is ▶ https://books.toscrape.com/catalogue/page-1.html.

When page = 2, the resulting URL is ▶ https://books.toscrape.com/catalogue/page-2.html.

HTTP GET Request: The *requests.get*() function is called with the dynamically generated URL as its argument. This sends an HTTP GET request to the server to fetch the webpage data for the specified page. The result of this request is stored in the variable *response*.

Result (the Excel file):

book_title	book_price	book_star_rating
A Light in the Attic	£51.77	3
Tipping the Velvet	£53.74	1
Soumission	£50.10	1
Sharp Objects	£47.82	4
Sapiens: A Brief History of Humankind	£54.23	5
... 400 books ...		

Step 3. Error Handling

9

Frequently, a web page fails to upload. Another potential problem is missing data—for instance, a certain book has no price information. A robust web scraping code should handle various errors and edge cases, such as missing data, network errors, and request rate limits.

USER:

Enhance the code by implementing error handling and rate limiting to no more than 1 request per second.

Assistant (the code is refactored for better understanding):

```
import requests
from bs4 import BeautifulSoup
import pandas as pd
import time

base_url = 'https://books.toscrape.com/catalogue/page-{}.html'
books = []

# Loop through the first 20 pages
for page in range(1, 21):
  try:
    response = requests.get(base_url.format(page))
    response.encoding = 'utf-8'  # Ensure UTF-8 encoding

    soup = BeautifulSoup(response.text, 'html.parser')

    # Extract data from each book on the page
    for book in soup.find_all('article', class_='product_pod'):
      book_title = book.h3.a['title']  # Book title
      book_price = book.find('p', class_='price_color').text
      book_star_rating = \
                book.find('p', class_='star-rating')['class'][1]

      star_rating_map = {
        'One': 1,
        'Two': 2,
        'Three': 3,
        'Four': 4,
        'Five': 5}

      book_star_rating = \
            star_rating_map.get(book_star_rating, 0)

      books.append({
        'book_title': book_title,
        'book_price': book_price,
        'book_star_rating': book_star_rating})

  except:
    print(f"An error occurred while processing page {page}")

  time.sleep(1) # Rate limiting: wait for 1 second

book_df = pd.DataFrame(books)

print(book_df)

book_df.to_excel("BooksData.xlsx", index=False)
```

Explanation:

The *time.sleep()* function is used to suspend the execution of a program for a specified number of seconds. For instance, many web servers enforce rate limits to prevent being overwhelmed by traffic. In such cases, we can use *time.sleep*(10) to introduce a 10-s delay between requests.

The *try-except* construct in Python is used for handling exceptions. Exceptions are runtime errors that occur during the execution of a program. This construct allows you to attempt (try) the execution of a block of code and handle any potential exceptions (errors) that might arise, preventing the program from crashing unexpectedly. The *try* block contains the code you want to execute, while the *except* block specifies how to handle any exceptions raised during the execution of the try block. In our code, the except block is used to catch exceptions, print error information, and then proceed to scrape the next page. This ensures the program can continue running even when an error occurs.

This basic exception handling can be enhanced to provide different handling for specific types of exceptions by including multiple except blocks for different error types. Additionally, the *else* block can be used to define actions to take if no exception occurs, while the *finally* block specifies code that should always be executed, regardless of whether an exception was raised or not. For instance, the try-except construct can be used to handle the extraction of all pages from a book catalog site by customizing the handling of specific exceptions, such as HTTP errors.

Step 4. Scrape All Book Data

In this section, we will make the final modifications to the code to increase its robustness further. Make the following modifications.

1. Import additional modules for customized exception handling:

```
from requests.exceptions import HTTPError, ConnectionError, Timeout
```

These modules allow differentiation among errors that may arise when sending a request to a web server.

2. Modify the request code:
   Replace

```
response = requests.get(base_url.format(page))
```

with

```
# Try to send an HTTP request
response = requests.get(base_url.format(page), timeout=10)

# Check if the response was successful, otherwise raise error
response.raise_for_status()
```

Explanation:

Requests.get() is a function provided by the requests library to send an HTTP GET request to the specified URL.

timeout = 10 is a parameter of the *requests.get()* function, set to an argument value of 10. It specifies the maximum time (in seconds) to wait for a server response. A Timeout exception will be raised if the server takes longer than this timeout.

response.raise_for_status() is a method of the response object returned by *requests.get*(). It raises an exception if the HTTP response indicates a client error (4xx) or a server error (5xx), such as a "404 Page Not Found" error.

The use of these tools ensures robust error handling when sending HTTP requests.

3. Modify exception handling:
   replace

```
except:
    print(f"An error occurred while processing page {page}")
```

with

```
except HTTPError as http_err: # No page found
    print(f"No more pages: {http_err}")
    break
except ConnectionError as conn_err: # Handle connection errors
    print(f"Connection error occurred: {conn_err}")
    break
except Timeout as timeout_err: # Handle timeout errors
    print(f"Timeout error occurred: {timeout_err}")
    break
except Exception as err: # Handle any other exceptions
    print(f"An error occurred: {err}")
    break
```

The new exception handling code will print "No more pages" or information about a specific error and then stop iterations of the *for* loop.

4. Increase the upper boundary of the *for* loop
   replace

```
for page in range(1,21):
```

with

```
for page in range(1,10000):
```

Hint:

The site has 50 pages. Scraping all 50 pages will take a long time. Start with scraping the last few pages, for example, setting the range of the loop variable to (49,10000):

```
for page in range(49,10000):
```

After making sure that the code works as intended, reverse the loop range.

Run the code and collect information on all 1000 books stored in the "online store".

Part 3. On Your Own

1. Use API to collect AMADEUS recommendations for travel destinations inside Spain for Madrid travel origin. You can see the API documentation at ▶ https://developers.amadeus.com/self-service/category/flights/api-doc/travel-recommendations/api-reference.
   Hint: The IATA code for Madrid is MAD.
2. Scrape all quotes from the site ▶ https://quotes.toscrape.com/.

Build Your Python Skills

Chapters 17 and 18 of the free book Learn Python with Jupyter by Serena Bonaretti. ▶ https://www.learnpythonwithjupyter.com/.

Additional Reading

Chapagain, A. (2019). *Hands-on web scraping with Python: Perform advanced scraping operations using various Python libraries and tools such as Selenium, Regex, and others*. Packt Publishing Ltd.

HTML scraping. From: The Hitchhiker's guide to Python. ▶ http://docs.python-guide.org/en/latest/scenarios/scrape/

▶ https://www.w3schools.com/: multiple tutorials for beginner Web developers.

▶ https://www.postman.com/: a platform for API learning and development.

Sentiment and Emotion Analysis

A. P. Kirilenko, *Practical Data Mining with AI for Social Scientists*,
Springer Texts in Social Sciences, https://doi.org/10.1007/978-3-031-89689-7_10

10.1 Introduction

Text mining (also known as text analytics) is the process of extracting meaningful patterns, insights, and knowledge from textual data. Its primary goal is to transform freeform text into structured, preferably numerical, data that can be analyzed using traditional quantitative methods, as discussed in earlier chapters of this book, or specialized techniques designed specifically for textual data. These specialized methods often account for linguistic principles, such as syntax, semantics, and pragmatics, to ensure a deeper and more accurate understanding of language.

Typical text mining data comes from social media, newspapers, abstracts from scientific publications, web pages, interviews and surveys, transcripts of audio records, and image recognition data. In social science applications, this data frequently originates from social media platforms, online forums, and digital archives. Because of that, text mining is commonly combined with a preceding web mining step, which we will discuss later in the book.

Typical text mining objectives include document clustering, classification, sentiment analysis, and topic modeling. For example, a sports scientist interested in the evolution of public discussion of the Olympic Games could **cluster** #Olympics2024 tweets according to their less common and more specific hashtags. A political scientist could use the same dataset to **classify** tweets into the ones discussing sports and those discussing politics for further in-depth analysis of the latter. The same data could also be used by a tourism scientist to investigate tourists' **sentiments** and **emotions** during their visit to Paris over the Olympics. They could apply **topic modeling** to understand major themes the Games' visitors were interested in.

In this chapter, text is understood as a combination of sentences written in one of the natural languages. The sentences are organized into paragraphs, typically dealing with one or a few underlying **topics**. In turn, the par-

Tokens and unigrams

Tokens are the smallest text analysis units. Depending on the goals,tokens can be words,subwords, characters,punctuation marks, etc.

Unigrams represent a single unit of text, such as a word. All unigrams are tokens, but not all tokens are unigrams.

agraphs form separate text units, typically called **documents**. Finally, a collection of documents is called a **corpus.**

Down this hierarchy, the sentences consist of separate words, often referred to as **unigrams**. Sequences of consecutive words are known as **n-grams.** In European languages (though not all languages), words are typically separated by whitespace, making it relatively straightforward to break sentences into individual words through a process called tokenization. These words can be further tokenized into smaller units, such as subwords or characters. In European languages, words (or n-grams) are often the primary unit of analysis in topic modeling because smaller units, like subwords or characters, are usually not meaningful on their own. However, this is not necessarily true for other languages, such as Chinese, where characters and subwords can carry significant meaning. This chapter focuses on analyzing texts written in English due to its widespread use. However, keep in mind that similar concepts can be applied to other languages, with adjustments to account for their unique linguistic structures.

The preceding chapters have already touched upon text clustering and classification. This chapter deals with analyzing sentiments and emotions expressed in written text. The next chapter will discuss topic modeling.

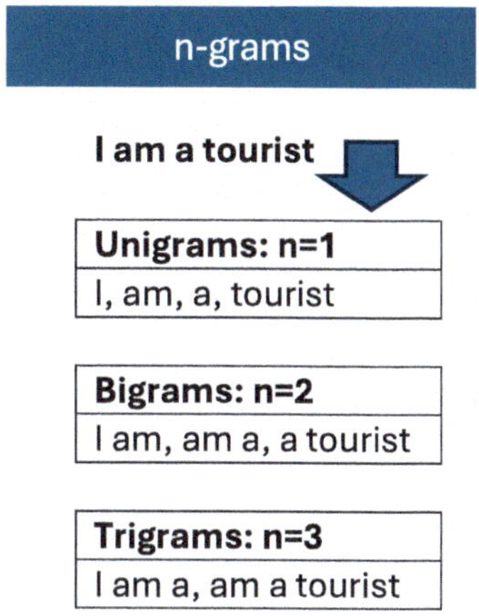

I am a tourist

Unigrams: n = 1

I, am, a, tourist

Bigrams: n = 2

I am, am a, a tourist

Trigrams: n = 3

I am a, am a tourist

10.2 Sentiment Analysis

Generally speaking, the texts can be categorized into facts, which contain objectively verified information (which still can be true or false), and opinions: subjective personal assessments, sentiments, and feelings directed toward this information. Sentiment analysis, also known as opinion mining, is the process of using computers to study people's opinions, feelings, and attitudes toward different things, like products, events, or topics. The word "sentiment" is similar but not identical to human emotions. Oxford Reference defines sentiment as "an attitude, in favour of or against people and their actions, which may involve both judgement and emotion." In text mining, **sentiment analysis** measures the emotional tone of a text on a scale from positive to negative. In comparison, **emotion analysis** goes further, aiming to recognize specific emotions defining the expressed sentiment. Joy, sadness,

Sentiment and emotions

Sentiment analysis measures the opinion sentiment on a scale from negative to neutral to positive.

Emotion analysis aims to recognize human emotions expressed in a written text,as well as in audio and video media.

amusement, excitement, horror, anger, and many others are examples of emotions.

Sentiments are generally characterized by their orientation (positive, negative, or neutral) and intensity (how strong the sentiment is). Depending on the instrument, the orientation and intensity both can use different scales, such as:

- A continuous scale from -1 (for very negative) to 0 (neutral) to + 1 (very positive) or
- A discrete scale: -2 (very negative); -1 (negative); 0 (neutral); 1 (positive); 2 (very positive).

Sentiment analysis can be applied to an entire document, individual sentences, or specific aspects of interest, like different product features ("hotel room was small but clean"). **Comparative sentiment analysis** can also compare sentiment estimates, such as when someone says, "Airbnb stay is better than a regular hotel's." The latter applications of sentiment analysis are frequently called **opinion mining**.

Consider the following hotel review:

10

> *HotelHopper2002*
> *I stayed at the Miracle Hotel for business. The location is good, but the room was nasty.*
> *June 2024*

Sentiment analysis levels

Sentence and document level: overall sentiment expressed in a sentence of the entire document.

Aspect-based: the sentiment of an opinion about a specific feature,object, or phenomenon.

From this review, one can identify several key elements of an expressed opinion:

- the **sentiment** reflecting the personal opinion of the reviewer about the hotel, expressed with the feeling-carrying words good and nasty;
- the target of the opinion telling about the object of the expressed opinion: Miracle Hotel
- the specific **aspects** being commented on: the location and room;
- the **person** holding the opinion: HotelHopper2002; and
- the **date** the opinion was expressed: June 2024.

The simplest form of sentiment analysis looks at the sentiment of an entire document, assuming it comes from one person and relates to one event or product. For example, the HotelHopper2002 review has one word carrying a highly negative sentiment (nasty) and one moderately positive sentiment word (good), from which we might conclude that the overall review is negative. A popular sentiment analysis program, VADER, estimates the negative component of this review as 0.584 and the positive component as 0.198, which adds up to an overall sentiment of -0.372 on a scale from -1 to 1.

Such **document-level sentiment analysis** works well for short texts such as product and service reviews or social media posts, but it can be limiting for more complex texts. For longer documents, a **sentence-level analysis** is more suitable, as it allows tracking sentiment change throughout the text. Such analysis would typically classify sentences into positive, negative, and neutral.

Even more specific **aspect-based (or feature-based) sentiment analysis** aims to learn opinions about a specific target. For example, an amusement park management may be interested only in sentiments directly related to the park; for instance, in a review, "We had a good time in the park, despite the terrible traffic on the way back," only the park aspect would be valuable. The aspect-based sentiment analysis can become very challenging. For instance, in a sentence, "Park driving was terrible," an analyst (human or algorithmic) should be able to differentiate between emotions directed at the park or driving conditions. Complementary to the aspect-based analysis, the **comparative sentiment analysis** seeks to *compare* sentiment towards several products or different aspects of the same product.

Emotions in animals

Many authors, including Charles Darwin, believed that the basic emotions have origins in animal world.

Chimpanzee disappointment and sulking. From: Darwin, C.R. 1872. The expression of the emotions in man and animals.London: John Murray. 1st edition.

10.3 Emotion Detection

Emotion detection is closely related to sentiment analysis. In practical terms, the sentiment is unidimensional and expressed using a single axis from negative to neutral to positive. As opposed to that, emotion detection seeks multidimensional text characterization. For example, one very popular emotional scale by Paul Ekman (1992) identified six "basic" emotions: Happiness, Sadness, Fear, Disgust, Anger, and Surprise.[1] Ekman's initial six basic emotions were later expanded to include additional emotions: contempt, embarrassment, amusement, and pride.

1 Ekman, P. (1992). "An argument for basic emotions". *Cognition and Emotion*. **6** (3): 169–200.

"My phone is broken!" Sad? Angry? Depressed?

Emoticons and emojis

Both emoticons and emotions are visual representations used to express emotions. Emoticons are simple textbased symbols created using keyboard characters, such as :-) for a smile or :-/ for skepticism. Emojis are more understandable graphical symbols: compare :-) to ☺ and :-/ to 😒. In contrast to emoticons, emojis represent a wider range of emotions, as well as objects and activities.

This seemingly small number of dimensions comes from a line of neurophysiological experiments supporting an argument that a small number of *basic* or *primary* emotions are hard-wired into human brains, possibly since birth. Activation of these emotions can be measured, e.g., through electroencephalography. It is hypothesized that these emotions benefit the evolutionary survival mechanisms by activating certain behavioral and physiological patterns (such as "fight or flight"). Ekman's research has shown that basic emotions are universally recognized through facial expressions, regardless of cultural differences. For example, a smile is recognized as a sign of happiness worldwide, while a frown is associated with sadness or displeasure.

A similar approach by Robert Plutchik suggested the "wheel of emotions" model with four pairs of primary emotions: joy—sadness, anger—fear, trust—disgust, and surprise—anticipation,[2] each affiliated with a specific stimulus, cognition, behavior pattern, and effect (Fig. 10.1). Notice that each dimension has multiple intensity levels. For example, joy can escalate to ecstasy or diminish to serenity. In addition, primary emotions may blend into complex feelings; for instance, joy combined with trust results in love.

Hence, emotion detection goes beyond sentiment analysis: the latter is only concerned with measuring the degree of objectivity and subjectivity in a written text and, especially, with text polarity (negativity or positivity). The former, however, is much more complex. In addition, there is no consensus on a set of basic emotions, further complicating emotion detection and making it less objective. Finally, recognition of emotions may be difficult for humans as well, affecting the effectiveness of emotion detection model training. The complexity of human emotions and their expression in language necessitates sophisticated linguistic and computational techniques. Moreover, the subtleties of emotional intensity and the interplay of multiple emotions in a single text add further layers of complexity.

The other side of this complexity is that emotion detection is better progressing in image and audio media, where facial expression and voice tone naturally com-

2 R. Plutchik, 1980. A general psychoevolutionary theory of emotion, pages 3–33. Academic press, New York.

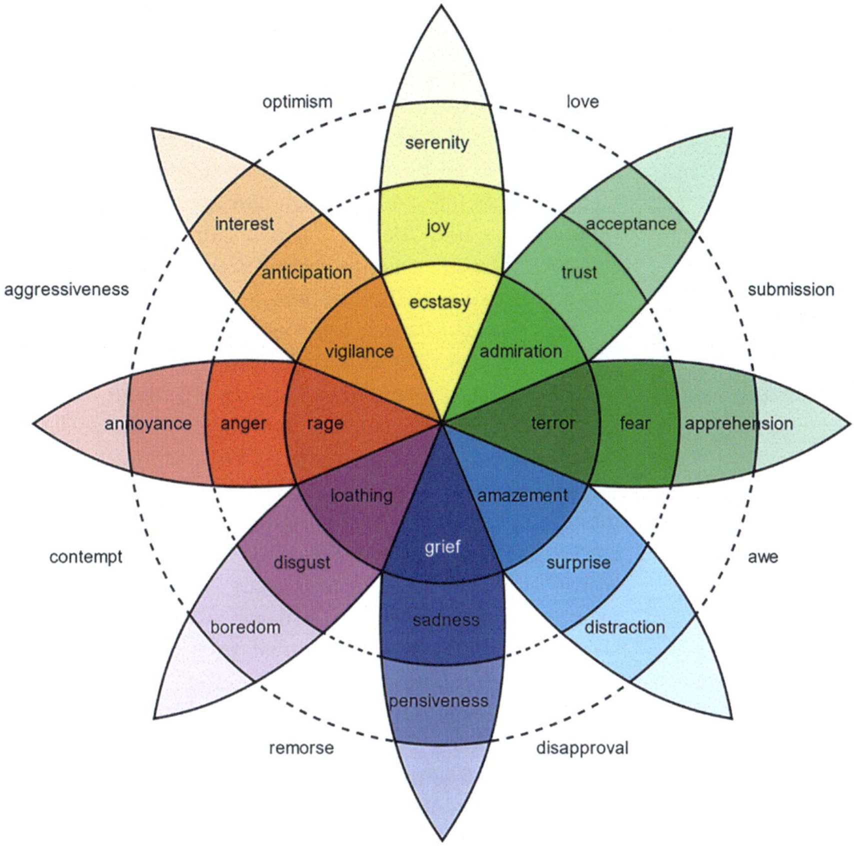

Fig. 10.1 Plutchik's wheel of emotions. By Machine Elf 1735—Own work, Public Domain, ▶ https://commons.wikimedia.org/w/index.php?curid=13285286

plement word semantics, as opposed to text analysis. For example, the text "I go for a business trip tomorrow" is emotion-free, yet it can express joy or sadness when spoken. Notice that social media, attempting to approach the richness of real-life human dialogs, employs **a multitude of emoticons and emojis**, convening the speaker's emotions.

10.4 Lexicon-Based Methods

The *lexicon-based* approach to sentiment analysis relies on a list of words and phrases that have been pre-assigned sentiment values associated with positive or negative sentiments, known as sentiment lexicons. These lexicons, together with a limited set of rules describing, e.g., how negation works, serve as a foundation for the algorithm to classify the sentiment of a given text. Essentially, the software compares each word in a document to this list to determine the overall sentiment. This sentiment score can be adjusted by the rules detecting the effect of words like "not," which can reverse the sentiment, or modifiers like "deeply," which can make a sentiment stronger. The document's overall sentiment is then calculated as the sum of these scores.

Opinion words in students' feedback sentiment lexicon[3]

Word	Score
Care	+2
Clear	+2
Complex	−3
Fast	−1
Sleepy	−2

The lexicon-based approach is simple, fast, and straightforward, but it is usually not very accurate. For instance, it may struggle to capture sentiment accurately in texts where the context significantly alters the sentiment associated with a word. For instance, the word "fight" is generally negative, but it carries a positive sentiment in sports reporting. Language semantics can change emotion-carrying word meaning substantially, like the word "cold" in "cold beer" versus "cold person," which have different sentiments. A more advanced method involves using a large collection of texts from the relevant field to understand the context better. Hence, the lexicon-based approach works best with lexicons developed for specific fields. For example, Aung and Myo created a lexicon of 745 sentiment words specifically targeting sentiment mining of teaching evaluations.[3] A very popular lexicon-based tool, VADER (Valence Aware Dictionary and Sentiment Reasoner), was specifically designed to analyze sentiment in social media texts, such as tweets or Facebook posts. Contrastingly, the SentiWordNet tool is trained on literary texts or news articles and, therefore, is more general-purpose.

10

Intensification words in students' feedback sentiment lexicon[3]

Word	Score
Slightly	−50%
Most	+100%
Really	+25%
Little	−50%
Easily	+25%

A systematic way of creating a sentiment lexicon for a specific field starts with a small set of seed words with known sentiments and then expands the list by finding synonyms and antonyms. After that, the cycle of searching for new words is repeated, with the expanded list serving as a new seed. After the search is saturated and no new words are found, a set of rules is developed to account for auxiliary words expressing negation (never,

3 Aung, K. Z., & Myo, N. N. (2017, May). Sentiment analysis of students' comment using lexicon based approach. In *2017 IEEE/ACIS 16th international conference on computer and information science (ICIS)* (pp. 149–154). IEEE.

no), soft negation (require, need), intensification (very, slightly), etc. The final list is manually edited to remove errors. Such an analytically developed lexicon is sometimes called a sentiment dictionary.

Lexicon-based methods can also be employed for emotion analysis. Emotion lexicons, similar to sentiment lexicons, consist of words associated with specific emotions and a set of rules defining how negation and intensification work. In addition, social media emotion lexicons include emoticons and emojis. Canada National Research Council (NRC) has published seven sentiment and emotion lexicons, including not only a general association between the words and emotions but also hashtags and emoticons.[4] Overall, however, the lexicon-based sentiment and emotion analysis has proven to be particularly effective in large-scale studies where the primary goal is to gauge the general sentiment trend rather than dissect individual texts' nuances. While lexicon-based analysis may not capture the full complexity of human emotions and sentiments, they are effective in providing a broad overview of sentiment trends and emotional patterns. A more detailed analysis that captures sentiments and emotions in a particular domain is provided by the machine learning methods, discussed in the next section.

Lexicon based sentiment analysis

The lexicon approach employs a list of words and phrases with positive and negative sentiment values preassigned by experts. It includes on a set of rules defining how the negation and intensification words modify the sentiment.

10.5 Machine Learning Sentiment Analysis

Algorithmically, sentiment analysis is a classification problem of mapping the documents or sentences to a limited set of sentiments, such as "Very negative," "Negative," "Neutral," "Positive," and "Very positive." The lexicon (or dictionary) based sentiment analysis is often considered "unsupervised" as it doesn't require any external information besides the dictionary itself. This interpretation, however, hints at the possibility of a "supervised" approach, working as follows. First, a representative sample of documents (for example, hotel reviews) is manually classified by experts in the field. Then, a machine learning classifier like those discussed in ► Chapter 5 is trained and validated on this sample, as discussed in ► Chapter 8. Finally, the trained model processes other documents belonging to the same domain (◘ Fig. 10.2).

Sentiment analysis is a classification problem

Sentiment analysis can be interpreted as a classification of a text to a sentiment "classes", such as "Negative", "Neutral", and "Positive". The lexicon approach is similar to unsupervised and machine learning to supervised classification. There is also a semi-supervised approach which takes the best from the lexicon and machinelearning classifications.

4 ► https://nrc.canada.ca/en/research-development/products-services/technical-advisory-services/sentiment-emotion-lexicons.

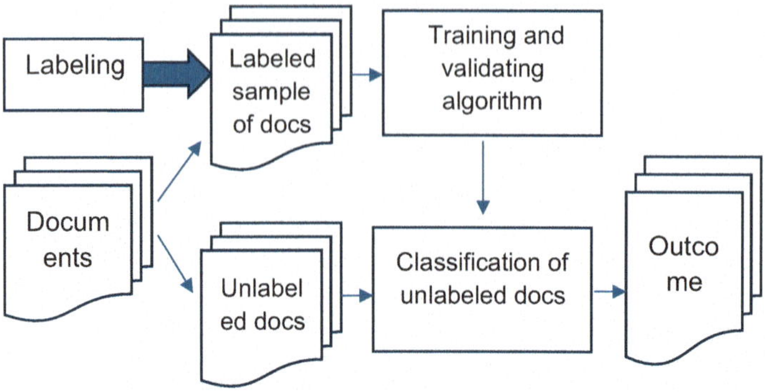

Fig. 10.2 Machine learning sentiment analysis

Notice that this supervised learning approach doesn't need a list of sentiment words or rules of negation and intensification, as the algorithm learns the sentiment directly from the training data. In machine learning sentiment analysis, the algorithm is trained on a labeled dataset of texts representative of the domain, where each text is associated with a sentiment label. The algorithm learns the relationship between the text features and the sentiment labels and applies this learned relationship to classify new, unseen texts. Conversely, extraction of main topics from the text, discussed in the next chapter, generally represents unsupervised learning and does not require labeled data. Instead, it identifies patterns and structures within the data, such as clusters of similar texts, to infer sentiment categories. Semi-supervised learning, as the name suggests, is a hybrid approach combining supervised and unsupervised learning elements. It leverages a small amount of labeled data to guide learning on a larger unlabeled dataset. This approach is particularly useful when labeled data is scarce or expensive to obtain, which is a common scenario in sentiment analysis.

We will study practical approaches to training the sentiment analysis machine learning models in the next chapter on text analysis. Meanwhile, for some popular domains, there are already pre-trained sentiment classifiers. For instance, RoBERTa classifier was trained on Twitter data. Comparisons show that properly trained machine learning methods perform better than lexicon-based ones. Firstly, they can capture complex linguistic features and their interactions, such as negations and intensifiers, often overlooked in lexicon-based approaches. Secondly, it can adapt to different domains and genres, as it learns from the data rather than relying on a predefined lexicon. However, lexicon-based methods have the advantage of being more transparent and don't require the intensive training that machine learning does. They are also less dependent on the quality of training data. Machine learning models, while powerful, can struggle when applied to documents from a different domain than the one they were trained on.

10.6 Popular Solutions

Many software packages capable of text sentiment analysis offer a wide range of functionalities. Among them are stand-alone applications, downloadable to personal computers; web-based applications; API (application programming interface) services, which allow access to online services; and software libraries providing specific functionality for custom computer programs. The stand-alone packages often provide a user-friendly graphical interface, allowing users to perform complex data mining tasks without writing code. Conversely, cloud-based services, such as Google Cloud Natural Language API, IBM Watson Natural Language Understanding, and Microsoft Azure Text Analytics, offer scalable, pay-as-you-go solutions that can handle large volumes of data. One example of a web-based application is OpenAI's ChatGPT. Finally, there are Python libraries that can provide sentiment analysis functionality.

MeaningCloud (▶ www.meaningcloud.com) provides a public API for sentiment analysis and can detect global, local, and aspect sentiment in texts in multiple European languages. Interestingly, this tool can also detect irony and polarity disagreement. While this tool cannot be used directly, it can be integrated with various programs such as Microsoft Excel and Google Sheets.

LIWC-22 (Linguistic Inquiry and Word Count, ▶ https://www.liwc.app/) is a lexicon-based package popular among social scientists for comprehensive text analysis. One advantage of the sentiment part of the package is its ability to compare the analyzed document with a "generic" writing in the same domain.

SentiStrength (▶ http://sentistrength.wlv.ac.uk/) is another example of a lexicon-based sentiment analysis tool aimed at academic users. It analyzes text sentiment and

Important!

Sentiment analysis tools should be used with texts from same domains they were trained on.
Sentiment analysis results should be validated
Machine learning algorithms are typically non-transparent; their outcomes are hard to explain and interpret.

provides a wide variety of sentiment classification types on document, sentence, and aspect levels. Another advantage of this tool is that multiple customized dictionaries are available for domains and various languages. Finally, SentiStrength can be used as a stand-alone package, integrated into another product, or with a custom computer program.

VADER (Valence Aware Dictionary for Sentiment Reasoning, ▶ https://github.com/cjhutto/vaderSentiment) is also a lexicon-based analysis directed at social media analysis. Due to this focus, it features not only a sentiment dictionary of common English words but also slang, initialisms (e.g., WTF), emojis, and acronyms. VADER was validated on tweets but works well with texts from other social media platforms. There are many versions of VADER, both free and commercial, stand-alone and cloud-based. There is also a VADER library, which can be called from a Python program; we will use it in this chapter's lab. Another tool, AFINN, is similar to VADER but less popular. It has also been trained on tweets and includes slang and multi-word tokens.

NLTK SentimentAnalyzer (▶ https://www.nltk.org/api/nltk.sentiment.html) Description: NLTK (The Natural Language Toolkit) is a large and very flexible software library that includes multiple sentiment analysis methods and is best used when integrated into computational linguistic software. I am unaware of a user-friendly version of the SentimentAnalyzer employing a graphical interface, but it is integrated into several other stand-alone products.

10.7 Conclusion

Sentiment analysis and closely related emotion detection are fast-developing fields of natural language analysis. Many of the packages listed above will be radically updated or disappear in a few years, and many new ones will appear. Among few considerations to keep in mind in selecting a particular package are:

- The domain used for developing the lexicon (for the lexicon-based tools) or for training (for the machine learning tools);
- Tool validation results. Generally, it is a good idea to validate a sentiment analysis tool every time it is used on a new dataset, even when this dataset belongs to

the same domain the tool developed for. One popular tour guide named Mr. Grim can ruin the outcomes of sentiment analysis of a popular destination (true story!)
- Trade-off between transparency of the lexicon-based tools and black box machine learning algorithms. In this respect, the large language models such as ChatGPT are the least transparent, and hence, the outcomes of their sentiment analysis are the hardest to explain and defend.

As with any tool, the key to effective use of sentiment analysis tools lies in understanding their strengths and weaknesses and adapting their use to the specific requirements of the task.

10.8 Review Questions

1. Define n-grams. Provide examples of unigrams, bigrams, and trigrams.
2. Explain the difference between sentiment analysis and emotion analysis.
3. Explain the difference between document-level sentiment analysis and sentence-level sentiment analysis. When might one be more appropriate than the other?
4. Describe the aspect-based sentiment analysis. Provide an example scenario where this type of analysis would be useful.
5. What challenges arise in emotion detection as compared to sentiment analysis?
6. Describe the "wheel of emotions" model proposed by Robert Plutchik. What are the primary emotions identified in this model?
7. How do emoticons and emojis enhance emotion detection in text mining?
8. What are the strengths and limitations of the lexicon-based approach to sentiment and emotion analysis?
9. How do machine learning methods for sentiment analysis differ from lexicon-based methods?
10. What is the significance of domain-specific training in the effectiveness of machine learning models for sentiment analysis?

11. Why is it important to validate sentiment analysis tools when applying them to new datasets, even within the same domain?
12. Discuss the trade-offs between using transparent lexicon-based tools and less transparent machine learning algorithms for sentiment analysis.

10.9 Lab

Objectives

1. Practice lexicon-based sentiment detection
2. Practice model evaluation and performance comparison

The overall goal of this assignment is to detect the sentiment expressed towards COVID-19 vaccines. We will use the lexicon-based sentiment analysis approach. Even though the lexicon-based models don't require labeled data (that is, human raters are not required to annotate the sentiment in a data sample manually), the lab is still using a labeled dataset to permit model performance evaluation. The lab will compare the performance of multiple models.

Data

The data for this chapter is modified from the Covid-19 Vaccine Tweets with Sentiment Annotation dataset (▶ https://www.kaggle.com/datasets/datasciencetool/covid19-vaccine-tweets-with-sentiment-annotation), provided under the CC BY-SA 4.0 license. The data is stored in a Microsoft Excel file and contains 6,000 records. The data represents tweets about the following widely distributed COVID-19 vaccines: Moderna (USA), Oxford-AstraZeneca (UK), Pfizer-BioNTech (Germany), Sinopharm BIBP (China), Sinovac (China), Covaxin (India), and Sputnik V (Russia).

The goal is to:

(1) Gauge public opinion towards COVID-19 vaccines.
(2) Using the labeled data, compare the performance of multiple sentiment analysis algorithms.

The data is stored as ▶ https://github.com/andykir/DataMiningLectures/raw/main/covid-19_vaccine_tweets_with_sentiment.xlsx.

Variables

Variable name	Explanation	Notes
tweet_id	Unique identifier for each tweet	String
label	Manually assigned sentiment	1: negative; 2: neutral; 3: positive
tweet_text	Tweet text	String

Part 1. Read, Describe, and Clean the Data

Step 1. Read the Data

Every lab starts with similar steps of reading and describing the data. Because of that, we reuse an older code with minimal modifications. Start with a new Markdown cell. Describe the purpose of the code. Then, describe the purpose of this step.

```
The code for Chapter 10 Lab: Sentiment analysis
Goal: practice text sentiment analysis
Data: Tweets on COVID-19

Part 1. Read, describe, and clean the data.
Step 1. Read the data
```

```
import pandas as pd

# Read the Excel file from the URL
url = \
"https://github.com/andykir/DataMiningLectures/raw/main/covid-
19_vaccine_tweets_with_sentiment.xlsx"
data = pd.read_excel(url, sheet_name="covid-19_vaccine_tweets")

# Print information about the data DataFrame
print("data info")
data.info()
print("\n\n data")
print(data.head())
print("\n\n data description")
print(data.describe())
```

Result:

Variables

```
<class 'pandas.core.frame.DataFrame'>
RangeIndex: 6000 entries, 0 to 5999
Data columns (total 3 columns):
 #   Column      Non-Null Count  Dtype
---  ------      --------------  -----
 0   tweet_id    6000 non-null   float64
 1   label       6000 non-null   int64
 2   tweet_text  6000 non-null   object
dtypes: float64(1), int64(1), object(1)
memory usage: 140.8+ KB
```

Data

```
       tweet_id  label  tweet_text
0  1.360342e+18      1  4,000 a day dying from the so called Covid-19 ...
1  1.382896e+18      2  Pranam message for today manifested in Dhyan b...
2  1.375673e+18      2  Hyderabad-based ?@BharatBiotech? has sought fu...
3  1.381311e+18      1  Confirmation that Chinese #vaccines "don't hav...
4  1.362166e+18      3  Lab studies suggest #Pfizer, #Moderna vaccines...
```

Descriptive statistics

```
           tweet_id        label
count  6.000000e+03  6000.000000
mean   1.372192e+18     2.246667
std    1.036489e+16     0.570856
min    1.337736e+18     1.000000
25%    1.366304e+18     2.000000
50%    1.373906e+18     2.000000
75%    1.380625e+18     3.000000
max    1.385237e+18     3.000000
```

There are no missing values, which simplifies the data cleaning process. Notice that the Tweet ID column has been interpreted as a floating-point number, which may truncate its values and result in multiple tweets appearing to have the same ID. Generally, these IDs should be stored as integers or strings to preserve accuracy. We can check for duplicate IDs using the *duplicated*() method of the pandas DataFrame object.

USER:

Count the number of duplicated values in column tweet_id of DataFrame "data".

Assistant:

```
import pandas as pd

duplicate_count = data['tweet_id'].duplicated().sum()
print(duplicate_count)
```

Explanation:

This code uses the *duplicated*() method to count duplicate entries in the *tweet_id* column and then sums them up to get the total count of duplicates.

Result:

0
Apparently, there are no duplicate IDs.

Step 2. Transform Textual Data

Prior to starting the analysis, it makes sense to extract mentions (another user's *handle* that appears after the @ symbol) and hashtags from the tweets and remove them from the sentiment analysis.

USER:

Use the column 'tweet_text' in DataFrame "data". Remove all alphanumeric strings preceded with '@' symbol and save them as a list in a new column, "handles". Print top rows of DataFrame "data".

Note: alphanumeric strings contain letters and numbers. The string "book15" is alphanumeric, while "book 15" is not alphanumeric as it contains a space.

Assistant:

```
import pandas as pd

data['handles'] = data['tweet_text'].str.findall(r'@\w+')
data['tweet_text'] = data['tweet_text'].str.replace(r'@\w+', '', \
regex=True)
print(data(head())
```

Explanation:

First, a *regular expression* is used to identify the strings that start with '@' and then store them in the new column called "handles", while also modifying the original 'tweet_text' column: *str.findall*(r'@\w+'). To learn more about regular expressions, consult ▶ https://www.regular-expressions.info/. To test a regular expression: ▶ https://regex101.com/.

Then, *str.replace*() method removes those strings from the 'tweet_text' column. We use an identical regular expression: *str.replace*(r'@\w+', '', *regex*=*True*).

Result (abridged):

	tweet_id	label	tweet_text	handles
0	1.360342e+18	1	4,000 a day dying from ...	[@DailyBeast]
1	1.382896e+18	2	Pranam message for today ...	[@meenapranam]
2	1.375673e+18	2	Hyderabad-based ...	[@BharatBiotech, @DBTIndia]
3	1.381311e+18	1	Confirmation that Chinese ...	[]
4	1.362166e+18	3	Lab studies suggest ...	[]

In a similar way, we can extract the hashtags. To do so, we can modify the previous code, replacing "@" with "#" and changing the column name to save the extracted hashtags.

Code:

```
data['hashtags'] = data['tweet_text'].str.findall(r'#\w+')
data['tweet_text'] = data['tweet_text'].str.replace(r'#\w+', '', \
regex=True)
data.head()
```

Result (abriged):

tweet_text	handles	hashtags
4,000 a day dying ...	[@DailyBeast]	[#vaccine, #PfizerVaccine ...
Pranam message ...	[@meenapranam]	[#truth, #love, #karm, ...
Hyderabad-based ?? ...	[@BharatBiotech, @DBTIndia]	[#Covaxin]
Confirmation that ...	[]	[#vaccines, #Sinovac]
Lab studies suggest ...	[]	[#Pfizer, #Moderna, ...

10

Part 2. Run Sentiment Analysis

Step 1. Use the Vader method to analyze the sentiment.

User:

Use the column tweet_text in DataFrame "data". Use the Vader method to extract the sentiment. Save sentiment in column "Vader" as follows: 1: negative; 2: neutral; 3: positive (using no context from Notebook).

Assistant:

```
import pandas as pd
from nltk.sentiment import SentimentIntensityAnalyzer
import nltk

# Ensure you have the VADER lexicon downloaded
nltk.download('vader_lexicon')

# Initialize the VADER sentiment analyzer
sia = SentimentIntensityAnalyzer()

# Function to classify sentiment
def classify_sentiment(text):
    score = sia.polarity_scores(text)['compound']
    if score < 0:
        return 1  # negative
    elif score == 0:
        return 2  # neutral
    else:
        return 3  # positive

data['Vader'] = data['tweet_text'].apply(classify_sentiment)
```

Explanation:

This code uses the *nltk* library, which includes the Vader sentiment analysis tool. Remember to install the *nltk* package if you havent already. The code works as follows. First, it imports the necessary modules from *nltk*. Then, Vader sentiment analyzer is initialized. Then, a function to classify the sentiment into the specified categories (1 for negative, 2 for neutral, and 3 for positive) is defined. Finally, this function is applied to the *tweet_text* column. The results are saved in a new column called *Vader*.

We can print the relevant columns of the dataset to compare the results of the automated analysis with those of manually assigned labels.

Code:

```
print(data[["tweet_text","label","Vader"]].head())
```

Result:

```
   tweet_text                                           label  Vader
0  4,000 a day dying from the so called Covid-19 ...        1      2
1  Pranam message for today manifested in Dhyan b...        2      2
2  Hyderabad-based ?? has sought funds from the g...        2      2
3  Confirmation that Chinese  "don't have very hi...        1      1
4  Lab studies suggest,  vaccines can protect ag...         3      3
```

Step 2. Use the SentiStrength method to analyze the sentiment.

User:

Replace the Vader method with TextBlob.

Assistant:

```
import pandas as pd
from textblob import TextBlob

# Function to classify sentiment
def classify_sentiment(text):
    polarity = TextBlob(text).sentiment.polarity
    if polarity < 0:
        return 1  # negative
    elif polarity == 0:
        return 2  # neutral
    else:
        return 3  # positive

data['TextBlob'] = data['tweet_text'].apply(classify_sentiment)
```

Explanation:

First, install the TextBlob package if you haven't already:

```
!pip install textblob
```

Then, the code imports the necessary modules from *textblob*. The rest of the code is identical to the Vader sentiment classifier. A function *classify_sentiment* is then defined. The function classifies the sentiment into the specified categories (1 for negative, 2 for neutral, and 3 positive). Then, *classify_sentiment* is applied to the *tweet_text* column, and the results are saved in a new column called *TextBlob*.

Finally, we can print sentiment analysis outcomes:

```
print(data[["tweet_text","label","TextBlob"]].head())
```

Result:

```
   tweet_text                                         label  TextBlob
0  4,000 a day dying from the so called Covid-19 ...      1         2
1  Pranam message for today manifested in Dhyan b...      2         2
2  Hyderabad-based ?? has sought funds from the g...      2         3
3  Confirmation that Chinese  "don't have very hi...      1         3
4  Lab studies suggest,  vaccines can protect ag...       3         2
```

TextBlob's results are noticeably inferior compared to Vader's. For instance, a human rater evaluated the fourth tweet (#3 in the table, "Confirmation that Chinese #vaccines don't have very high protection rates, according to the director of the China Centers for Disease Control, Gao Fu. #Sinovac effectiveness found to be as low as 50.4% by researchers in Brazil is not useful.") as negative, while TextBlob evaluated it as positive. There are several reasons for this failure.

Let us start with sentiment analysis of the sentence "My mood isn't very high":

```
TextBlob("My mood isn't very high").sentiment.polarity
```

Result:

```
0.20800000000000002
```

TextBlob and many other methods require prior text cleaning. For instance, they don't understand word contractions. To deal with this problem, we can use library **contractions,** which can expand the contractions common in the English language, such as "don't" or "can't". To install:

```
!pip install contractions
```

New Code:

```
import contractions
mySentence=contractions.fix("My mood isn't high")
print(mySentence)
TextBlob(mySentence).sentiment.polarity
```

Result:

```
My mood is not very high
-0.0615384615384615 35
```

TextBlob is also confused with complex negations. Compare, for example:

```
print(TextBlob("I do not have very good life").sentiment.polarity)
print(TextBlob("My life is not very good").sentiment.polarity)
```

```
0.9099999999999999
-0.26923076923076916
```

This problem, unfortunately, is not easily corrected. The final code:

```
from textblob import TextBlob

data['tweet_text'] = data['tweet_text']. \
    apply(lambda x:contractions.fix(x))

# Function to classify sentiment
def classify_sentiment(text):
    polarity = TextBlob(text).sentiment.polarity
    if polarity < 0:
        return 1  # negative
    elif polarity == 0:
        return 2  # neutral
    else:
        return 3  # positive

data['TextBlob'] = data['tweet_text'].apply(classify_sentiment)

print(data[["tweet_text","label","TextBlob"]].head())
```

Part 3. Model Performance Evaluation

Step 1. Evaluate Vader and TextBlob.

Vader

User:

Use columns “label” and “Vader” of DataFrame “data”. The column “label” is labels; the column “Vader” is outcomes of data classification into 3 classes: 1, 2, and 3. Compute: accuracy, precision, recall, and F1-score of classification.

Assistant

```
import pandas as pd
from sklearn.metrics import accuracy_score, precision_score,
recall_score, f1_score

y_true = data['label']  # True labels
y_pred = data['Vader']  # Predicted labels

# Compute metrics
accuracy = accuracy_score(y_true, y_pred)
precision = precision_score(y_true, y_pred, average='weighted')
recall = recall_score(y_true, y_pred, average='weighted')
f1 = f1_score(y_true, y_pred, average='weighted')

# Print the results
print(f"Accuracy: {accuracy}")
print(f"Precision: {precision}")
print(f"Recall: {recall}")
print(f"F1 Score: {f1}")
```

Explanation:

This code uses the the *scikit-learn* library methods that compute accuracy, precision, recall, and F1-score. In this code, *y_true* contains the true labels from the *label* column, and *y_pred* contains the predicted labels from the *Vader* column. The metrics are computed using the respective functions from sklearn.metrics, with *average='weighted'* to account for the multi-class nature of the classification. Finally, the results are printed out.

Result:

Accuracy: 0.464.
Precision: 0.5904990967019625.
Recall: 0.464.
F1 Score: 0.4843217570789823.

Text Blob

We can reuse this code for TextBlob performance evaluation:

```
# TextBlob performance evaluation

y_true = data['label']   # True labels
y_pred = data['TextBlob']  # Predicted labels

# Compute metrics
accuracy = accuracy_score(y_true, y_pred)
precision = precision_score(y_true, y_pred, average='weighted')
recall = recall_score(y_true, y_pred, average='weighted')
f1 = f1_score(y_true, y_pred, average='weighted')

# Print the results
print(f"Accuracy: {accuracy}")
print(f"Precision: {precision}")
print(f"Recall: {recall}")
print(f"F1 Score: {f1}")
```

Result:

```
Accuracy: 0.46
Precision: 0.5558996271342894
Recall: 0.46

F1 Score: 0.4724631923058448
```

The sentiment analysis model and human raters agree in 46% of cases. Comparing Vader with TextBlob, Vader's sentiment evaluation is slightly more reliable.

Step 2. Repeat Model Performance Evaluation Using Misclassification Costs

Intuitively, the cost of mistaking a positive review for a negative one is greater than mistaking a positive or negative review for a neutral one. For example, we can subjectively assign the former cost to 1 and the latter to 0.5. Weighted evaluation allows the incorporation of relative costs of misclassifications, adjusting model performance metrics to reflect the real-world impact of errors.

There are two useful metrics. The *total cost of misclassification* metrics summarizes the costs of all misclassifications. A more interpretable variation, the *relative total cost of misclassification*, is calculated by dividing the total cost of misclassification by the maximum possible cost when every prediction was incorrect with the highest error cost. The relative total cost of misclassification varies between 0 and 1, where 0 corresponds to a perfect classification, and 1 corresponds to the worst possible misclassification. When the cost of all mistakes is identical, the relative total cost of misclassification is similar to the accuracy measure.

User:

Use columns "label" and "Vader" of DataFrame "data". The column "label" is labels; the column "Vader" is outcomes of data classification into three classes: 1, 2, and 3. Compute: *relative total cost of misclassification*. The cost of mistaking classes 1 and 2 is 0.5. The cost of mistaking classes 2 and 3 is 0.5. The cost of mistaking classes 1 and 3 is 1.

Assistant:

```
true_labels = data['label']
predicted_labels = data['Vader']

# Define the costs of misclassification
costs= {(1, 2):0.5,(2, 1):0.5,(2, 3):0.5,(3, 2):0.5,(1, 3):1,(3, 1):1}

# Calculate the confusion matrix
cm = confusion_matrix(true_labels, predicted_labels, labels=[1, 2, 3])

# Calculate total cost of misclassification
total_cost = 0
for i in range(len(cm)):
    for j in range(len(cm)):
        if i != j:  # Only consider misclassifications
            total_cost += cm[i, j] * costs.get((i + 1, j + 1), 0)

# Calculate the relative total cost of misclassification
total_instances = cm.sum()
relative_total_cost = total_cost / total_instances if total_instances
> 0 else 0

# Print the results
print(f'Total Cost of Misclassification: {total_cost}')
print(f'Relative Total Cost of Misclassification:
{relative_total_cost}')
```

Explanation:

First, the true labels and predicted labels are extracted from the DataFrame *data*. Then, misclassification costs are defined. For instance, the cost of classifying neutral sentiment as positive is 0.5: {(1, 2):0.5}. Then, the confusion matrix is calculated using the *confusion_matrix* method from the *scikit-learn* library. The confusion matrix shows the counts of true positives, false positives, true negatives, and false negatives. This matrix is then used compute the total cost of misclassification. Finally, the relative total cost of misclassification is calculated by dividing the total cost of misclassification by the total number of instances in the dataset.

Result:

```
Total Cost of Misclassification: 1812

Relative Total Cost of Misclassification:
0.302
```

Part 4. On Your Own

1. Make a table that compares the relative total cost of misclassification for Vader and TextBlob sentiment analysis models using three different weight matri-

ces. In total, there will be six model runs, but some of those you have already done.

In the first matrix, the weights of misclassification should be the same.

The second matrix should represent 1/2 the cost of mistaking neighboring classes (for example, predicting negative sentiment for a neutral-sentiment document) as compared to non-neighboring classes (for example, predicting negative sentiment for a positive-sentiment document).

The third matrix should represent zero costs of mistaking neighboring classes.

Elaborate on model evaluation outcomes. How frequent is it for sentiment analysis models to confuse positive and negative sentiments?

2. For those who are following the free Python course. Convert the code into a function that computes a relative total cost of misclassification given a sentiment analysis model output and a weight for specific misclassification errors.

» Function parameters:
model (str): sentiment analysis model (Vader or TextBlob);
weight (int): The cost of mistaking 0 for 1 or 2 for 0.

Use this function in a loop to calculate all measures required for the previous problem without repeating the code.

Build your Python Skills

Chapter 19, 20 of the free book Learn Python with Jupyter by Serena Bonaretti. ▶ https://www.learnpythonwith-jupyter.com/

Additional Reading

Liu, B. (2010). Sentiment analysis and subjectivity. *Handbook of natural language processing, 2*, 627–666.

Liu, B. (2020). *Sentiment analysis: Mining opinions, sentiments, and emotions*. Cambridge University Press.

Kim, A. (2021). Sentiment Analysis: Extracting emotion through machine learning. TED talk (Video). ▶ https://www.youtube.com/watch?v=n4L5hHFcGVk

Text Mining Essentials

A. P. Kirilenko, *Practical Data Mining with AI for Social Scientists*,
Springer Texts in Social Sciences, https://doi.org/10.1007/978-3-031-89689-7_11

11.1 Introduction

IBM defined text mining as "the process of transforming unstructured text into a structured format to identify meaningful patterns and new insights", noting that "since roughly 80% of data in the world resides in an unstructured format … text mining is an extremely valuable practice within organizations".[1] In the previous chapter, we discussed sentiment analysis, one of the typical tasks of text mining. That gave us a chance to explore the challenges of text mining. Unlike numerical data, text data is inherently noisy, messy, and unstructured. It is laden with typos, jargon, dialects, nuances, ambiguities, and subtleties that can easily be lost or misinterpreted in the computer-based text interpretation process. Furthermore, text data is high-dimensional, meaning it contains a large number of variables, which complicates computations.

Exacerbating the problem, humans have trouble understanding texts, too. For instance, studies show that two human raters evaluating the sentiment of the same text have only a fair agreement.[2] This raises an interesting question of properly defining the ground truth for text analysis validation. For example, does a sentiment of a given piece of text objectively exist, or is it subject to human interpretation?

Text encoding

In natural language processing (NLP), text encoding transforms text into a numerical format. Text encoding allows machine learning algorithms to analyze, understand, and make predictions based on text data.

Text mining (also called text analytics) frequently employs methods originally developed to analyze numerical data, such as regression analysis, correlation analysis, classification, and cluster analysis. However, that necessitates a prior conversion of texts into numbers. Because of that, text mining involves elaborate preprocessing such as tokenization, stemming, or lemmatization, as well as feature extraction: bag-of-words, N-gram, and so on. These steps aim to transform unstructured text data into structured data that can be analyzed using traditional numerical techniques. These methods are discussed in the next section.

1 ► https://www.ibm.com/topics/text-mining.

2 See for instance Imtiaz, N., Middleton, J., Girouard, P., & Murphy-Hill, E. (2018, June). Sentiment and politeness analysis tools on developer discussions are unreliable, but so are people. In Proceedings of the 3rd International Workshop on Emotion Awareness in Software Engineering (pp. 55–61).

11.2 Text Pre-processing

11.2.1 Text Cleaning

The first step in text analysis is cleaning. In the previous lab, before analyzing the sentiment of tweets on COVID-19 vaccines, we removed mentions and hashtags. We also replaced text contractions. These are the typical tasks performed during text cleaning. This step aims to standardize text as much as possible without changing its meaning and typically includes:

- URL extraction and/or removal from the main text;
- Emoji and emoticon extraction and/or removal;
- Handles (mentions) extraction and/or removal;
- Spelling correction;
- Contraction replacements.

Text cleaning

The first step in text analysis is cleaning the data. Fundamentally, the purpose of text cleaning is to remove or replace parts of documents that are irrelevant to the analysis or just incorrect.

There are also optional cleaning steps: stop word removal and punctuation removal. The purpose of these steps is to exclude words irrelevant to the intentof text processing. Stop words are commonly occurring words that do not contribute to text "understanding" by the text analysis algorithm. This reference to the *algorithmic understanding* is important. Even though the common stop words, such as the definite and indefinite articles, are important for *human* text understanding, they may interfere with the performance of a *computer* text mining algorithm such as the Latent Dirichlet Analysis (LDA). There are multiple stopword datasets developed for different languages; in addition, each particular project may require a separate list of stopwords. For instance, in a project dealing with newspaper publications about the Olympic Games one might want to remove the words "Olympic" and "Games" as they are not contributing anything to data understanding (we know already it's about the Olympics).

11.2.2 Spelling Correction

Text data is noisy. The social media text data is extremely noisy, especially due to misspellings, both intentional and unintentional. Because of that, spelling correction is a necessary step in text pre-processing. One popular method employs *Levenshtein distance* (also known as *Edit distance*) to detect misspelled words. This metric measures the distance between two strings of symbols, such as two different words, as the number of edits (substitutions, deletions, or insertions of a single symbol) required to transform one string to the other. For instance, the distance between the words "Carribbean" and "Caribbean" is 1 because one edit is required to correct the misspelled word. Similarly, the distance between "Carribbian" and "Caribbean" is 2, and the distance between "Carribian" and "Caribbean" is 3. When an algorithm encounters a word missing in its dictionary, it uses the Levenshtein distance to search for the closest word from the dictionary.

Typo correction approaches

Syntactic similarity: how the words are written. This method is based on the edit distance.

Phonetic similarity: how the words are pronounced. This method is based on phonetic dictionaries.

Semantic similarity: prediction of misspellings based on context.

There are more advanced algorithms that use similar approaches. For instance, the Damerau-Levenshtein distance handles common typing errors where characters are swapped ("**hte**" → "**the**"). Norvig's algorithm handles cases with multiple substituting candidates by selecting the substituting word based on its relative frequency. The Bayesian methods predict the substituting words from the sentence's context. The latter approach is even further developed in the new generation of transformer-based models like BERT or GPT.

11

11.2.3 Standardization

The goal of standardization is to reduce variability in the text while preserving its meaning. One common technique is lowercasing, which eliminates case distinctions by transforming all words to lowercase, such as *Tourist* → *tourist*. This step reduces the corpus vocabulary size and memory requirements for tokenization, which are discussed further in this section.

Text standardization approaches

Lowercasing
Punctuation removal
Lemmatization
Stemming
Synonym replacement

A more radical reduction in text variability is achieved through lemmatization and stemming. Lemmatization is the process of reducing the words to their base or dictionary form, called a *lemma*. For example, the word *tourists* would be reduced to *tourist*, *traveling* to *travel*, and *betterment* to *better*. Lemmatization can also handle irregular forms: *better*→*good; teeth*→*tooth,* and *was*→*be*.

In comparison, stemming goes even further, reducing the words to their roots: Helpful→Help, Happiness→Happi, and Fried→Fry. Occasionally, stemming transforms semantically different words into identical base forms; for instance, the words caring and car would be transformed into the same base form *car*. Stemming is a fast and popular method of text variability reduction, yet it is prone to producing nonsensical or linguistically incorrect results.

Another useful technique in text variability reduction is synonym replacement. A common approach is to use a thesaurus to join words with similar meanings. For instance, the phrases "I was furious", "I was enraged", and "I was mad" would all be replaced with "I was angry".

11.2.4 POS Tagging

In text analysis, one important problem is *polysemy*, the phenomenon where a single word or expression has multiple related meanings. Compare the sentiment expressed in the following sentences:

» *I like playing football* (positive sentiment).
  and
  I play football like a professional (neutral sentiment).

In these two sentences, the word "like" functions as different parts of speech, changing the sentiment accordingly. To alleviate the polysemy problem, it makes sense to transform the text, tagging each word with the part-of-speech (POS) tag:

» *I/PRP like/VBP Playing/VBG Football/NN.*
  I/PRP Play/VBP Football/NN like/IN a/DT Professional/NN.

POS Tagging

POS tagging tags each word of the text with a respective part-of-speech tag. POS tagging helps with the polysemy problem and also allows focused analysis on a specific part of speech.

▫ Table 11.1 presents some common POS tags, but there are many more. For example, aside from the verb tag "VB", there is also a gerund (VBG), present (VBP), past (VBZ), and so on. A well-build sentiment analyzer then would be able to differentiate between the word "like" used as a verb (VBP) and a preposition (IN) and assign appropriate sentiment (positive and neutral, respectively).

POS tagging can help in focusing only on those text features that are relevant to the analysis. For instance, to extract the keywords, we may focus only on nouns (NN tag). In sentiment analysis, the adjectives (JJ) and adverbs (RB) are frequently the most useful:

» *I am* ***really sad****: the travel agency canceled the trip because of the weather.*

Table 11.1 Common POS (part-of-speech) tags

POS Tag	Description	Example
CC	Coordinating conjunction	And, but, or
DT	Determiner	The, a, an
IN	Preposition/Subordinating conjunction	In, of, like, after
JJ	Adjective	Small, happy, easy
NN	Noun, singular or mass	cat, computer, love
PRP	Personal pronoun	I, he, she, it
RB	Adverb	Quickly, softly
RP	Particle	Up, off, down
VB	Verb, base form	Go, run, play

On the opposite, in text summarization, the nouns and verbs carry the most important information:

> *I am really sad: the* ***travel agency canceled*** *the* ***trip*** *because of the* ***weather****.*

In conclusion, POS tagging may be useful for many types of text analysis, including sentiment analysis, named-entity recognition (NER), text summarization, and others.

11.2.5 Word Tokenization

Tokenization

Word tokenization is the process of splitting sentences into tokens, which then serve as the basic units of text analysis.

We have already mentioned tokenization in an earlier chapter. Word tokenization is the process of splitting sentences into individual words or tokens, which will then serve as the basic units of analysis in natural language processing (NLP). Tokenization typically involves breaking down text based on spaces, punctuation, or special characters but can be more complex depending on the language or the specific task at hand. For example, in English, spaces usually separate words, but in languages like Chinese or Japanese, tokenization requires more advanced techniques to segment characters into meaningful tokens. Beyond individual word tokens, *n-grams* are also used to capture sequences of n words, where *unigrams* refer to single words, *bigrams* refer to two-word combinations, and *trigrams* refer to three-word combinations. N-grams provide context, capturing word dependencies and relationships across sequences. For instance, in a sentence "I am not happy," a sentiment analysis not using bigrams would miss the importance of negation and classify the sentence as positive.

In the *bag-of-words* text parsing model, each document is represented with an unordered collection of tokens, disregarding grammar and word order but keeping track of frequency. For instance, the sentence "*I like playing football*" would be represented as a bag of words: {I, like, playing, football}. This approach is simple but popular and powerful. However, this model's simplicity is also its limitation, as it fails to capture the context and semantic relationships between words. In addition, the method generates highly inefficient, high-dimensional text representations.

The reason for high dimensionality is that each token generates a new dimension in a word space. A sentence "*I like playing football*" will then require four dimensions for its representation. A new sentence, "*I play football like a professional,*" will add three more dimensions representing the words *play*, *a*, and *professional*. In this word space, the sentences will be represented as the vectors (0111010) and (1110101) (◘ Table 11.2). When the number of documents is large, it is not uncommon to have over 10,000 dimensions, each representing a unique word. Most documents contain only a small subset of the total vocabulary, so most of the elements in the vector are zero, making it a *sparse* vector. Despite having only a few non-zero values, the entire vector with all its zeroes must still be stored in memory.

The first limitation, which deals with a lack of context, is partially addressed in the n-gram moving window approach, where the text is separated into tokens with a moving n-word long window. For example, the 2-g (bigram) moving window approach represents the sentence "*I like playing football*" as {I_like, like_playing, playing_football}. This method captures more context than the bag-of-words model. Still, it radically increases the di-

Methods' comparison

Bag-of-words (BoW): the order of words is disregarded. The syntax can be preserved as POS tags.

N-grams: word order is partially captured through a moving window.

Word2vec: words are represented as vectors in a continuous space. The context is captured in vector representation.

◘ **Table 11.2** Bag-of-word representation of two sentences

	A	Foot-ball	I	Like	Play	Playing	Professional
I like playing football	0	1	1	1	0	1	0
I play football like a professional	1	1	1	0	1	0	1

mensionality of the data: while the number of word-space dimensions generated by the bag-of-words approach equals the number of unique words, the n-gram moving window approach also adds combinations of these words. Because of that, in practical applications, the most frequent and the least frequent tokens are removed from consideration.

11.2.6 Embedding

A breakthrough in natural language processing came with the development of the word2vec approach by Google in 2013, which introduced the concept of word embeddings. Word embeddings are dense vector representations of words in a continuous vector space, where words with similar meanings are positioned closer together. Unlike traditional models that require a separate dimension for each word in the vocabulary, word2vec employs an artificial neural network (ANN) to generate these embeddings. In this model, the ANN is trained to predict words in a document based on their context (i.e., the surrounding words). The resulting word embeddings are learned as the weights between the input and hidden layers of the neural network, capturing relationships between words based on their co-occurrence patterns in the training data.

Word2vec

Word2Vec employes an artificial neural network (ANN) predicting the words from the context.

The Continuous Bag-of-Words (CBOW) architecture predicts a target word based on the surrounding words.

The Skip-Gram architecture predicts surrounding words based on the target word.

The dimensionality of the word embeddings is determined by the size of the ANN's hidden layer rather than the size of the vocabulary. This dimensionality is typically much smaller, usually in the range of a few hundred, enabling a significant reduction in computational complexity. One key property of word embeddings is that semantically similar words are represented by vectors that are close to each other, as measured by cosine similarity (discussed in ► Chapter 5).

The word2vec model not only reduces the dimensionality of word representations but also captures meaningful semantic relationships between words, offering a significant advantage over traditional methods such as the bag-of-words approach. The introduction of word embeddings by word2vec laid the foundation for more advanced natural language processing (NLP) models, including the transformer-based architectures that underpin today's state-of-the-art language models.

11.2.7 Visualization of the Word Space

An ever-popular way of text visualization is a word cloud (tag cloud), popularized in 2004 by the photo-sharing platform Flikr cofounder Stewart Butterfield. Word clouds represent the text as a cloud of words so that more frequent words are displayed in larger, bolder fonts and/or brighter colors than less frequent words. Since word frequency distribution in a text typically follows a power law, font type assignment is usually preceded by a logarithmic transformation of word frequencies. It also makes sense to complete text pre-processing before its visualization to exclude stopwords, typos, duplicated low- and upper-case words, and synonyms. ◘ Figure 11.1 provides an example of word clouds visualizing a set of interviews with hotel guests asked about their perspective on the importance of the tone and manner of hotel personnel communication. Notice that text preprocessing (right plot) has removed words unimportant for text understanding given our knowledge of the context (left plot).

While word cloud is arguably an overused and not very informative way of text summarization, a similar Scattertext method is more informative. Scattertext targets visualization of the relative frequency of word use in two groups: Democrats and Republicans, climate change activists and skeptics, or males and females. Similar to the word cloud, the Scattertext visualization should be preceded by text preprocessing and a logarithmic transformation of word frequencies. ◘ Figure 11.2, for instance, presents a scatterplot comparing USA and Swiss hotel guest interview responses. From the plot, we can conclude that US guests were much more likely to use the

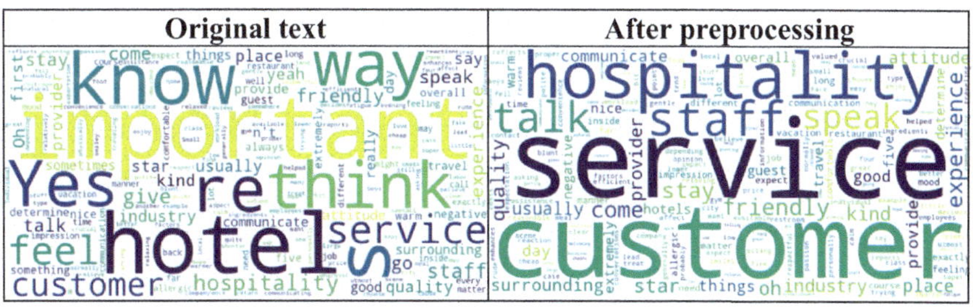

◘ **Fig. 11.1** Word cloud before and after text preprocessing. Notice that preprocessing made visualization more informative

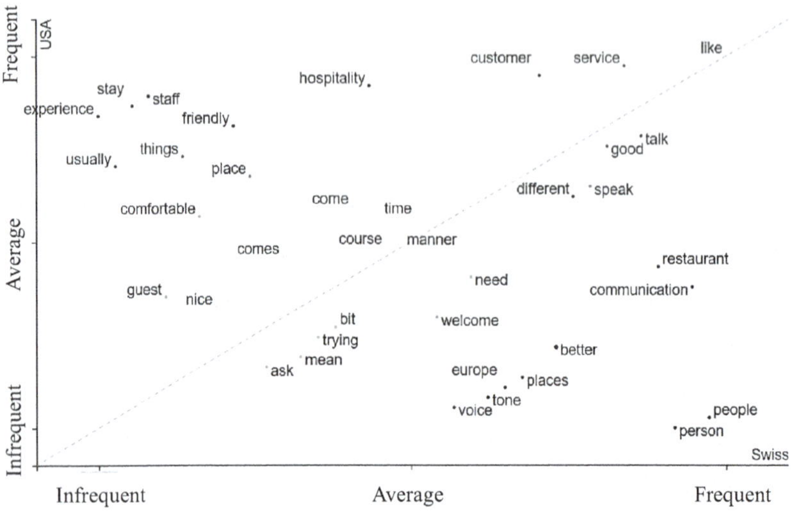

Fig. 11.2 Scattertext plot comparing word frequency use in the USA and Swiss samples of hotel guests

words "hospitality", "friendly", "comfortable", and "customer service" when describing factors important in conversation with hotel staff. Comparably, Swiss guests were more likely to discuss staff voice tone and communication. They also used the words "Europe" and Swiss, hinting at a comparison of hotel experience in Switzerland and the rest of Europe.

11.3 Text to Numbers

Many of the methods originally developed for processing numerical data can also work with texts. However, that frequently requires the conversion of textual data into numbers. Some algorithms, such as the ones used in large language models (LLMs), would apply this conversion automatically; others require the user's attention. In this section, we will cover the most popular approaches.

The popular bag-of-words approach, introduced in the previous section, presents documents as a "bag" of words contained in the document. In this way, the entire corpus can be represented with a matrix of words (Table 11.3).

The one-hot representation of this corpus converts these three documents into the presence/absence matrix (Table 11.4). This representation works well with the

Table 11.3 Bag-of-word representation of a corpus of three documents. Only unigrams are used for simplicity

I like playing football	Football	I	Like	Playing	–	–
I play football like a professional	A	Football	I	Like	Play	Professional
I play football with friends	Football	Friends	I	Like	With	–

Table 11.4 One-hot representation of the corpus

ID	A	Football	Friends	I	Like	Play	Playing	Professional	With
1	0	1	0	1	1	0	1	0	0
2	1	1	0	1	1	1	0	1	0
3	0	1	1	1	0	1	0	0	1

analysis of short documents such as tweets. One may, for example, use factor analysis with the dice distance measure or hierarchical cluster analysis to understand how the tweets group around different topics.

For longer documents, it is important to account for repeated use of the same words. Instead of the one-hot, the term frequency (TF) conversion would be more useful. Consider the following documents:

Text 1: – I like playing football. Playing football is something I really like. I enjoy playing football because football is my favorite sport. Like many, I like playing football every weekend.

Text 2: – I play football like a professional. Playing football like professionals do takes time and practice. Professionals play football with great skill, and I play football to get better like them. Football is something I always play, and playing football like a pro is my goal.

Text 3: – I play football with friends often. When I play football with friends, we have a lot of fun. Playing football with friends is a great way to bond, and I love playing football with friends every chance I get. With friends, playing football is always exciting.

The TF representation converts these documents into a one-hot-like matrix with cells counting the number of times each term appears in a document divided by the total number of terms in the document:

Table 11.5 Term frequency (TF) representation of the corpus

ID	always	and	because	...	football	friends	...	with
1	0	0	0.04	...	0.20	0	...	0
2	0.03	0.08	0	...	0.15	0	...	0.03
3	0.03	0.03	0	...	0.13	0.13	...	0.13

Table 11.6 Term frequency-inverse document frequency (TF-IDF) representation of the corpus

ID	always	and	because	...	football	Friends	...	with
1	0	0	0.019	...	0	0	...	0
2	0.004	0.013	0	...	0	0	...	0.004
3	0.004	0.004	0	...	0	0.060	...	0.022

$$\mathrm{TF}(t,d) = \frac{\text{Number of times term } t \text{ appears in document } d}{\text{Total number of terms in document } d}$$

For example, football appears frequently in all three documents giving it a high weight. Other terms like friends have high TF only in specific documents (Table 11.5).

The TF representation gives higher weight to terms that appear more frequently in documents. This representation, however, is not always useful for differentiating between the documents. For instance, the word "football" is frequent in all three documents, giving it a high weight in analyses, but it does not represent the differences among the document topics well. In comparison, the word "friend" is frequent in only one document, reflecting the document's uniqueness. The term "frequency-inverse document frequency" (TF-IDF) reflects this uniqueness, multiplying the TF value by an inverse of the logarithm of the frequency of the documents containing the particular term:

$\mathrm{TF\text{-}IDF}(t, d) = \mathrm{TF}(t,d) * \mathrm{IDF}(t, D);$

$$\mathrm{IDF}(t, D) = \log\left(\frac{N}{|\{d \in D : t \in d\}|}\right),$$

where N is the total number of documents and $|\{d \in D: t \in d\}|$ is the number of documents where the term t appears. The TF-IDF representation dives a high weight to the term "friends" in the third document, underscoring its uniqueness in discussing playing with friends (Table 11.6).

Text representations

One-hot encoding is a simple representation of text where each unique word in the corpus is assigned a unique vector of binary values (0 s and 1 s).

TF encoding accounts for how often a word appears in a document.

TF-IDF improves upon TF by weighing words based on how common they are across the entire document corpus.

In all of these encodings, the length of the vector is equal to the number of unique words in the texts being analyzed.

On the other hand, the term football has zero weight as it appears in all documents.[3]

The TF-IDF reduces the impact of commonly occurring words (like "the" or "is"), improving distinguishing between documents. It is a robust method for converting text to numbers, enabling further statistical analysis and data mining techniques such as document classification, clustering, and topic extraction.

11.4 Measuring Documents' Similarity

In the previous chapters, we discussed the concepts of distance and similarity several times. These concepts are applicable to text mining as well. For instance, many word spelling correction methods (▶ Sect. 11.2.2) are based on *syntactic similarity* (how the words are written) or *phonetic similarity* (how the words are pronounced). ▶ Section 11.2.5 introduced the *cosine* distance, defined as the angle between the vectors representing two words in a multi-dimensional space. The cosine distance measures the *semantic similarity* between words based on the context. The assumption behind the algorithms that use semantic similarity is that words with similar meanings tend to appear in the same context; in other words, syntactic similarity implies semantic similarity. The TF and TF-IDF encoding of documents frequently use the "normal" Euclidean distance.

One-hot document encoding requires a special similarity measure appropriate for measuring the distance between binary vectors. Two frequently used measures are Jaccard or Dice similarity.[4] These measures are specifically created to measure the similarity between two binary vectors. The Jaccard similarity is equal to the relative number of 1's shared between two encoded documents, d_1 and d_2:

$$J(d_1, d_2) = \frac{d_1 \cap d_2}{d_1 \cup d_2}$$

Here, $d_1 \cap d_2$ is the total number of unique words common to both documents and $d_1 \cup d_2$ is the sum of the number of unique words when two documents are merged.

Frequently used text similarity measures

Jaccard and **Dice** similarity both require one-hot encoding.

Jaccard measure is less sensitive when two documents have a large number of shared and a small number of unique words.

Dice similarity is more preferable for measuring commonality between documents rather than uniqueness.

Euclidean distance is frequently used with TF and TF-IDF encodings.

3 There are several variations of the TF-IDF encoding.

4 The Jaccard and Dice distance equal to 1 – the respective similarity.

Table 11.7 Comparison of different measures of distance between three "football player" texts (▶ Sect. 11.3)

Distance between …	One-hot encoding		TF encoding	TF-IDF encoding
	Jaccard	Dice	Euclidean	
Text 1 & Text 2	**0.813**	**0.684**	**0.227**	**0.068**
Text 2 & Text 3	0.737	0.583	0.236	0.087
Text 1 & Text 3	0.882	0.789	0.307	0.093

The Dice similarity measure is closely related to the Jaccard index but gives more weight to shared elements:

$$Dice(d_1, d_2) = 2\frac{d_1 \cap d_2}{d_1 + d_2}$$

Here, $d_1 + d_2$ is the sum of the number of unique words in both documents.

Table 11.7 presents an example of different similarity measures applied to three "football playing" text documents from ▶ Sect. 11.3. Notice how the distance between the texts changes with the way documents are encoded and the distance is measured. Yet, the distance between Text 1 and Text 3 is still the greatest regardless of the measure and encoding, which suggests that these two documents are the most dissimilar.

11

11.5 Moving Further: Grouping the Texts

Document clustering

Document (text) clustering is a process of categorizing similar documents into the same groups based on a similarity measure.

Once documents have been encoded and the distance measure is defined, they can be organized into groups according to the between-document distance. When these groups are undefined, cluster analysis (Chapters 5 and 6) can be used to search for groups of similar documents. As new documents appear, classification methods (▶ Chapter 7) can be used to search for the existing group of similar documents.

The process of categorizing similar documents into the same group is referred to as document clustering. Document clustering is frequently used with large volumes of text data and as a preliminary step in more detailed text analysis. For instance, Table 11.8 pre-

Table 11.8 Example of cluster analysis applied to grouping the documents. Major topics of articles published in three main academic tourism journals, 1974–2017[5]

Subject area	Word cluster
Tourism as a social phenomenon	Culture, social, place, politics, role, authentic, identity, historical, power, relationship
Image and risk	Image, tourism destination, risk, behavior, perception, trip, family, response, destination image, media
Attractions	Visit, attraction, nature, site, management, national, area, park, heritage, value
Tourism industry	Industry, management, tourism industry, sector, business, organization, hospitality, competition, firm, organizational
Service quality and satisfaction	Service, satisfaction, quality, value, relationship, attribute, custom, intention, brand, airline
Modeling and forecasting	Model, method, forecast, application, technique, methodology, measure, series, efficient, system
Conferences	Hotel, state, event, city, publication, copyright, cost, association, conference, rate
Tourist experience and motivation	Experience, factor, motivation, response, scale, knowledge, tourist experience, group, dimension, framework
Market segmentation	Market, segment, tour, operator, strategy, product, agency, tourism market, travel agency, agent
Decision making process	International, decision, choice, vacation, pattern, holiday, product, domestic, accommodation, spatial
Tourism demand	Country, demand, price, expenditure, income, growth, season, period, foreign, show
Governing tourism development	Economy, region, policy, case, government, problem, resort, island, area, tourism development
Sustainable tourism	Environment, active, sustainable, interest, participation, leisure, recreation, life, involvement, action
Local communities	Community, resident, attitude, benefit, local, rural, perception, negative, positive, tourism development

sents the results of the bibliographical analysis of tourism scholarship. Here, the texts of 6,110 papers published in the top three tourism scientific journals starting from 1974 were clustered according to the word distance between the papers.

5 Kirilenko, A. P., & Stepchenkova, S. (2018). Tourism research from its inception to present day: subject area, geography, and gender distributions. PLoS One, 13(11), e0206820.

11.6 Conclusion

In this chapter, we have found that once the texts are cleaned and converted into numbers, they can be analyzed even with methods not specifically designed for text mining. In the next chapter, we will move further, introducing the algorithms specifically designed for working with texts. These algorithms will allow looking deeper into the text structure and moving from clustering separate documents to clustering document parts, that is, to topic mining.

11.7 Review Questions

1. Explain the dimensionality challenge associated with processing and analyzing text data. How does text data differ from numerical data in this regard?
2. What are the main steps involved in preprocessing text data?
3. Discuss the purpose of removing stop words during text preprocessing. Why might stop words interfere with certain text mining and visualization algorithms?
4. Discuss using distance measures in spelling correction.
5. Compare stemming and lemmatization. How do they reduce variability in text data, and what are the potential risks of using stemming?
6. What is the purpose of part-of-speech (POS) tagging? How the POS tagging address the problem of polysemy?
7. Explain the bag-of-words (BoW) text encoding. What are its advantages and limitations when it comes to capturing the context of text?
8. Describe how the word2vec model represents text data and how it differs from traditional methods like the bag-of-words.
9. How can text preprocessing improve the quality of text visualizations?
10. What are the one-hot, TF, and TF-IDF text representations? How does TF-IDF compare to TF in representing the importance of terms in documents?
11. Explain the difference between different similarity measures. What are the appropriate uses of these measures?

11.8 Lab

Objectives

1. Practice making the output more readable
2. Practice text preprocessing and visualization
3. Practice comparative analysis of texts

The overall goal of this assignment is to learn how to conduct comparative lexical analysis using text visualization.

Data

The data for this chapter is abridged from hotel guest interviews conducted in the USA and Switzerland.[6] The interviews aimed to learn guests' opinions on the importance of authentic (genuine, honest, and respectful) communication with the hotel staff. The data is stored in a Microsoft Excel file and contains 40 interviews. We will use the answers to the first interview question:

> *When you communicate with providers of hospitality services, is the way they talk important to you? Apart from being outright rude and unhelpful. And why?*

The objectives are:
(1) Clean the data.
(2) Visualize the data.

The data is stored as ▸ https://github.com/andykir/DataMiningLectures/raw/main/authenticityData.xlsx

The Excel file contains sheets titled "USA" (USA interviews), "Swiss" (Switzerland interviews), and "Questions" (Interviewer's questions).

6 The author is thankful to Dr. Girardin (EHL Hospitality Business School, Switzerland) and Dr. Stepchenkova (University of Florida, USA) for providing the data.

Variables

Variable name	Explanation	Notes
ID	Unique identifier for each interview	Number
Question 1	Answer to Interview Question 1	Text
...	...	...
Question 5	Answer to Interview Question 5	Text

Part 1. Read and Clean the Data

Step 1. Read the Data

Every lab starts with similar steps of reading and describing the data. Because of that, we reuse an older code with minimal modifications. Start with a new Markdown cell. Describe the purpose of the code. Then, describe the purpose of this step.


```
The code for Chapter 11 Lab: Text analysis
Goal: practice text cleaninig and vizualization
Data: Hotel guest interviews

Part 1. Read and clean the data.
Step 1. Read the data.
```

```
import pandas as pd

URL="https://github.com/andykir/DataMiningLectures/raw/main/authenticityData.
xlsx"

# Read data
dataUSA=pd.read_excel(URL, sheet_name="USA")
dataSwiss=pd.read_excel(URL, sheet_name="Swiss")
dataQuestions=pd.read_excel(URL, sheet_name="Questions")

#Print questions of the interviews
print(dataQuestions)

#print information and the first line of the USA interview data
print(dataUSA.info())
print(dataUSA.head(1))

#print information and the first line of Swiss interview data

print(dataSwiss.info())
print(dataSwiss.head(1))
```

Execute the code and observe the results. Notice that the output is hard to read: Swiss and USA outputs are merged, text lines are truncated, and questions are not well formatted. To take care of these issues, we need to (1) print additional information on the output, (2) insert an additional line between the output related to different kinds of data, and (3) stop line trancating. The first two issues can be resolved by printing the titles:

```
print("\n USA:")
```

Notice the special "new line" symbol "\n" included for skipping one line in the output.

The second issue is related to truncating pandas strings to 50 symbols by default. One way of dealing with this is to change the data type to String before printing:

```
print(dataUSA.head(1).to_string())
```

Another option is to change the default truncation option in pandas. The following code removes line truncation:

```
# Keeping pandas from truncating long strings
pd.set_option('display.max_colwidth', None)
```

The corrected code:

```
import pandas as pd

URL="https://github.com/andykir/DataMiningLectures/raw/main/authenticityData.
xlsx"

# Read data
dataUSA=pd.read_excel(URL, sheet_name="USA")
dataSwiss=pd.read_excel(URL, sheet_name="Swiss")
dataQuestions=pd.read_excel(URL, sheet_name="Questions")

# Keeping pandas from truncating long strings
pd.set_option('display.max_colwidth', None)

#Print questions of the interviews
print("Questions:")
print(dataQuestions)

#print information and the first line of the USA interview data
print("\n USA:")
print(dataUSA.info())
print(dataUSA.head(1))

#print information and the first line of Swiss interview data
print("\n Swiss:")
print(dataSwiss.info())
print(dataSwiss.head(1))
```

Run the code. The output becomes significantly more concise.

Notice missing data in the respondents' answers columns:

```
USA:
<class 'pandas.core.frame.DataFrame'>
RangeIndex: 20 entries, 0 to 19
Data columns (total 6 columns):
 #   Column     Non-Null Count  Dtype
---  ------     --------------  -----
 0   ID         20 non-null     int64
 1   Question1  20 non-null     object
 2   Question2  18 non-null     object
 3   Question3  20 non-null     object
 4   Question4  13 non-null     object
 5   Question5  19 non-null     object
dtypes: int64(1), object(5)
memory usage: 1.1+ KB
```

Missing data are typical for interviews: respondents may have no opinion on the discussed topic, be unwilling to share their thoughts, or be short on time.

Step 2. Visualize Question 1 Data as a Word Cloud

USER:

Use dataframe dataUSA. Build a word cloud from the column "Question1".

Assistant:

```
import pandas as pd
import matplotlib.pyplot as plt
from wordcloud import WordCloud

# Combine all text from "Question1" and preprocess
text = ' '.join(dataUSA['Question1'].dropna()).lower()  # Combine and convert
to lowercase

# Generate the word cloud
wordcloud = WordCloud(width=800, height=400,
background_color='white').generate(text)

# Display the word cloud
plt.figure(figsize=(10, 5))
plt.imshow(wordcloud, interpolation='bilinear')
plt.axis('off')  # Turn off axis
plt.title('Word Cloud from Question1')
plt.show()
```

Explanation:

Import necessary libraries: we need *pandas* for data manipulation, *wordcloud* for generating the word cloud, and *matplotlib* for displaying it.

Preprocess the text: Combine all the text from the *Question1* column, convert it to lowercase, and remove any unwanted characters.

Generate the word cloud: Use the *WordCloud* class from the *wordcloud* library to create the word cloud.

Display the word cloud: Use *matplotlib* to visualize the generated word cloud.

Notice that the code uses the libraries *matplotlib* and *wordcloud*. We have already installed *matplotlib*, but the library *wordcloud* will likely be missing. To install the missing library:

```
!pip install wordcloud
```

Result:

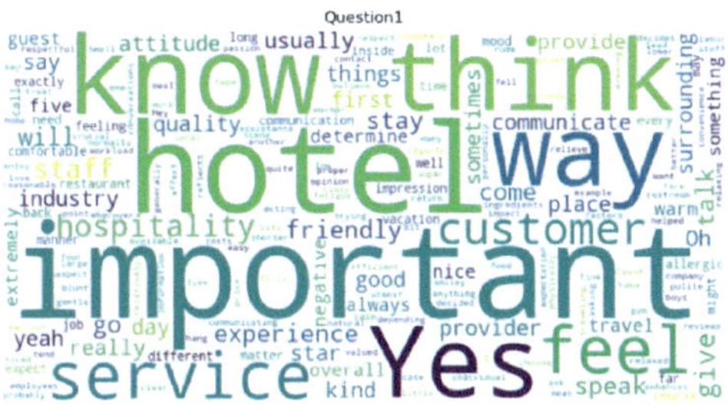

One noticeable feature of this word cloud is that it contains no stopwords. The wordcloud library includes a list of stopwords eliminated by default. It also includes lemmatization. If you are wondering what the wordloud would look like without stopword removal, use the following option:

```
# Generate the word cloud
wordcloud = WordCloud(width=800, height=400, background_color='white',
stopwords=[]).generate(text)
```

Result:

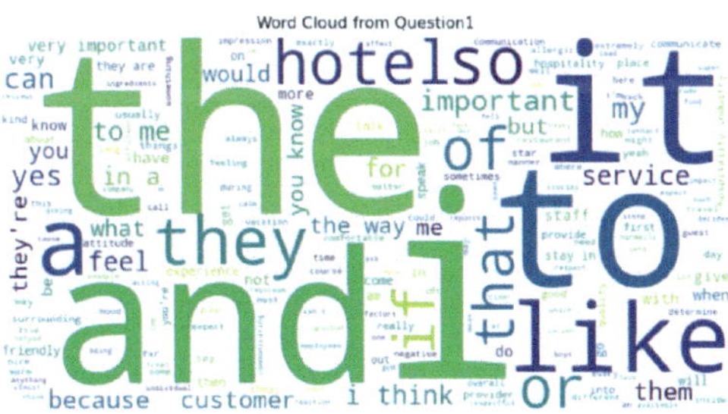

The words "the", "a", "and", and similar frequent English words make the word cloud unintelligible. However, extending the standard stop words, including those specific to the analyzed text, is advisable. For instance, the words "hotel", "important", "Yes", "know", "feel", "way", "think", and "know" are frequent in our text, but not very informative. Survey respondents are asked what they *know*, *think*, or *feel* about the *way* the *hotel* staff communicates and whether this is *important* for them, so these words are expected to be present in their replies. We can also expect to have filler words such as "yes", "yeah", or "uhm". These words can be included in the stop word list.

You can explore other options of the wordcloud package here: ▶ https://amueller.github.io/word_cloud.

Step 3. Clean the Data

In this chapter, we outlined the following components of text cleaning:
1. URL extraction and/or removal from the main text;
2. Emoji and emoticon extraction and/or removal;
3. Handles (mentions) extraction and/or removal;
4. Spelling correction;
5. Contraction replacements.

The first three steps do not apply to the interview data. Hence, we will start with step 4.

USER:

Use column "Question1" of the DataFrame dataUSA. Correct the spelling and word contractions.

Assistant:

```
...
# Define a dictionary for contractions
contractions = {
    "can't": "cannot",
    "won't": "will not",
    "it's": "it is",
    "I'm": "I am",
    "you're": "you are",
    "they're": "they are",
    "he's": "he is",
    "she's": "she is",
    "that's": "that is",
    "what's": "what is",
    "where's": "where is",
    "who's": "who is",
    "let's": "let us",
    "I've": "I have",
    "we're": "we are",
    "you're": "you are",
    "they've": "they have",
    # Add more contractions as needed
}
...
```

Explanation:

Notice that the code suggested by the Assistant includes a small custom list of contractions that is far from complete; for instance, common contractions such as “don't”, “didn't,” and “can't” are not included. Instead of manually expanding the contraction dictionary, we can use a popular library, “contractions”.

Assistant Code:

```
import pandas as pd
import contractions
from textblob import TextBlob

# Function to expand contractions using the contractions library
def expand_contractions(text):
    return contractions.fix(text)

# Function to correct spelling
def correct_spelling(text):
    return str(TextBlob(text).correct())

# Apply the functions to the "Question1" column
dataUSA['Question1'] =
dataUSA['Question1'].dropna().apply(expand_contractions).apply(correct_spelli
ng)

# Display the corrected DataFrame
print(dataUSA['Question1'])
```

Explanation:

This code uses two additional libraries that must be installed: *textblob* and *contractions*. As usual, we have to use *pip* for library installation.

```
!pip install contractions
!pip install textblob
```

Let us also print one interview before and after the cleaning.

The Final Code:

```
import pandas as pd
import contractions
from textblob import TextBlob

#print data before cleaning. Use [0] to output only the first cell
print("Before cleaning")
print(dataUSA['Question1'][0])

# Function to expand contractions using the contractions library
def expand_contractions(text):
    return contractions.fix(text)

# Function to correct spelling
def correct_spelling(text):
    return str(TextBlob(text).correct())

# Apply the functions to the "Question1" column
dataUSA['Question1'] =
dataUSA['Question1'].dropna().apply(expand_contractions).apply(correct_spelli
ng)

#print data after cleaning. Use [0] to output only the first cell
print("\n After cleaning")
print(dataUSA['Question1'][0])
```


Result:

Before cleaning

I say yeah. Because **I thnik** how they talk is like it decides the quality of service. You know. If they're kind, I would be like, "Oh, it's very nice here". But it's they're being rude, I feel like "Oh so low quality here". My costs fell or something. But you know, if you go to like, like five-star or four-star hotels they're always kind you know, always been nice.

After cleaning

I say yeah. Because **I think** how they talk is like it decides the quality of service. You know. Of they are kind, I would be like, "Oh, it is very nice here". But it is they are being rude, I feel like "Oh so low quality here". By costs fell or something. But you know, if you go to like, like five-star or four-star hotels they are always kind you know, always been nice.

Notice how the typo "I thnik" was corrected and the contractions were expanded:

» *it's they're being rude → it is they are being rude.*

Part 2. Standardize the Text

To standardize the text, we will (1) remove stopwords, (2) transform the text to lowercase, and (3) replace words with their base forms (lemmas).

Step 1. Remove Stopwords

There are many different stopword dictionaries, typically including a few hundred words. We will use the stopword dictionary from a natural language processing library, *SpaCy* (► https://spacy.io/), which is relatively aggressive compared to others.

USER:

Use column Question1 of the DataFrame dataUSA. Use the library spaCy to remove stopwords. Include additional stop words: ["hotel", "important", "Yes", "know", "feel", "way", "think", "know", "yeah", "uhm"].

Assistant:

```
import pandas as pd
import spacy

# Load the spaCy English model
nlp = spacy.load('en_core_web_sm')

# Add custom stop words
custom_stop_words = ["hotel", "important", "yes", "Yes", "know", "feel",
"way", "think", "yeah", "uhm"]
for word in custom_stop_words:
    nlp.vocab[word].is_stop = True

# Function to remove stopwords
def remove_stopwords(text):
    doc = nlp(text)  # Process the text with spaCy
    return ' '.join([token.text for token in doc if not token.is_stop])  #
Join tokens that are not stopwords

# Apply the function to the "Question1" column
dataUSA['Question1'] = dataUSA['Question1'].dropna().apply(remove_stopwords)

# Display the modified DataFrame
print(dataUSA['Question1'])
```

Make sure that *spaCy* is installed before running this code. Additionally, *spaCy* requires the English language model *en_core_web_sm*:

```
# Load the spaCy English model
nlp = spacy.load('en_core_web_sm')
```

In fact, *spaCy* offers several English language models, with *en_core_web_sm* being the smallest one (see ▶ https://spacy.io/models/en). Instead, we will use a medium-sized model, *en_core_web_md*. Whichever model to use, it must be downloaded and installed. The installation and download can be performed with the following code. You only need to run this code once; afterward, you can delete or comment it out to improve processing speed.

```
!pip install spacy
!python -m spacy download en_core_web_sm
!python -m spacy download en_core_web_md
```

Finally, we can modify the Assistant's code to observe text modifications.

The final code:

```
import pandas as pd
import spacy

print("Before stopword removal ")
print(dataUSA['Question1'][0])

# Load the spaCy English model
nlp = spacy.load('en_core_web_md')

# Add custom stop words
custom_stop_words = ["hotel", "important", "yes", "know", "feel", "way",
"think", "yeah", "uhm"]
for word in custom_stop_words:
    nlp.vocab[word].is_stop = True

# Function to remove stopwords
def remove_stopwords(text):
    doc = nlp(text)  # Process the text with spaCy
    return ' '.join([token.text for token in doc if not token.is_stop])  #
Join tokens that are not stopwords

# Apply the function to the "Question1" column
dataUSA['Question1'] = dataUSA['Question1'].dropna().apply(remove_stopwords)

# Display the corrected DataFrame
print("\n After stopword removal ")
print(dataUSA['Question1'][0])
```

11

Result:

```
Before stopword removal
I say yeah. Because I think how they talk
is like it decides the quality of service.
```

```
You know. Of they are kind, I would be like,
"Oh, it is very nice here". But it is they
are being rude, I feel like "Oh so low qua-
lity here". By costs fell or something. But
you know, if you go to like, like five-star
or four-star hotels they are always kind you
know, always been nice.

After stopword removal
. talk like decides quality service . . kind
, like , " Oh , nice " . rude , like " Oh low
quality " . costs fell . , like , like - star
- star hotels kind , nice .
```

Step 2. Lemmatize the Text

USER:

Use column Question1 of the DataFrame dataUSA. Use the library spaCy to lemmatize the text.

Assistant:

```
import pandas as pd
import spacy

# Load the spaCy English model
nlp = spacy.load('en_core_web_md')

# Function to lemmatize text
def lemmatize_text(text):
    doc = nlp(text)  # Process the text with spaCy
    return ' '.join([token.lemma_ for token in doc])  # Join the lemmatized
tokens

# Apply the function to the "Question1" column
dataUSA['Question1'] = dataUSA['Question1'].dropna().apply(lemmatize_text)

# Display the modified DataFrame
print(dataUSA['Question1'])
```

Similar to the previous steps, we will update the code to observe the changes (printing the first four interviews instead of just the first one) and replace the 'small' English model with the 'medium' one. Additionally, we will remove any unnecessary lines of code.

The final code:

```
print("Before lemmatization")
print(dataUSA['Question1'][0:4])# Load the spaCy English model
nlp = spacy.load('en_core_web_md')

# Function to lemmatize text
def lemmatize_text(text):
    doc = nlp(text)  # Process the text with spaCy
    return ' '.join([token.lemma_ for token in doc])  # Join the lemmatized
tokens

# Apply the function to the "Question1" column
dataUSA['Question1'] = dataUSA['Question1'].dropna().apply(lemmatize_text)

# Display the corrected DataFrame
print("\n After lemmatization")
print(dataUSA['Question1'][0:4])
```

Result:

```
Before lemmatization
0 . talk like decides quality service . . kind , like , " Oh , nice " . rude , like " Oh low quality " . costs fell . , like , like - star - star hotels kind , nice .
1 opinion , extremely . employees hospitality industry communicating guest crucial factors determine quality service .
2 service provider speaks tend speak calm negative , friendly warm , affect travel experience overall impression place . usually stay hotels travels , staff contact local scene , speak determines experience place mood place , .
3 clear communication helped exactly need exactly available . example , restaurant staff asking food allergic ask like , ingredient ? inside inside meal ? allergic reaction like . , communication , . like , love restroom boys grm . surrounding .
Name: Question1, dtype: object

After lemmatization
0 . talk like decide quality service . . kind , like , " oh , nice " . rude , like " oh low quality " . cost fall . , like , like - star - star hotel kind , nice .
1 opinion , extremely . employee hospitality industry communicate guest crucial factor determine quality service .
2 service provider speak tend speak calm negative , friendly warm , affect travel experience overall impression place . usually stay hotel travel , staff contact local scene , speak determine experience place mood place , .
```

```
3 clear communication help exactly need exactly available
. example , restaurant staff ask food allergic ask like ,
ingredient ? inside inside meal ? allergic reaction like
. , communication , . like , love restroom boy grm . sur-
round .
```

Notice that all words are now transformed into the lowercase base word forms: decides → decide, surrounding → surround.

Step 3. Visualize Question 1 Survey Responses as a Word Cloud

We can reuse the Part 1 *word cloud* code to visualize the outcomes of text cleaning.

Modified Code:

```
# Combine all text from "Question1" and preprocess
text = ' '.join(dataUSA['Question1'].dropna()).lower()  # Combine and convert
to lowercase

# Generate the word cloud
wordcloud = WordCloud(width=800, height=400,
background_color='white').generate(text)

# Display the word cloud
plt.figure(figsize=(10, 5))
plt.imshow(wordcloud, interpolation='bilinear')
plt.axis('off')  # Turn off axis
plt.title('Word Cloud from Question1')
plt.show()
```

Result:

Compare this visualization with the word cloud generated in Part 1. Significant changes highlight key terms related to hotel-guest interactions, such as *customer, staff, service, hospitality, friendly, and experience*. However, some filler words, like mood, could be removed to refine the message further. Additionally, replacing words with their synonyms can help standardize the text even more. While the *spaCy* model does not provide this functionality, it can be achieved using the *Wordnet* library.

Part 3. Compare the Interviews with the US and Swiss Customers

Step 1. Clean and Standardize Swiss Question 1 Interviews

We can reuse the code developed for the USA customers.

```
#------- Wordcloud function and initial word cloud ---------
def draw_wordcloud(text):
    # Generate the word cloud
    wordcloud = WordCloud(width=800, height=400, \
        background_color='white', stopwords=[]).generate(text)
    # Display the word cloud
    plt.figure(figsize=(10, 5))
    plt.imshow(wordcloud, interpolation='bilinear')
    plt.axis('off')  # Turn off axis
    plt.title('Word Cloud from Question1')
    plt.show()

text = ' '.join(dataSwiss['Question1'].dropna()).lower()
draw_wordcloud(text)

#--------------expand contractions and correct spelling-------------
print("Before cleaning")
print(dataSwiss['Question1'][0])

# Function to expand contractions using the contractions library
def expand_contractions(text):
    return contractions.fix(text)

# Function to correct spelling
def correct_spelling(text):
    return str(TextBlob(text).correct())

# Apply the functions to the "Question1" column
dataSwiss['Question1'] = dataSwiss['Question1'].dropna(). \
    apply(expand_contractions).apply(correct_spelling)

#--------------remove stopwords-------------
nlp = spacy.load('en_core_web_md')

custom_stop_words = ["hotel", "important", "yes", "know", "feel", \
    "way", "think", "yeah", "uhm", "Yes"]
for word in custom_stop_words:
    nlp.vocab[word].is_stop = True

def remove_stopwords(text):
    doc = nlp(text)  # Process the text with spaCy
    return ' '.join([token.text for token in doc if not token.is_stop])

# Apply the function to the "Question1" column
dataSwiss['Question1'] = dataSwiss['Question1'].dropna(). \
    apply(remove_stopwords)

#--------------Lemmatization-------------
def lemmatize_text(text):
    doc = nlp(text)  # Process the text with spaCy
    return ' '.join([token.lemma_ for token in doc])

# Apply the function to the "Question1" column
dataSwiss['Question1'] = dataSwiss['Question1'].dropna(). \
    apply(lemmatize_text)

#--------------Final word cloud-------------
text = ' '.join(dataSwiss['Question1'].dropna()).lower()
draw_wordcloud(text)
```

Result: Swiss dataset word cloud before and after preprocessing.

Before preprocessing	After preprocessing
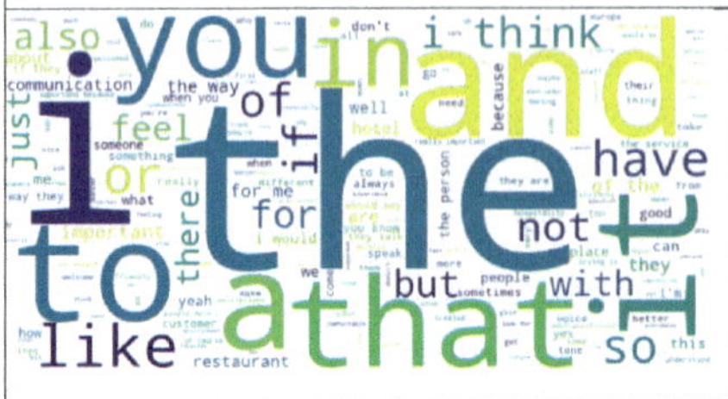	

A comparison of Swiss and USA word clouds shows a considerable difference in answers to the survey questions. To see these dissimilarities in expectations from hotel customer service, we will use a *scattertext* visualization.

Step 2. Compare Word Frequencies in the US and Swiss Interviews

USER:

Use column Question1 of the DataFrame dataUSA. Use column Question1 of the DataFrame dataSwiss. Use the library scattertext to compare word distributions in these columns.

Assistant:

```
import pandas as pd
import scattertext as st

# Preprocess the text: drop NaN values and reset index
dataUSA = dataUSA[['Question1']].dropna().reset_index(drop=True)
dataSwiss = dataSwiss[['Question1']].dropna().reset_index(drop=True)

# Add a column to identify the source
dataUSA['Source'] = 'USA'
dataSwiss['Source'] = 'Swiss'

# Combine the two DataFrames
combined_data = pd.concat([dataUSA, dataSwiss], ignore_index=True)

# Create a Corpus object
corpus = st.CorpusFromPandas(
    combined_data,
    category_col='Source',
    text_col='Question1').build()

# Generate the scattertext visualization
html = st.produce_scattertext_explorer(
    corpus,
    category='USA',  # Choose the category to focus on
    category_name='USA',
    not_category_name='Swiss',
    width_in_pixels=1000,
    metadata=combined_data['Source'])

# Save the visualization to an HTML file
with open('scattertext_visualization.html', 'w', encoding='utf-8') as f:
    f.write(html)
```

Explanation:

The code is using a new library, *scattertext*, which must be installed:

```
!pip install scattertext
```

The results are saved as an HTML file:

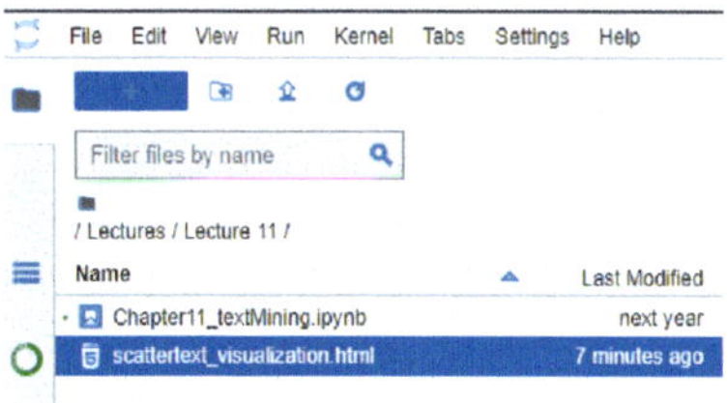

Double-clicking the file opens a new notebook window. Click on "Trust HTML" to see the plot.

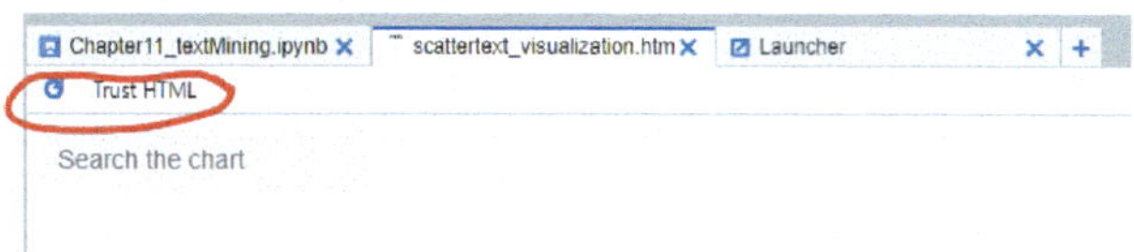

The scatter text plot presents a comparative frequency of word use among the USA and Swiss customers.

This scatter text visualization displays words based on their frequency of use in two datasets: USA and Swiss. Words in the top-left corner are frequently used in the USA dataset but rarely in the Swiss dataset, while words in the bottom-right corner are frequent in the Swiss dataset but rare in the USA dataset. Words along the diagonal have similar frequencies in both datasets.

To better interpret this plot, we will draw a diagonal line and adjust the minimum word frequency threshold required for a word to appear in the visualization.

USER:

Modify the code. Draw a diagonal. Set the minimal word frequency required to include a word in a visualization to 3.

Assistant:

```
import scattertext as st

# Preprocess the text: drop NaN values and reset index
dataUSA = dataUSA[['Question1']].dropna().reset_index(drop=True)
dataSwiss = dataSwiss[['Question1']].dropna().reset_index(drop=True)

# Add a column to identify the source
dataUSA['Source'] = 'USA'
dataSwiss['Source'] = 'Swiss'

# Combine the two DataFrames
combined_data = pd.concat([dataUSA, dataSwiss], ignore_index=True)

# Create a Corpus object
corpus = st.CorpusFromPandas(
    combined_data,
    category_col='Source',
    text_col='Question1').build()

# Generate the scattertext visualization with a diagonal line and
minimum word frequency
html = st.produce_scattertext_explorer(
    corpus,
    category='USA',  # Choose the category to focus on
    category_name='USA',
    not_category_name='Swiss',
    width_in_pixels=1000,
    metadata=combined_data['Source'],
    minimum_term_frequency=3,  # Set minimum word frequency to 3
    show_diagonal=True  # Draw a diagonal line
)
```

Result:

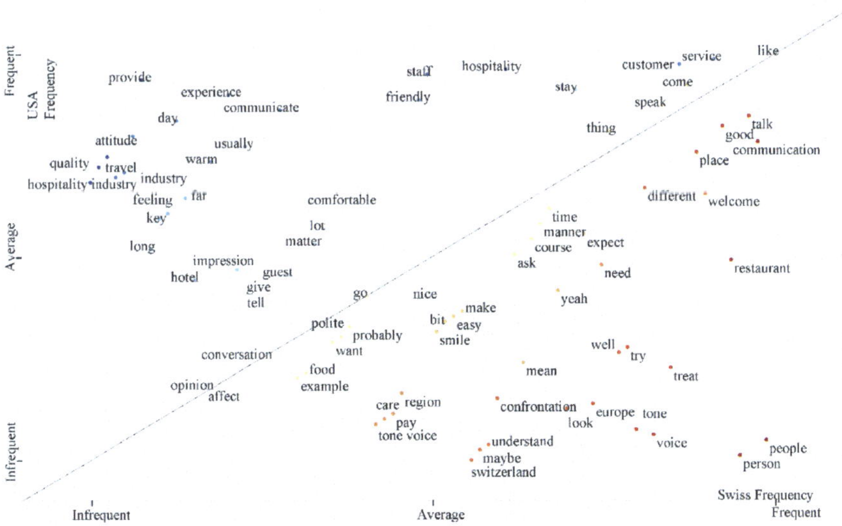

Certain words, such as *like* and *polite*, occur with similar frequency in the USA and Switzerland datasets, while others show notable differences. For example, Swiss study participants often commented on the voice and tone used in communication with the hotel staff. In contrast, USA hotel guests emphasized aspects like attitude, quality, hospitality, and overall experience. *Scattertext* offers many other useful features for exploring relationships between texts. For more information about the package, refer to its GitHub repository: ▶ https://github.com/JasonKessler/scattertext?tab=readme-ov-file.

Part 4. On Your Own

1. Compare survey participants' replies to questions 2–5. Try to run all computations in a loop rather than creating a "spaghetti code" by copy-pasting the same code with minimal modifications for Questions 2, 3, 4, and 5.
2. Interpret the scattertext visualization for questions 2 – 5. What are the differences and similarities between the Swiss and the USA hotel guests?

Build Your Python Skills

Chapter 21 of the free book Learn Python with Jupyter by Serena Bonaretti. ▶ https://www.learnpythonwithjupyter.com/

Addtional Reading

Singh, J. (2023). *Natural language processing in the real world: Text processing, analytics, and classification*. Chapman and Hall/CRC (Chapters 3 and 4).

Topic Modeling: Latent Dirichlet Allocation

A. P. Kirilenko, *Practical Data Mining with AI for Social Scientists*,
Springer Texts in Social Sciences, https://doi.org/10.1007/978-3-031-89689-7_12

12.1 Introduction

In ▶ Chapter 11, we explored various methods for clustering texts into meaningful and coherent groups. Similar methods allow the classification of texts into existing groups so that new documents would join the "most suitable" group, that is, the group with similar documents. These methods work well with short texts that typically discuss only one topic. Examples of such texts include tweets, Facebook status updates, a social media influencer's followers' comments, and similar documents. Notably, not all short texts belong to this category. For instance, product reviews, while short, frequently cover multiple topics. Consider the following hotel review:

> *Everything in the room was super clean, including the frig, microwave & bathroom. Decor was fresh & the room was big…perfect for a longer stay! Pools and facilities were super clean, too! Drains were a little slow.*

In just 38 words, the review author covered room feeling, appliances, hotel facilities, cleanliness, and suitability for a long stay. Allocating this review to any particular group of texts might create a problem.

Long texts, as well as some types of short texts, contain many topics, which makes clustering, classification, dimension reduction, and many other methods successfully used for text analysis unsuitable for finding these topics. One early approach was to cut long texts into smaller records, e.g., 50 to 100 words, which could be analyzed separately. Nowadays, however, there are more effective methods specifically designed for the analysis of longer texts.

Topic modeling

Topic modeling is an unsupervised machine learning technique for discovering hidden thematic structures (topics) within a large collection of documents.

The umbrella name for these methods is *topic modeling*. This term covers the unsupervised machine learning methods for discovering the hidden thematic structure in a large collection of documents. Topic modeling views each document as a *mixture of various topics*, where a topic is defined as a *distribution* over a fixed vocabulary. This approach allows the discovery of the main themes in a large and otherwise unstructured set of documents without any prior annotations or labeling.

As discussed in ▶ Chapter 11, text mining mostly deals with measures based on word frequency counts, keyword extraction, and sentiment analysis. Meanwhile, topic modeling moves deeper into the semantic structure of the text, providing a more holistic view of the text by uncovering the latent topics common to the analyzed collection of documents. The purpose of this chapter is to review the most popular and effective topic modeling approaches.

12.2 Latent Dirichlet Allocation Idea

Fundamentally, automated topic modeling is based on the *distributional hypothesis* of the linguistic theory. In the famous words of British linguist J. R. Firth, "You shall know a word by the company it keeps," the distribution of words in text relates to text meaning. In topic modeling context, it means that words that frequently co-occur within the same documents tend to belong to the same topic. For instance, from an observation of the words "ticket", "museum", "price", "hours", "open", and/or "admission" frequently co-occurring in a corpus of TripAdvisor reviews, one can conclude that one of the topics discussed in these documents is museum visitation logistics.

Latent Dirichlet Allocation (LDA) is a highly popular method that introduced topic modeling to many social science researchers. The LDA algorithm was first published over 20 years ago in David Blei's publication[1] which has been cited over 50,000 times. Since then, the method has been integrated into multiple software libraries and easy-to-use stand-alone programs, making it well recognized even among mostly qualitative disciplines. LDA ideas are relatively transparent and easy to comprehend compared to more advanced transformer-based methods, making its results more easily accepted.

Latent Dirichlet Allocation (LDA) is a *generative statistical model*. That means the algorithm tries to "understand" the probabilistic process by which the data (words in documents) was generated. Specifically, it assumes that (1) each document is generated by a mixture of topics, and (2) each topic is represented as a distribution over words. In other words, LDA posits that each document in a collection of a (large number) of documents is a mixture of a (small) number of *hidden* topics and that the *observed* word's presence is attributable to one of these unobserved topics. The algorithm aims to model the joint

Important!

LDA cannot be applied to extract topics from a single document! Typically, a very large number of texts (>1,000) are required to obtain satisfactory results.

LDA is probabilistic

LDA results can be interpreted as the likelihood of specific words appearing in a topic and the likelihood that a document discusses a specific topic.An interpretation that a document contains proportions of certain topics is incorrect!

1 Blei, D. M., Ng, A. Y., & Jordan, M. I. (2003). Latent Dirichlet Allocation. Journal of machine learning research, 3(Jan), 993–1022.

probability distribution *P*(words, topics). From this joint distribution, we can infer which topics generated a given document and also generate new documents by sampling from the distribution.

Consider three short documents containing the following words:
Doc 1: Fishing, Sea, Beach
Doc 2: Beach, Sea, Vacation, Hotel
Doc 3: Vacation, Hotel, Fishing, Sea

An LDA model might explain the observed distribution of the words fishing, sea, beach, vacation, and hotel over three documents with the underlying existence of three hidden topics: sea fishing, beach-going, and holiday hotel (◘ Fig. 12.1).

12.3 LDA Details

12.3.1 Assumptions

Similar to methods discussed in ▶ Chapter 11, LDA has a bag-of-words assumption. That is, the order of words in a document does not affect topic extraction; only word frequency is important. This assumption simplifies the computational process of topic modeling but disregards

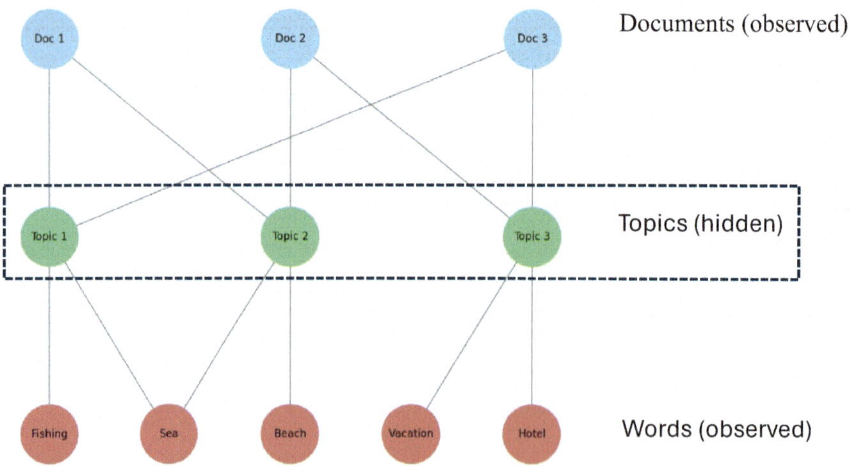

◘ **Fig. 12.1** LDA framework example. We can interpret the topics as sea fishing, beach-going, and holiday hotels

the importance of syntactic and semantic structures for text understanding. Partially, this drawback is alleviated by using bi-grams and tri-grams instead of single words. In addition, in large text corpora, the loss of local word order is offset by the gain in statistical power.

The second assumption is the exchangeability of documents and exchangeability of words: the order in which the documents appear in the corpus and the order in which the words appear in the documents does not affect the topic model. The model assumes that the documents are generated independently, and their order does not provide any additional information for understanding the topics these documents contain.

The third assumption is that each document in a corpus is a mixture of a fixed number of topics, and each topic is a distribution over a fixed vocabulary. The probability of a particular word appearing in a particular topic depends only on the topic; it is irrelevant to the document type, the time a document was published, the author, or even the language (if the corpus is multilingual). An important derivative of this assumption is that the total number of topics should be known prior to modeling.

The final assumption is the *sparsity* of topics and words. This implies that each document is associated with only a small number of topics, and each topic is associated with only a small number of words. This assumption is crucial for the computational feasibility of topic modeling, as it reduces the dimensionality of the problem and allows for efficient inference algorithms.

LDA assumptions

Word order in texts is irrelevant for their topics

Document order is irrelevant

The vocabulary is fixed; no new words may appear

A document contains only a few topics

For a topic, there are only a few important words that describe it.

12.3.2 Parameters

The LDA uses the Dirichlet distribution to describe the distributions of topics and words. The Dirichlet distribution describes the "probability of probabilities". Suppose you have a corpus of documents containing three topics: sea fishing, beach-going, and holiday hotel. For each document, LDA assumes that it is generated by a combination of these three topics, but how much each topic contributes to the document varies. The distribution of topics in each document is a *probability distribution*. For instance, Document 1 could be 70% likely to be about a holiday hotel, 20% about beach-going, and 10% about sea fishing, expressed as [70%, 20%, 10%]. In Document 2, the probabilities of topics could differ: [10%, 40%,

50%]. Now, if we look at the distribution of these topics over all documents, this corpus-level *distribution of topic distributions* is described in LDA by the Dirichlet distribution. Similarly, the LDA uses a Dirichlet distribution to describe the distribution of words over the topics.

Hence, the LDA uses two Dirichlet distributions: one for the distribution of topics within a document and another for the distribution of words within a topic. Two key LDA hyperparameters, α and β (in place of β, some authors use symbols θ or δ), describe our "initial belief" of the distribution of topics within a document and the distribution of words within topics, respectively. A higher α value leads to more uniform distributions, meaning that documents are more likely to contain a mixture of many topics. A lower α value results in a "spikier" distribution with documents focusing on fewer topics.

Similarly, a higher β value implies that each topic is more evenly spread over words, allowing more words to appear in each topic. Conversely, a lower β results in topics concentrating on a smaller subset of words, making topics more distinct.

The final parameter of the LDA model is the number of topics T. The values of α, β, and T affect model results, yet in practice, authors usually use default values set by the software, such as $\alpha=0.1$ and $\beta=0.01$. These small values lead to "spiky" distributions that work well for real-world textual datasets, where each document focuses on a small number of topics, and each topic is defined by a small number of distinct words.

LDA parameters

α: controls the number of topics per document. A smaller value of α results in fewer topics per document;

β: controls the number of words per topic. A smaller value of β results in topics with fewer "important" (high likelihood) words.

Number of topics

Before applying LDA to data, an analyst has to decide on the number of topics that describe the corpus of documents.

12

12.3.3 Interpreting Results

LDA iteratively adjusts the initial distributions of words over topics over documents to fit the observed data best, thus providing a meaningful decomposition of documents into topics and words into thematic groups. LDA results are represented by two matrices:

- Document-Topic Matrix, which represents the distribution of topics for each document. Each row corresponds to a document, and each column represents a topic, with the values indicating the likelihood of the document being associated with a particular topic. Note that any particular document can be associated with multiple topics.
- Word-Topic Matrix, which represents the distribution of words across topics. Each row of the matrix corre-

LDA results

Document-Topic Matrix:for each document, the table provides a likelihood of topics that the document describes.

Word-Topic Matrix: for each topic, the table provides the highlikelihood words that constitute this topic.

Fig. 12.2 Word cloud describing reviews of Arenal National Park in Costa Rica

sponds to a topic, and each column represents a word (or an n-gram) from the corpus vocabulary. The values in this matrix indicate the likelihood of each word appearing in a specific topic. The words with the highest likelihood are used in topic interpreting.

Consider a dataset of reviews from visitors to Arenal National Park in Costa Rica (Fig. 12.2). The word cloud hints at the major park attractions: La Fortuna waterfall, hot springs, animals, and active volcano, as well as the main activities: tours, hikes, and walks. This representation, however, is very generic.

The topic-word and topic-document matrices generated by LDA present a much better option for understanding documents' content. In this chapter's lab, we will explore a dataset of reviews written by visitors to Arenal Volcano National Park in Costa Rica. A preliminary LDA run with an arbitrary number of topics ($T = 5$) and default values for parameters α and β (in genism, the default value for α and β is $1/T$) returns the following five topics (Table 12.1).

Arenal Volcano National Park

From this table, we can conclude that Topic 0 is likely to deal with waterfalls, animals, and swimming, Topic 1 with volcanoes, views, and tours, Topics 2 and 3 with hot springs and pools, and Topic 4 with waterfall hiking. We can also conclude that more topics are needed to describe data.

La Fortuna waterfall ladder

To understand how these topics relate to specific reviews, we can refer to the topic-document matrix (Table 12.2). The first review is likely to contain topics 1 (vol-

Table 12.1 Five topics from the Arenal Volcano National Park dataset. The columns "word" represent a word from the dictionary, and the columns "p" represent the likelihood of this word appearing in a topic

Topic	Word 1		Word 2		Word 3		Word 4		Word 5	
	Word	p	Word	p	Word	p	Word	P	Word	p
Topic 0	Waterfall	0.029	Animal	0.018	Swim	0.014	Worth	0.011	Climb	0.009
Topic 1	Volcano	0.017	View	0.016	Guide	0.016	Tour	0.014	Bridge	0.014
Topic 2	Pool	0.026	Hot	0.02	Water	0.018	Spring	0.017	Time	0.011
Topic 3	Hot	0.015	Great	0.014	Pool	0.014	Spring	0.013	Place	0.012
Topic 4	Waterfall	0.042	Water	0.023	Step	0.022	Worth	0.02	Beautiful	0.019

Table 12.2 Topic-document matrix. The numbers represent the likelihood of a particular topic being present in a review. Notice that review texts are cleaned, standardized, and abridged to fit two lines of the table

Review	Topic 0	Topic 1	Topic 2	Topic 3	Topic 4
amazing waterfall big area base swim see howler monkey tree come	0.017	**0.503**	0.017	0.017	**0.446**
pretty low impact loop walk maintain take 90 min take time hire guide see couple bird fact		**0.979**			
lake 40 min travel la fortuna monteverde famous jeep boat jeep route beautiful view lake	**0.966**				

cano) or 4 (waterfall). The review's text confirms it is, in fact, about waterfall swimming. The second review likely contains topic 1 (volcano), confirmed by the review's text discussing a mountain hike. Finally, the third review is clearly about the waterfall La Fortuna (Topic 0). Notice that reviews 2 and 3 were assigned correct topics even though review 2 does not contain the word "volcano" and review 3 does not contain the word "waterfall". This shows that the decision on the document' topic is not based on a single important word but on the document's similarity to other documents belonging to this topic.

12.3.4 Finding the Optimal Number of Topics

In ▶ Chapter 5, we introduced the Elbow method to find the optimal number of clusters in the K-means method. Recall that the method helps to determine the "optimal" value of a K-means parameter (the number of clusters K) by measuring model error at different values of K. To apply the same idea to find the optimal number of LDA topics, we need to define a relevant error function that reliably quantifies a particular model's error.

A widely popular method of LDA performance evaluation is *topic coherence*. This method assesses the quality of the topics by considering how semantically similar the words within a topic are to one another. In other words, topic coherence is high when the words grouped in a topic are related in a *meaningful* way. To define what the word *meaningful* means, the method uses the *distributional hypothesis* of the linguistic theory we discussed in

Topic coherence

Topic coherence is a metric used to evaluate the quality of topics produced by LDA and other topic modeling algorithms. It measures how frequently the words in a topic *generally* appear together in documents.

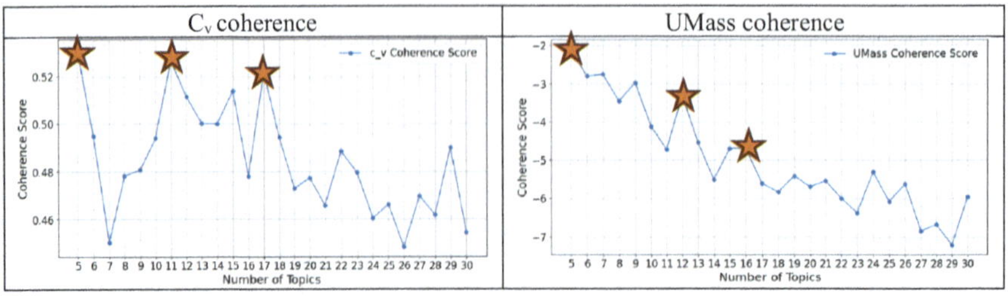

Fig. 12.3 Coherence scores as functions of the LDA number of topics. Stars designate the "sweet spots" to investigate for interpretable topics

Sect. 12.2.1: a topic is coherent if the words grouped in the topic are also frequently grouped in the corpus's documents.

There are several coherence measures that realize this idea in practice. Among those methods, the most popular are the Word Co-occurrence Coherence C_V (frequently written as c_v) and UMass Coherence (frequently written as U_Mass). One important difference between these two measures is that C_V uses an external reference dataset of texts while u_mass is based exclusively on the texts of the analyzed documents. That is, C_V is an external and u_mass is an internal measure.

Coherence measures: which one to use?

Some analysts advise UMass as a preferable measure of topic coherence. In my view,using multiple measures provides better guidance on the number of topics to choose from.

C_V is positive and distributed from 0 to 1, with higher values indicating better coherence. The UMass measure is *generally* negative, with higher values indicating better coherence.

In our case, five topics seem to be an appropriate number (Fig. 12.3).

It makes sense to test LDA with a large range of topics, for instance, changing the number of topics from three to thirty to find the "sweet spot" intervals and then investigating the interpretability of these "optimal" topics. Notice other coherence spikes at around 11 or 12 topics and around 16 or 17. This is typical to have multiple topic number "sweet spots" as larger, more general topics split into smaller and more precise ones. The final number of topics can be decided on during the topic interpretation step.

12.4 Implementation

12.4.1 Best Practices

This chapter's lab follows the established pipeline of topic modeling. Generally, this pipeline includes the following steps.

Data collection and initial preparation. This step includes optical character recognition (OCR) of printed materials and data conversion to a uniform plain text format. One important decision in this step is dealing with words in different languages. You may decide to remove them, translate everything to the prevalent language of the text, or leave foreign words as is. The risk here is that the same topic will be described with different words and hence may appear twice in topic modeling outcomes. This risk, however, may also be an advantage, allowing to investigate differences in interests of people from different cultures.

Data cleaning. This step is especially important for the social media data that must be cleaned of hyperlinks and similar artifacts present in online discussions and passed through a spellchecker. Note that an incorrect spelling by itself is irrelevant for topic extraction, but different spelling variants of the same word affect topic modeling outcomes. For the same reason, it is highly advisable to harmonize different spelling standards, such as in American and British English (color–colour), and replace word contractions (don't–do not).

Data standardization. The goal of this step is to reduce the word-space. The common practice is to remove stop words, that is, the words common in English and not informative, for the purpose of topic modeling. Aside from removing the words appearing in one of the common stop word lists, many researchers also opt to remove the most and the least frequent words. The reasoning is that the most frequent words are unhelpful in differentiating between the documents. Contrasting, the unfrequent words are unhelpful in showing similarity between the documents. There are multiple informal guidelines, such as removing words found in over half of the documents or in fewer than 5–10 documents. The same rule applies to n-grams if they are used. To reduce the number of words even more, the researcher may decide to pass the text through POS recognition with the goal of leaving

Important!

LDA is very sensitive to data quality. Inadequately cleaned and standardized data result in topics that are hard to interpret.

LDA steps

Data collection & harmonizing

Data cleaning

Data standardization

Selecting values of α and β

Selecting the optimal number of topics

Interpretation of topics

Analysis of topic distribution over documents

only nouns and/or verbs, depending on the analysis goals. Finally, it is advisable to convert all words of the documents to their base forms (moved–move; travelers–traveler). To further standardize the text, all words are converted to lowercase, and punctuation is removed.

Selection of hyperparameter values. Some software packages automatically generate the "optimal" values for α and β hyperparameters. However, in practical text analysis, a common practice seems to set $\alpha=0.1$ and $\beta=0.01$. The reasoning is that small β values lead to higher differentiation among the topics in terms of contributing words, making the topics easier to interpret. In turn, small values of α lead to higher differentiation among the documents in terms of their topics.

Selection of the number of topics. Fundamentally, there is no objective way to define the number of topics in the text. However, there are multiple measures that guide the decision-making process. Some measures, such as perplexity, result in hundreds of topics, making them hard to interpret. The coherence measures discussed in this chapter return fewer topics, which can be interpreted realistically. Whichever number is selected, the final litmus test is the interpretability and usefulness of the generated topics.

12.4.2 LDA Software

LDA is a mature algorithm with multiple software libraries and stand-alone applications implementing it. The lab exercise accompanying this chapter uses a *genism* implementation of LDA. This particular package leaves all data preprocessing steps to users, making it ideal for learning the method's best practices.

Another popular package, *pyLDAvis,* has a powerful data visualization and analysis interface that allows interactive exploration of topic distribution. This makes pyLDAvis a very attractive option for topic modeling. The downside is that installing the package may become a challenge.

An interesting LDA extension, hierarchical LDA (hLDA), organizes topics in a hierarchical tree-like structure rather than as a fixed list of topics. This hierarchy reflects a relationship between topics with broader topics at higher levels of the tree and their subtopics at lower levels. For instance, an hLDA model may discover a topic "waterfall" at a higher level with subtopics "hiking" and "swimming" at the lower levels. Notably, hLDA does not require a pre-assigned number of topics.

In ▶ Chapter 11, we introduced the word2vec method of numerical representations of words in vector space. LDA2vec combines the LDA method with word embedding of word2vec, partially solving the high dimensionality problem of LDA. LDA2vec is very efficient in accounting for related terms such as "car," "vehicle", "truck", and "SUV". Since these terms are unlikely to appear in the same document, LDA might struggle to group them into the same topic. In contrast, LDA2vec is able to recognize similarities in how these terms are used, hence producing more coherent and meaningful topics. The drawback is that LDA2vec results are lesser interpretable compared to LDA.

12.5 Conclusion: Why LDA?

In the previous chapter, we discussed problems associated with high dimensionality of the word space. In LDA, the number of dimensions equals the size of the dictionary, that is, the number of unique words after preprocessing. One major goal of text standardization is the reduction of the word space dimensionality. Still, even after text cleaning and standardization, the number of dimensions is too high to process large sets of texts. This is the problem that LDA2vec has attempted to solve by using a (relatively) low-dimensional hidden layer of an artificial neural network to represent the words.

This idea has progressed spectacularly in the past few years with the development of a crop of new approaches generally classified as Large Language Models. The text analysis capabilities of these models are significantly more advanced compared to LDA, creating a viable alternative, particularly for topic modeling. We will cover one of these models in detail in the next chapter. However, while losing in consistency and efficiency of topic modeling, LDA is winning in explainability of the process leading to topic generation.

Overall, LDA is very successful in extracting highly detailed topics when the dataset is large and cohesive, and the documents are reasonably long. Contrasting, document clustering, as discussed in ▶ Chapter 11 is likely to be successful when documents are short, such as in tweets or followers' comments. However, when texts

LDA: when to use?

Despite the recent development of LLMs capable of topic modeling, fast and easy-to-use LDA has one major advantage: its mechanism is easy to explain, and topic interpretation is transparent. Hence, LDA is preferable when trust in model results is a decision-making factor.

are noisy and/or the dataset size is inadequate, there is insufficient context for successful LDA topic extraction. Approaches such as BERTopic will be more successful in the summarization of such texts, even though the result's explainability will suffer.

12.6 Review Questions

1. What is the primary purpose of topic modeling in text analysis?
2. How does topic modeling differ from clustering in text mining?
3. Explain the "distributional hypothesis" and its role in topic modeling.
4. What assumptions does LDA make about word order and document order?
5. Latent Dirichlet Allocation (LDA) is a probabilistic model. What are the implications for interpreting topics based on their word distribution?
6. Why can't LDA be applied to a single document?
7. What are the roles of the hyperparameters α and β in LDA?
8. How do α and β affect the number of topics and words per topic in LDA?
9. What is the "bag-of-words" assumption in LDA?
10. Describe the purpose of the Document-Topic Matrix and the Word-Topic Matrix in LDA.
11. What is topic coherence, and how is it used to evaluate LDA models?
12. How can you determine the optimal number of topics in an LDA model?
13. Explain the main steps of LDA data cleaning.
14. What are some advantages of LDA compared to more advanced transformer-based topic modeling methods?

12.7 Lab

Objectives

1. Practice text preparation for the analysis
2. Develop an LDA topic model
3. Practice topic visualization

The overall goal of this assignment is to extract topics from a set of travelers' reviews.

Data

The data for this chapter is abridged from reviews of visitors to Volcano Arenal National Park in Costa Rica. The data is stored in a Microsoft Excel file and contains over 10,242 reviews. The objectives are:
(1) Clean the data.
(2) Visualize the data
(3) Extract main topics using LDA
(4) Visualize the topics

The data is stored as ▶ https://github.com/andykir/DataMiningLectures/raw/refs/heads/main/ArenalReviews.xlsx

The Excel file contains one sheet titled "reviews".

Variables

Variable name	Explanation	Notes
ID	Unique identifier for each interview	Number
Title	Review title	Text
Review	Review text	Text

Part 1. Read, Clean, and Standardize the Data

Step 1. Read the Data

Every lab starts with similar steps of reading and describing the data. Because of that, we reuse an older code with minimal modifications. Start with a new Markdown cell. Describe the purpose of the code. Then, describe the purpose of this step.

```
The code for Chapter 12 Lab: Text analysis
Goal: Develop an LDA topic model
Data: Arenal National Park reviews

Part 1. Practice text preparation for the analysis
Step 1. Read the data
```

```
import pandas as pd

URL="https://github.com/andykir/DataMiningLectures/raw/refs/heads/main/Arenal
Reviews.xlsx"

# Read data
data=pd.read_excel(URL)

#print information and the first line of data

print(data.info())
print(data.head(1))
```

Result:

```
<class 'pandas.core.frame.DataFrame'>
RangeIndex: 10242 entries, 0 to 10241
Data columns (total 3 columns):
 #   Column  Non-Null Count  Dtype
---  ------  --------------  -----
 0   ID      10242 non-null  int64
 1   Title   10241 non-null  object
 2   Review  10242 non-null  object
dtypes: int64(1), object(2)
memory usage: 240.2+ KB
None
   ID              Title                                             Review
0   1  Be ready to hike.  The hike going down is okay with no problem. I...
```

Step 2. Clean Data

Previously, we outlined the following components of data cleaning:

1. URL extraction and/or removal;
2. Emoji and emoticon extraction and/or removal;
3. Handles (mentions) extraction and/or removal;
4. Spelling correction;
5. Contraction replacement.

We can skip steps 3 and 4. Step 3 is not applicable because there are no @mentions in reviews, but skipping step 4 (spelling correction) is more involved. The reason for skipping spellcheck is specific to the destination. Many reviews include Spanish words, which the spelling correcting software would attempt to "correct". For instance, the word "Pinzón" (finch) would be corrected to "Pirate". In addition, spelling correction is not very important in review-type data as customer reviews are typically well-written compared with, for example, Facebook status updates.

Let us implement the rest of the data cleaning.

URL and emoticon extraction and/or removal (items 1 and 2).

USER:

Use existing DataFrame data. Use the column "Review". Apply the following transformations to the column: remove the URLs and emoticons.

Assistant:

```
import pandas as pd
import re

# Function to remove URLs
def remove_urls(text):
    # Check if the input is a string
    if isinstance(text, str):
        return re.sub(r'http[s]?://\S+', '', text)
    return text  # Return unchanged if not a string

# Function to remove emoticons
def remove_emoticons(text):
    # Check if the input is a string
    if isinstance(text, str):
        return re.sub(r'[\U0001F600-\U0001F64F]', '', text)  # Range

    return text  # Return unchanged if not a string

# Apply transformations to the "Review" column
data['Review'] = data['Review'].apply(remove_urls).apply(remove_emoticons)

# Display the transformed DataFrame
print(data)
```

Explanation:

To remove URLs and emoticons from the *Review* column, we use regular expressions. First, the regular expressions library *re* is imported. Then, the function *remove_emoticons* is defined. This function uses a regular expression to remove emoticons from the text. The Unicode block from U+1F600 to U+1F64F covers common emoticons.[2] Similarly, the function remove_urls, which uses a regular expression to find and remove any URLs that start with `http://` or `https://`, is defined. The *apply* method is used to apply these functions to each entry in the *Title* and *Review* columns.

Note that the initial code may return an error due to missing data (empty reviews). The code above is corrected for this type of error.

2 ▸ https://en.wikipedia.org/wiki/Emoticons_(Unicode_block).

Contraction replacement (item 5)

For contraction replacement, we can reuse the code from ► Chapter 11's lab:

```
import contractions
from textblob import TextBlob


print("Before cleaning")
print(data['Review'][0])

# Function to expand contractions using the contractions library
def expand_contractions(text):
    return contractions.fix(text)

# Function to correct spelling
def correct_spelling(text):
    return str(TextBlob(text).correct())

# Clean the data. We are not correcting spelling this time.
for column in ["Title","Review"]:
    data["Review"] = data["Review"].dropna().apply(expand_contractions)

# Display the corrected DataFrame
print("\n After cleaning")
print(data['Review'][0])
```

Result:

Before cleaning

The hike going down is okay with no problem. It gets crowded specially in the morning with all the tour bus from san jose. The water is cold and you can't really swim as it was not that deep. Not allowed to go behind the falls. Nice view from the top. And you will see people doing zip line. Going up is a bad hike as it was steep and the floor was wet so you will make a big effort. You can stop once in a while as they have resting area and chair. It is worth the visit.

After cleaning

The hike going down is okay with no problem. It gets crowded specially in the morning with all the tour bus from san jose. The water is cold and you cannot really swim as it was not that deep. Not allowed to go behind the falls. Nice view from the top. And you will see people doing zip line. Going up is a bad hike as it was steep and the floor was wet so you will make a big effort. You can stop once in a while as they have resting area and chair. It is worth the visit.

Notice how the contraction "can't" was expanded.

Step 3. Standardize the Text

To standardize the text, we will (1) remove stopwords, (2) transform the text to lowercase, and (3) replace words with their base forms (lemmas). We can reuse the code from the previous lab. We need to apply some modifications, for example, replacing "dataUSA" with "data" and "Question1" with "Review". In addition, we remove the custom list of stopwords; it can be restored later if needed. After modifications, the code is as follows:

```
import spacy

print("Before standardization: ")
print(data['Review'][0])

# Load the spaCy English model
nlp = spacy.load('en_core_web_md')

# Add custom stop words
custom_stop_words = []
for word in custom_stop_words:
    nlp.vocab[word].is_stop = True

# Function to remove stopwords
def remove_stopwords(text):
    doc = nlp(text)  # Process the text with spaCy
    return ' '. \
    join([token.text for token in doc if not token.is_stop])


# Function to lemmatize text
def lemmatize_text(text):
    doc = nlp(text)  # Process the text with spaCy
    return ' '.join([token.lemma_ for token in doc])

# Apply the function to the "Review" column
data['Review'] = data['Review'].dropna(). \
    apply(remove_stopwords).apply(lemmatize_text).str.lower()


# Display the corrected DataFrame
print("\n After standardization: ")
print(data['Review'][0])
```

Note: lemmatization is computationally expensive. Expect to wait few minutes to complete.

Result

Before standardization:

```
The hike going down is okay with no prob-
lem. It gets crowded specially in the morning
with all the tour bus from san jose. The wa-
ter is cold and you cannot really swim as it
was not that deep. Not allowed to go behind
the falls. Nice view from the top. And you
```

```
will see people doing zip line. Going up is a
bad hike as it was steep and the floor was wet
so you will make a big effort. You can stop
once in a while as they have resting area and
chair. It is worth the visit.
```

After standardization:

```
hike go okay problem. get crowd specially
morning tour bus san jose. water cold swim
deep. allow fall. nice view. people zip line.
go bad hike steep floor wet big effort. stop
rest area chair. worth visit.
```

The final operation is to remove the punctuation. Recall that LDA has the "bag of words" approach: all words in a document, regardless of their position or syntactic meaning, are treated the same. In this context, punctuation is irrelevant to outcomes and should be treated the same as stop words. Because of that, many LDA packages, but not *genism*, remove the punctuation automatically, by default.

USER:

Use column "Review" of DataFrame "data". Remove punctuation.

Assistant:

```
import pandas as pd
import string

data['Review'] = data['Review'].str. \
    replace(f'[{string.punctuation}]', '', regex=True)
```

Explanation:

- Regular Expression: The expression *f'[{string.punctuation}]'* creates a pattern that matches any punctuation character defined in the string.punctuation constant.
- *str.replace*: This method replaces all occurrences of the matched punctuation characters with an empty string, effectively removing them from the *Review* column.

After running this code, the *Review* column in your DataFrame will have all punctuation removed.

To observe the results of this operation, we can use the following code:

```
# Display the corrected DataFrame
print("\n After punctuation removal: ")
print(data['Review'][0])
```

Results (after punctuation removal):

`hike go okay problem get crowd specially morning tour bus san jose water cold swim deep allow fall nice view people zip line go bad hike steep floor wet big effort stop rest area chair worth visit`

At this point, it is a good idea to save the cleaned data in a separate file. If you have to restart your development environment, you can restore cleaned data without the need to run the previous code:

```
#save cleaned data
data.to_excel("dataCleaned.xlsx")
```

```
# To restore data start from here:
Import pandas as pd
data=pd.read_excel("dataCleaned.xlsx")
```

Step 4. Visualize Data as a Word Cloud

We can reuse the word cloud code from the previous lab to visualize the outcomes of text cleaning and standardization. After modifying the code, such as replacing *dataUSA* with *data* and *Question1* with *Review*, importing the libraries, and making other adjustments, the code looks like follows:

```
#import libraries
from wordcloud import WordCloud
import matplotlib.pyplot as plt

# Combine all text from "Review" and preprocess
text = ' '.join(data['Review']) # Combine data

# Generate the word cloud
wordcloud = WordCloud(width=800, height=400, \
        background_color='white').generate(text)

# Display the word cloud
plt.figure(figsize=(10, 5))
plt.imshow(wordcloud, interpolation='bilinear')
plt.axis('off')  # Turn off axis
plt.title('Word Cloud from Review column')
plt.show()
```

Result:

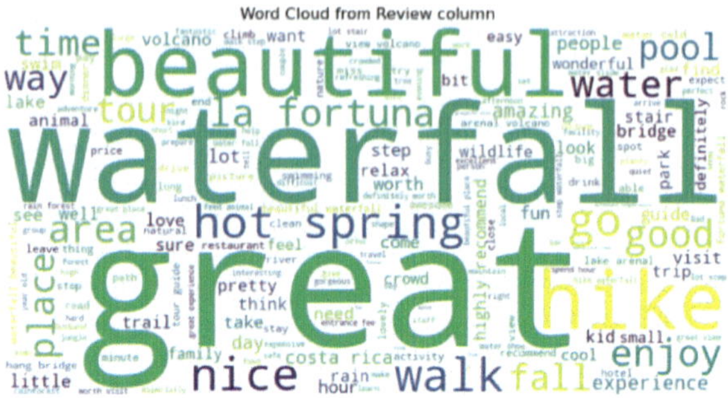

Observe that the word cloud adequately relates the content of reviews for Arenal National Park. The major attractions of the park are La Fortuna waterfall, natural hot springs, nature walks and hikes, wildlife, and the volcano Arenal, all present in the word cloud. In the next part of this lab, we will extract separate topics from visitors' reviews.

Part 2. Topic Extraction

Step 1. Initial LDA Analysis

We will use LDA implementation from the genism library. We will start by extracting five topics. The first step is data sampling. LDA is computationally expensive. To reduce our waiting time, we will start by sampling 1000 records of data:

```
# Sample the data to debug the code
data=data.sample(n=1000).reset_index()
```

Explanation:

We have already encountered the method *reset_index*(). Imagine an Excel table with lines numbered from 1 to 10,000. Samping 1,000 lines might leave only the rows numbered 5, 18, 23, etc. The function *reset_index*() would re-number the rows in the sample to 1, 2, 3, … 1000.

USER:

Use column Review of DataFrame "data". Each row represents one document. Extract main topics from the documents using the gensim library. The number of topics = 5. Save the document-topic and word-topic matrices to an Excel file.

Note that parts of the code provided by the Assistant are unnecessary. For instance, we have already converted the documents to lowercase and removed the stop words.

Modified Assistant Code:

```
import pandas as pd
from gensim import corpora
from gensim.models import LdaModel

documents = data['Review'].tolist()

# Create a dictionary and corpus
dictionary = corpora.Dictionary([doc.split() for doc in documents])
corpus = [dictionary.doc2bow(doc.split()) for doc in documents]

# Apply LDA model
lda_model = LdaModel(corpus, num_topics=5, id2word=dictionary, passes=15)

# Print topics
topics = lda_model.print_topics(num_words=10)
for topic in topics:
    print(topic)
```

Explanation:

Dictionary and Corpus: The *corpora.Dictionary* creates a mapping of words to their integer ids, and the corpus is a list of bag-of-words representations of the documents.

LDA Model: The *LdaModel* is used to fit the model to the corpus with the specified number of topics.

Print Topics: The *print_topics* method retrieves the topics along with the most significant words associated with each topic, which are then printed.

Make sure that the *genism* library is installed.

The parameter *num_words* controls the number of words in the topic output. For better topic understanding, we set the parameter's value to 10:

```
# Print topics
topics = lda_model.print_topics(num_words=10)
```

Result:

(0, '0.029*"waterfall" + 0.018*"animal" + 0.014*"swim" + 0.011*"worth" + 0.009*"climb" + 0.009*"hike" + 0.009*"fall" + 0.009*"tour" + 0.009*"guide" + 0.009*"beautiful"').

(1, '0.017*"volcano" + 0.016*"view" + 0.016*"guide" + 0.014*"tour" + 0.014*"bridge" + 0.012*"good" + 0.011*"see" + 0.011*"great" + 0.010*"nice" + 0.010*"lake"').

(2, '0.026*"pool" + 0.020*"hot" + 0.018*"water" + 0.017*"spring" + 0.011*"time" + 0.011*"lake" + 0.011*"good" + 0.010*"enjoy" + 0.010*"area" + 0.009*"beautiful"').

(3, '0.015*"hot" + 0.014*"great" + 0.014*"pool" + 0.013*"spring" + 0.012*"place" + 0.011*"love" + 0.011*"experience" + 0.010*"food" + 0.010*"animal" + 0.009*"nice"').

(4, '0.042*"waterfall" + 0.023*"water" + 0.022*"step" + 0.020*"worth" + 0.019*"beautiful" + 0.019*"walk" + 0.015*"hike" + 0.015*"swim" + 0.013*"fall" + 0.012*"great"').

Step 2. Topic Interpretation

12

LDA results include five topics numbered from 0 to 4. From each topic, we printed out the five most representative words. For example, in Topic 0, the following five words are the most representative: *waterfall, animal, swim, worth, climb*. The numbers indicate the likelihood of a specific word appearing in this topic and can be interpreted as the word relevance. Observe that the most relevant word is "waterfall". There is a popular La Fortuna waterfall in this park, which is "worth a climb". We can interpret this topic (as well as Topic 4) as a "waterfall hike". Topic 1 can be interpreted as a "volcano", and Topics 2 and 3 as "hot springs and pools". Notice an apparent mixture of topics. For instance, Topic 1 also includes the words "animal" and "swim". Topic 3 includes "place" and "food", which hints at a hot spring resort, differentiating Topic 3 from Topic 2. Meanwhile, Topic 2 includes "lake" [Arenal, famous for its beauty].

To learn how the topics are distributed across the documents, we should analyze the topic-document matrix.

USER:

Join the document texts stored in DataFrame data['Review'] to "topic_documents". Save topic_documents to an Excel file topic_documents.xlsx.

Assistant:

```
# Get document-topic matrix
doc_topic_matrix = lda_model.get_document_topics(corpus)

# Convert document-topic matrix to DataFrame
doc_topic_df = pd.DataFrame(doc_topic_matrix)

# Add document texts to the DataFrame
doc_topic_df['Document'] = data['Review'].values

# Save to Excel
doc_topic_df.to_excel('topic_documents.xlsx', index=False)
```

Explanation:

Document-Topic Matrix: The *get_document_topics* method retrieves the distribution of topics for each document.

DataFrame Conversion: The document-topic matrix is converted to a DataFrame *doc_topic_df*.

Joining Document Texts: The document texts from *data*['*Review*'] are added as a new column named *Document* in the DataFrame *doc_topic_df*.

Excel Output: The combined DataFrame *doc_topic_df* is saved to an Excel file named topic_documents.xlsx.

Note that the original snippet generated by the Assistant included the code for LDA topic generation. These lines can be safely excluded since we have already trained the LDA. You must download the Excel file *topic_documents* to your computer to see the results.

0	1	2	3	4	Document
(0, 0.011)	(1, 0.572)	(2, 0.393)	(3, 0.011)	(4, 0.011)	great experience long hike beautiful night arrive large tour bus pull locker room crowd individual lockers belonging
				(4, 0.975)	fall pretty easy park car walk step start fine old step bit tricky lookout walk view amazing stay swam swimming hole fall fine young daughter walk fall suck bench break totally worth

0	1	2	3	4	Document
(0, 0.010)	(1, 0.010)	(2, 0.229)	(3, 0.01)	(4, 0.74)	bit hike fall lot step end beautiful lagoon swim base waterfall water cold refreshing contrast humidity great photo opportunity

The first five columns in this output represent the likely topic for specific documents. The first field is the topic's index, and the second is the likelihood that this topic is present in the document. For instance, the first document likely contains Topics 1 (volcano tour, likelihood 0.57) and 2 (hot springs and pools, likelihood 0.39). Similarly, the second document contains Topic 4 (waterfall, likelihood 0.98), and the third document is likely to contain Topics 4 (waterfall, likelihood 0.74) and 2 (hot springs and pools, likelihood 0.29). From observation of the documents' text, this topic designation seems reasonable.

Note: LDA is a probabilistic model. Every time you run the LDA, the topics may differ from those in the previous run. Because of that, refrain from re-training a model with already interpreted topics.

Part 3. Find the Optimal Number of Topics.

We will use the *u_mass* and *c_v* coherence scores to find the optimal number of topics. First, we will approximate the optimal number of topics by running LDA with the number of topics from 3 to 30 with the step of 3. Then, we will improve our approximation by running LDA in a narrower range of topics.

Select the cell with the LDA code. Then, in the Assistant window, type the following directions.

USER:

Refactor the selected code. Run the LDA model with the number of topics changing from 3 to 30 with the step of 3 to make two plots with u_mass and c_v coherence scores.

Assistant (the code is modified for brevity):

```
import matplotlib.pyplot as plt
from gensim import corpora
from gensim.models import LdaModel
from gensim.models import CoherenceModel

documents = data['Review'].tolist()

# Create a dictionary and corpus
dictionary = corpora.Dictionary([doc.split() for doc in documents])
corpus = [dictionary.doc2bow(doc.split()) for doc in documents]

# Lists to hold coherence scores
umass_scores = []
cv_scores = []
num_topics_range = range(3, 31, 3)  # From 3 to 30 with step 3

# Run LDA model for different numbers of topics
for num_topics in num_topics_range:
    lda_model = \
        LdaModel(corpus, num_topics=num_topics, id2word=dictionary, passes=15)

    # Compute UMass coherence score
    umass_coherence_model = CoherenceModel(model=lda_model, \
        texts=[doc.split() for doc in documents], \
               dictionary=dictionary, coherence='u_mass')
    umass_coherence = umass_coherence_model.get_coherence()
    umass_scores.append(umass_coherence)

    # Compute c_v coherence score
    cv_coherence_model = CoherenceModel(model=lda_model, \
        texts=[doc.split() for doc in documents], \
               dictionary=dictionary, coherence='c_v')
    cv_coherence = cv_coherence_model.get_coherence()
    cv_scores.append(cv_coherence)

# Plotting the coherence scores
for scores in [umass_scores, cv_scores]:
    plt.figure(figsize=(12, 6))
    plt.plot(num_topics_range, scores, marker='o')
    plt.xlabel('Number of Topics')
    plt.ylabel('Coherence Score')
    plt.xticks(num_topics_range)
    plt.grid()
    plt.show()
```

Result:

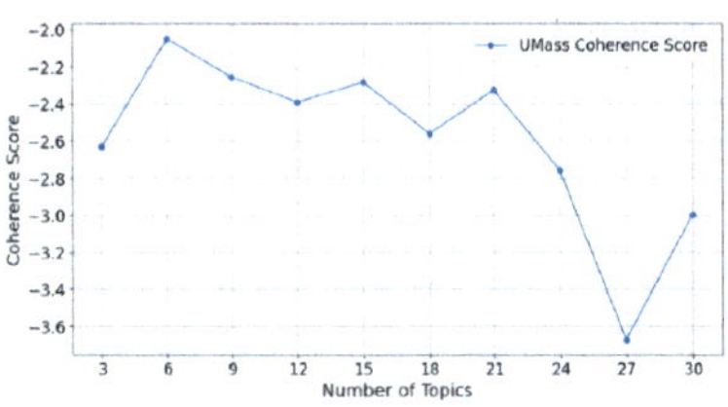

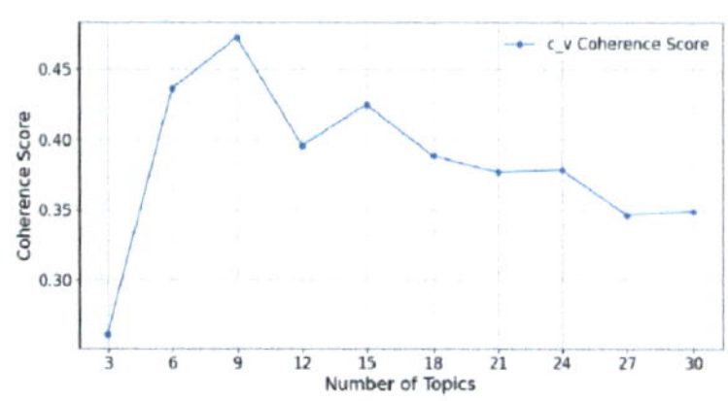

The plots indicate that the maximum *u_mass* and *c_v* coherence values are achieved somewhere between 5 and 11 topics. To determine the optimal number of topics (based on coherence scores), we will evaluate the model's coherence for each number of topics within this range.

Find the following line of the code:

```
num_topics_range = range(3, 31, 3)  # From 3 to 30 with step 3
```

Change the code as follows:

```
num_topics_range = range(5, 12)  # From 5 to 11 with step 1 (default step)
```

In addition, increase the number of iterations for the LDA model by replacing the line

```
lda_model = LdaModel(corpus, num_topics=num_topics, id2word=dictionary,
passes=15)
```

With

```
lda_model = LdaModel(corpus, num_topics=num_topics, id2word=dictionary,
passes=30)
```

Result:

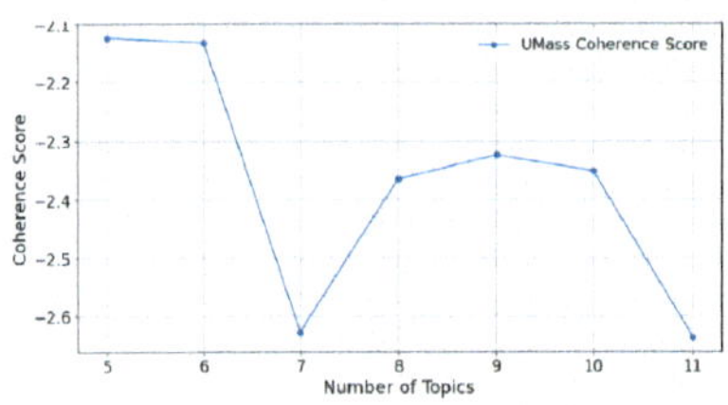

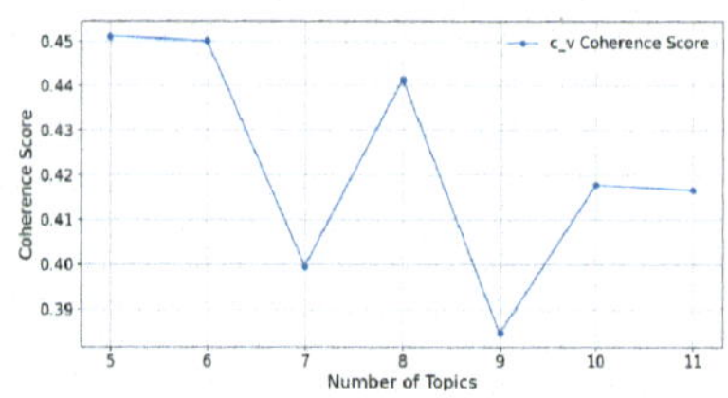

Extracting five or six topics seems reasonable for our sample of 1000 reviews. Note that a different sample may change the coherence plot.

Part 4. On Your Own

Find the optimal number of topics for the entire dataset instead of a sample. Note that it will take a considerable amount of time. Then, run the LDA model with the entire dataset of reviews for the selected "optimal" number of topics. Interpret the obtained topics and find representative documents for each of the topics.

For each topic, print the list of words constituting this topic, the words' likelihood scores, and a representative document.

Build Your Python Skills

Chapter 22 of the free book Learn Python with Jupyter by Serena Bonaretti. ▶ https://www.learnpythonwithjupyter.com/

Additional Reading

Blei, D. M., Ng, A. Y., & Jordan, M. I. (2003). Latent Dirichlet allocation. Journal of Machine Learning Research, 3(Jan), 993–1022.

Text Analysis with Large Language Models (LLMs)

A. P. Kirilenko, *Practical Data Mining with AI for Social Scientists*,
Springer Texts in Social Sciences, https://doi.org/10.1007/978-3-031-89689-7_13

13.1 Introduction

Text modeling techniques we have covered so far were based on the bag-of-word approach. This approach assumes that the document word order does not affect semantics; that is, an analysis of word distribution gives an understanding of the text content. This assumption still allows the generation of well-interpreted topics in many cases, especially when combined with techniques such as text tokenizing by n-gram or part-of-speech (POS) tagging. Still, accounting for the word order in automated topic analysis would be clearly advantageous.

Important concepts

Word embedding
Contextual embedding
Attention mechanism
Encoding and decoding
Transfer learning

Recall how the LDA functions. First, the words present in the documents are combined in a dictionary. Then, each word is represented as a vector with a length of 1 in a multidimensional space with the number of dimensions equal to the size of the dictionary. For instance:

- the word "home" could be represented with a vector (1,0,0,0…0.0),
- the word "house" could be represented with a vector (0,1,0,0,…0.0), and
- the word "alligator" could be represented with a vector (0,0,1,0,…0.0).

The distance between each pair of these three vectors would be the same,[1] even though it is intuitively clear that the words "home" and "house" are closer one to the other, as compared to the words "home" and "alligator". The other drawback is that the number of dimensions required for storing a real-world dictionary is huge; this is why complex text cleaning and standardization are required to reduce the size of the word space by removing the least frequent words.

This chapter introduces several innovations that overcame the limitations of the early topic models and serve as the foundation for the current crop of large language models (LLMs). One of these innovations is *word embedding*, which radically improves the format of vector word representation. In this format, the distance between the words "home" and "house" is much shorter than the distance between the words "home" and "alligator". In ad-

1 The cosine distance (Chapters 5 and 11) between each pair of words in this representation is 1, indicating no similarity.

dition, the number of dimensions required to store the entire dictionary of a corpus is drastically smaller compared with the "one-hot" representation.

The word embedding mechanism was briefly covered in ▶ Chapter 11 when we discussed the word2vec model, and in ▶ Chapter 12, when we examined improvements to the original LDA topic model. The LLMs go even further than traditional word embedding by introducing contextual word embedding. *Contextual embedding* allows differentiation between identically written words with different meanings (polysemy), such as "flight" (to Florida) vs. "flight" (of stairs). The contextual embedding would map such words to different vectors, depending on the context in which they are used.

Another important innovation is the *attention mechanism*. In the LDA, all tokens have the same weight, with the assumption that all tokens are equally important for text understanding. This assumption is clearly unhelpful. For instance, we had to spend significant efforts on data preparation in ▶ Chapter 12 Lab to partially compensate for assigning equal weights to words in the input sequence. The attention mechanism is designed to account for tokens' relative importance, allowing the model to focus on relevant parts of an input sequence while generating the output. Notably, this focus is changing as the model progresses.

Cocktail party effect

Similar to how the artificial neural networks (introduced in ▶ Chapter 7) were inspired by the structure and function of the human brain, where interconnected neurons process and transmit information, allowing the network to learn and adapt by adjusting the strengths of these connections, the attention mechanism was inspired by the brain's selective focus process. This biological process enables humans to prioritize certain stimuli over others, focusing on the most relevant information at any moment while filtering out distractions. For instance, the "cocktail party effect" allows a party visitor to attend a conversation with a particular person while not attending to all other conversations in the room.

A foundational 2017 paper authored by eight scientists at that time working at Google[2] introduced a novel

2 Vaswani, A., Shazeer, N., Parmar, N., Uszkoreit, J., Jones, L., Gomez, A.N., Kaiser, L. and Polosukhin, I., 2017. (2017). Attention is all you need. Advances in Neural Information Processing Systems. arXiv:1706.03762, 10, p.S0140525X16001837.

transformer model architecture that incorporated these ideas to radically improve the computational efficiency of language models. This allowed the development of LLMs pre-trained on billions- and trillions-word datasets extracted from variable media sources. Current LLMs with hundreds of millions or even billions of parameters have *emergent properties* not present in smaller-scale models, that is, the skills to perform tasks for which the models have no instructions and have not been trained to perform. One example is *analogical reasoning*: an ability to identify similarities between two different concepts and use this to draw inferences or solve problems. Raven's Progressive Matrices test measures the analogical reasoning ability by presenting a person with a matrix of various elements and asking to identify a missing element. Another example is the ability to perform language translation of novel metaphors and memes correctly.

Analogical reasoning example. Identify the missing element. https://en.wikipedia.org/wiki/Raven%27s_Progressive_Matrices.

Yet another emergent ability is in-context learning. The in-context learning is the ability to learn how to perform a new task from a limited number of examples provided in LLM's input (*context*). This ability is used in the *few-shot prompting* technique. In a *zero-shot prompting*, the model is given its task without any examples or demonstrations of a similar task. Meanwhile, better performance may be achieved by providing the LLM with a few examples of completed similar tasks. For instance, in a classification task that requires an LLM to group tourists' comments, just a few examples of correct group allocation may radically improve classification outcomes.

In ▶ Chapter 12, we used the text of the analyzed documents to train the LDA model to "understand" the hidden topics. In essence, the model had no prior information about the language. One implication is the large amount of data required for model training. It would be very beneficial to pre-train the model on a large set of generic texts. Then, the documents of interest could be used to fine-tune the model. This would relax the requirement for a large amount of input data while reducing the model training time. The mechanism allowing for fine-tuning a pre-trained model is called *transfer learning*. LLMs, such as OpenAI's GPT and Google's BERT, are trained on vast corpora of diverse text data, including Wikipedia, books, and social media. This allows the models to capture the patterns and structures of written language.

Zero-shot and fewshot prompts

Zero-shot prompting: an LLM is given a task without any prior taskspecific examples. Few-shot prompting: an LLM is provided with a small number of examples or demonstrations to guide it in completing a task.

13.2 Language Models

A model able to predict the likelihood of a word or sequence of words appearing in a given context is called a *language model*. The early language models were primarily *rule-based*; that is, they relied on a complex set of rules governing the generation of the output word sequence. For instance, a *rule-based machine translation* model purposed for translation from English to Spanish would have an English-to-Spanish dictionary and a set of rules capturing the morphological structure of English and Spanish languages. However, languages are fluid and hard to capture in a rigid system of rules, making the rule-based language models extremely complex yet limited in their ability to operate on real-world tasks. A story attributed by some to a pioneer machine translation scientist, Antony Oettinger, exemplifies this limitation by telling how a Bible verse, "The spirit is willing, but the flesh is weak," after being machine-translated into Russian and then back to English transformed into "The vodka is ready, but the meat is rotten".

A radically different approach is to use *data-driven* probabilistic language models. In machine translation, a model could be trained on a large parallel corpus of English–Spanish sentence pairs, such as translations of the European Parliament or the United Nations documents, to return the most likely Spanish word based on the English input. Early *statistical language models* based on this approach targeted learning probability distributions of words and word sequences. For instance, n-gram models would learn probabilities for predicting the word n + 1 based on n previous words. By examining small sequences of words (e.g., trigrams), these models could capture simple patterns, yet they could not account for long-distance dependencies. One critical limitation here is an over-simplistic internal structure of the statistical models.

Rule-based and data-driven language models

A rule-based language model generates its output from the input sequence based on a rigid set of rules.
A data-driven model instead rely on statistical analysis and machine learning without being explicitly programmed with linguistic rules.

The internal complexity of the contemporary LLMs allows for capturing much more nuance patterns in human language compared to just a few preceding words in the statistical models. However, the overall principle is identical: the model is trained to predict the output word sequence based on the input. While the methods and capabilities have evolved significantly, the foundational task of learning probabilistic relationships between words or sequences is shared with earlier statistical models. For instance, OpenAI's GPT is designed and trained for the ability to predict the next word in a sequence, hence capturing the preceding context. This design contributes toward GPT's capability to generate a coherent and contextually relevant text.

In comparison, Google's BERT (Bidirectional Encoder Representations from Transformers) LLM is trained to predict a missing word in a sequence, capturing the preceding and the following context, hence the word "bidirectional" in BERT's acronym. This capability makes BERT versatile in text analysis tasks rather than text generation. Examples of these tasks include:

- Text Classification: grouping the documents into predefined categories based on their topics or sentiment;
- Named Entity Recognition (NER): identification of the names of people, places, brands, organizations, and similar information in the texts;
- Extraction of event dates and monetary amounts;
- Answering questions; and
- Text summarization.

The following section illustrates LLMs' abilities in some of these tasks.

13.3 Applications of LLMs for Text Analysis

Transformer architecture. https://en.wikipedia.org/wiki/Transformer_(deep_learning_architecture)

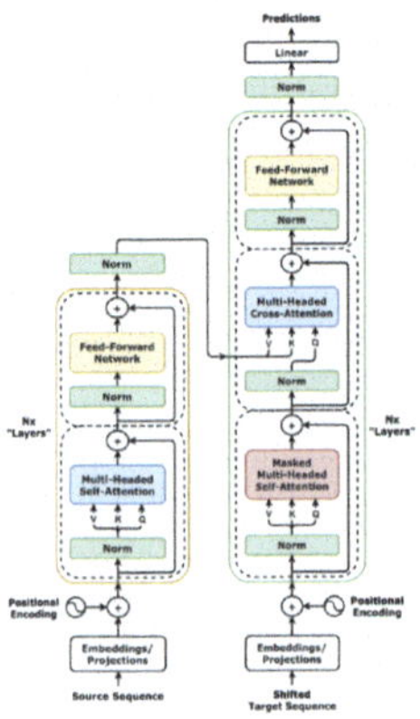

The transformer architecture outlined in the Introduction was used to develop dozens of LLMs over the past few years. Some of these models, such as Meta's BART and Google's T5, cater to tasks that require a comprehensive understanding of an input sequence and generation of a respective output sequence, such as in translation between different languages. Other models, such as Google's BERT, target predictive analytics, text classification, sentiment analysis, named entity recognition (NER), and similar. Yet other models, such as OpenAI's GPT family, target the generation of coherent and contextually appropriate text.

On top of these models, there are multiple applications that address specific needs. For instance, ChatGPT is an application built on top of a GPT LLM to allow for conversational interactions. ChatGPT is fine-tuned to maintain dialog context over multiple exchanges and to follow instructions. Other models, such as Jasper AI or Writesonic, use GPT capabilities with their own models.

Similarly, BERTopic uses BERT LLM but is optimized for topic modeling. This tool can analyze customer feedback, organize documents, or cluster news articles based on underlying themes. This chapter will focus on typical text analysis tasks: topic modeling, classification, and sentiment analysis.

13.3.1 Sentiment Analysis with BERT

In ▶ Chapter 10, we introduced two approaches to sentiment analysis. The lexicon-based sentiment analysis involves the development of a dictionary covering the majority of the sentiment-carrying words of the language. The machine learning approach requires a large, manually labeled sample of texts to train the algorithm. Neither of these approaches is perfect. Lexicon-based algorithms tend to fail on texts from specific fields where word sentiment may differ from general usage. For instance, the word "fight" in sports news usually carries a positive sentiment. Additionally, lexicon-based methods often struggle with context, sarcasm, and negation, where sentiment shifts based on subtle language cues.

Meanwhile, machine learning-based approaches require large datasets representative of the specific domain and language variation. The model may not generalize well without a large enough labeled dataset and may misinterpret sentiment in unfamiliar contexts. Moreover, machine learning models can be computationally expensive to train and deploy, while simple models fail to recognize the context in which sentiment words are used.

Deep learning models like BERT largely lift the latter consideration by capturing the general structure of the language during the model pretraining process. The complex structure of the model allows recognition of nuanced expressions and complicated sentence structures affecting the sentiment as well as handling sarcasm and negation. To refine the model for a specific field, the model is typically fine-tuned on a small, labeled sentiment dataset, such as sports news. This allows the LLM to learn the specific patterns associated with each sentiment class in this field.

The base version of BERT is trained on a large number of texts from books and English Wikipedia with a total size of several billion words. The model was trained on two tasks: prediction of masked words and prediction of the next sentence. Randomly masking words accomplished the first task in a sentence, training the model to predict the masked words from their context (the words to the left and right of the word). In the second task, the model was trained to recognize if two text spans logically follow one after another. For example, given the text "The hotel room was small", the text "Yet the room price was unbeatable" would be marked as "next," while the text "Yet the taste of persimmons was excellent" would be marked "not next".

The trained BERT model is then fine-tuned to perform specific tasks. For instance, a popular BERT implementation, bert-base-multilingual-uncased-sentiment, was trained for sentiment analysis of product reviews in five European languages. Specifically, the model was trained to use review text to predict the number of stars (1 to 5) in a product's online review. The authors claim a 67% accuracy score for reviews in English and a 95% likelihood that star predictions would be off by no more than one star.

Resources for NLP models

Hugging Face: ▶ https://huggingface.co/models

Kaggle: ▶ https://www.kaggle.com/models

Google Colab: ▶ https://colab.research.google.com

Note that while review star rating correlates with review sentiment, the correlation is not perfect. In addition, the model suitable for classifying customer reviews may be deficient for other corpora. When selecting the model suitable for your data, it is important to select a model fine-tuned on texts from a similar domain. For instance, a specific BERT implementation, twitter-roberta-base-sentiment-latest, was fine-tuned on 124 million tweets to predict the likely sentiment (0 for negative, 1 for neutral, and 2 for positive). This makes it a strong prospective candidate for social media sentiment analysis. In an emotion analysis domain, the BERT implementation emotion-english-distilberta-base was fine-tuned to predict Ekman's six basic human emotions (see ▶ Chapter 10) plus a neutral class.

13.3.2 Text Classification and Summarization with BART

While BERT is a powerful text analysis model, it cannot be directly used for text generation. The autoregressive models, such as GPT, excel in text generation but lack the right-side context because they are trained to predict the next word based only on the preceding sequence. This unidirectional approach means they do not fully leverage information in a sentence during training. The model BART (Bidirectional and Auto-Regressive Transformers) developed by Facebook AI (Meta) combines the BERT bidirectional approach and the GPT text generative capabilities. This makes the BART model well-suited for tasks such as summarization, translation, and classification of texts.

Like BERT, BART was pre-trained on a large volume of data from books, Wikipedia, news, and other high-quality texts. This base model can then be fine-tuned for various tasks. For instance, the BART implementation bart-large-mnli is trained to evaluate hypotheses like "This text is about {…}", where the {…} contains a series of alternatives. For example, for hotel review classification, the relevant alternatives could be "front desk", "room", and "hotel restaurant". The relevant hypotheses then would be "This text is about the front desk", "This text is about the room", and "This text is about the restaurant". Such an approach proved to be very effective not only for classification but also for emotion recognition and similar problems of mapping the texts into pre-assigned categories.

This approach to classification is known as *zero-shot* text classification (or 0Shot-TC). Note the difference between the 0Shot-TC and the supervised classification discussed in ► Chapter 4. In supervised classification, the model was trained to match the data (numerical or textual) with classes known *prior* to model training. The zero-shot text classification algorithm learns the classes *after* the model training. That means the model is not provided with samples of the correct task completion. This is an example of transfer learning where an LLM model originally trained for performing a different task, such as restoring a corrupt sentence, is applied to a dissimilar problem. Note similar concepts of a single or a few-shot classification, where one or a few samples of correctly performed tasks are provided to further improve model performance.

Similarly, in text summarization tasks, fine-tuned versions of BART enable the distilling of information expressed in texts into a few sentences. Meanwhile, a multi-shot text summarization returns improved summarization results by allowing samples of summarized texts. For instance, healthcare providers could provide examples of summarized medical reports.

Text summarization

Text summarization reduces the text length while keeping its main informative part intact.

Extractive summarization extracts sample key sentences from the text.

Abstractive summarization generates text abstract.

13.3.3 Topic Modeling with BERTopic

Topic modeling heavily relies on understanding words context. For instance, the LDA bag-of-words (BoW) model (Chapters 11 and 12) has a rudimentary understanding of the context by considering words that tend to appear in the same documents belonging to the same topics. The LLMs have greatly improved the ability to capture the contextual meaning of words in a sentence by considering not only word co-occurrence but also the distance between the words. This makes BERT highly suitable for topic modeling. However, BERT doesn't have a built-in ability to identify clusters of related terms. Instead, BERT outputs represent high-dimensional words, sentences, or document embeddings that capture contextual information in the text. These embeddings can be further utilized in topic modeling. The main idea is to cluster the embeddings so that similar textual units would be joined in the same clusters and then interpret these clusters in terms of common themes.

For instance, the BERTopic algorithm operates in four steps. First, the textual data is processed with the BERT embedding model to represent sentences (by default, while paragraphs or entire documents can also be used) as high-dimensional vectors of the same length. These vectors capture the semantic information of the text units with semantically similar texts represented by spatially close vectors.

BERTopic process

Input text is provided. Noisy text may be preprocessed.

BERT generates text embeddings.

The embedding vector dimensionality is reduced.

The vectors are clustered making potential topics

Words specific for each of the clusters are extracted.

The topics are interpreted and visualized.

13

The high dimension of the embedding vectors (default BERT base word embeddings have 768 dimensions) presents a problem for clustering. Because of that, a dimension reduction algorithm is used to reduce the number of dimensions to a manageable size, for instance, 5. Various dimension reduction methods can be used in this step, including the Principal Component Analysis (PCA). However, the default method is UMAP (Uniform Manifold Approximation and Projection), which is able to capture non-linear data relationships. After this step, each sentence, paragraph, or document is represented by a low-dimensional vector.

In the third step, these vectors are grouped with a clustering algorithm. By default, clustering employs the HDBSCAN (Hierarchical Density-Based Spatial Clustering of Applications with Noise) algorithm, an enhanced version of DBSCAN that we covered in ► Chapter 5. Recall that DBSCAN groups data points based on density;

data points outside the densely populated regions are treated as outliers. In this chapter's lab, you will observe a significant number of documents marked as outliers.

The words common in texts that are grouped in the same cluster represent topics. The final step of the algorithm is extracting word representations for each of the topics. This step uses the TF-IDF approach (introduced in ▶ Chapter 11), modified to work with clusters. Recall that TF-IDF is designed to find words, which are typical for a limited number of documents but rare for the corpus as a whole. Similarly, the class-based c-TF-IDF is designed to search for words, typical for documents in a small number of clusters, but rare for the corpus as a whole.

The outcome of BERTopic is similar to LDA, with each topic represented by a list of words together with the likelihood of their appearance in that topic. The distance between the topics can be further explored numerically using a similarity matrix or various built-in visualization techniques. Some of these methods will be covered in the lab. BERTopic documentation can be found online at ▶ https://maartengr.github.io/BERTopic.

13.4 Text Analysis with ChatGPT

13.4.1 Examples: Sentiment Analysis and Topic Modeling

In contrast to the BERT-type LLMs, the autoregressive models, such as OpenAI's ChatGPT, are trained to predict the next word based on the left-side context only. That is, the model forecasts future words based only on past content. This makes ChatGPT a conversational partner able to generate coherent texts; consequently, multiple authors explored ChatGPT's analytic abilities, including the ability to perform typical NLP tasks: sentiment analysis and topic modeling. For instance, sentiment analysis can be accomplished with a single ChatGPT prompt:

- Which one of the attributes, "negative sentiment", "neutral sentiment", or "positive sentiment," characterizes the sentence: "I hate pizza"?
  - The sentence "I hate pizza" is characterized by negative sentiment. The word "hate" conveys a strong

GPT features

Decoder-only transformer.

Trained on unidirectional context.

Used mainly in language generation tasks.

ChatGPT is a GPT version fine-tuned for conversational tasks.

negative emotion toward pizza, which makes the sentiment negative.

For a large number of documents, a user is able to upload a file with data to be analyzed. For instance, sentiment analysis of the Arenal Volcano National Park reviews can be completed with the following dialog:

- I will give you an Excel file.
  - Please upload the Excel file, and let me know how I can assist you with it!
- *Upload the file with reviews.*
- Which one of the attributes, "negative sentiment", "neutral sentiment", or "positive sentiment," characterize each review? Provide the table in a downloadable format.
  - The table with sentiment classifications has been saved. You can download it using the link below.

ChatGPT can be further prompted to explain analysis results. For instance:

- Explain your first answer.
  - Review: "The hike going down is okay with no problem. It's the coming back up that's a killer. Make sure you are physically able."
  - Explanation:
    The phrase 'okay with no problem" indicates a neutral tone for the hike going down.
    The phrase "a killer" is a strongly negative term, highlighting the difficulty of the return hike.
    Overall Sentiment: Negative sentiment, as the focus is on the challenging aspect.

Similarly, ChatGPT can be prompted for topic modeling. Notice that we prevent ChatGPT from using Python for text analysis.

- Describe the main topics encountered in reviews. Do not use Python programming.
  - Based on the reviews, the main topics encountered include:
    1. Hiking Difficulty. Many reviews discuss the physical challenge of the hike, particularly the climb back up. Terms like "killer," "difficult," and "not for the faint of heart" emphasize the strenuous nature of the experience. This topic reflects the physical effort required and how it impacts the overall perception of the destination.

2. Scenic Beauty. Several reviews mention the destination's natural beauty, such as the "waterfall" or "falls." Descriptions like "amazing" and "nice" highlight the location's visual appeal and serene environment.
3. Accessibility. Some reviews discuss the ease or difficulty of reaching the site, including parking and steps. Comments about "500 steps" or "short walk" indicate varying levels of accessibility and how it influences visitors' experiences.
4. Adventure and Experience. Several reviews reflect on the adventure aspect of visiting the site. Phrases like "an adventure in itself" and "way more than we planned on" suggest that the visit can feel thrilling, though sometimes unexpectedly challenging.
5. Effort vs. Reward. A recurring theme is the balance between the effort required (e.g., climbing) and the perceived reward (e.g., the waterfall). Some reviewers mention whether the experience was "worth it," indicating how personal expectations align with the actual visit.

The main problem with this approach is that, in contrast with models covered in ▶ Sect. 13.3, ChatGPT has never been trained to perform these tasks. The tests of ChatGPT on various tasks have shown that while ChatGPT was able to successfully perform a wide variety of analyses, in all of these tasks, its performance was inferior to more appropriate LLMs fine-tuned for these types of problems.[3] This is why the study authors called ChatGPT a "jack of all trades, master of none." This conclusion should not discourage using ChatGPT for text analytics but rather caution analysts not to take the outcomes for granted and always evaluate model performance to validate the results.

13.4.2 Prompt Engineering

The performance of conversational LLMs such as ChatGPT can be improved by providing carefully crafted

3 Kocoń, J., Cichecki, I., Kaszyca, O., Kochanek, M., Szydło, D., Baran, J., … & Kazienko, P. (2023). ChatGPT: Jack of all trades, master of none. *Information Fusion*, *99*, 101,861.

Prompt engineering

A set of guidelines for designing and optimizing input prompts to that guide the behavior of LLMs and other AI models.

instructions. For instance, the topic modeling example in the previous section applied the following ChatGPT prompt:

- Describe the main topics encountered in reviews. **Do not use Python programming**.

This instruction forced ChatGPT to search for recurring patterns in the review texts as opposed to writing a Python code that would, for example, use LDA for topic extraction. The new discipline of crafting LLM input prompts to elicit an expected and accurate response from the model is called prompt engineering. One may interpret prompt engineering as reprogramming an LLM which guides it towards a specific task. Multiple prompt engineering lists have been developed for various fields, but the following guidelines are common for all of them:

Instruction: A clear instruction on the model's goal.

- Describe the main topics encountered in reviews.

Steps to achieve the goal: A detailed list of steps.

- To achieve this goal, follow the following steps: …

Context: The context provides background information that helps the model in understanding the data.

- The data represents reviews of travelers to the Arenal Volcano National Park in Costa Rica.

13

Domain: The specific area or audience for which the model response is intended: general public, medical domain, technical report, etc.

- The results are intended for a government report.

Output format: The desired format for model response.

- Provide the table in Excel format.

Constraints: Additional constraints on model response.

- The output should be no longer than one paragraph.

Pretending: Instructing the model to assume a particular role.

- Act as a public health official.

Other useful techniques that help in changing LLMs behavior include:

Iterating and prompt chaining: refining the prompt or even repeating the same prompt tends to improve LLM response gradually.

Setting the LLM temperature: the temperature parameter controls the measure of randomness in model output. Usually, this parameter is set automatically. For analytical tasks, to a low value for more deterministic output; for text generation, to a high value for more creativity. While GPT allows a direct temperature setting through the API interface, the ChatGPT default temperature can be changed with prompts like "be creative" or "be deterministic".

Few-shot learning: providing the model with a few examples of the completed task. For instance:

- Use the following example for the output: 'Topic 1: Hotels. Representative document: "We stayed in a nice lodge"'.

The Further Reading section lists publications providing deeper insides on prompt engineering techniques.

13.5 Conclusion

We have explored several text analysis methods, ranging from traditional approaches like bag-of-words to advanced language models powered by neural networks. Selecting the "best" approach for text analysis depends on various factors, including the nature of the data, the objectives of the analysis, and the available computational resources.

In text analysis, when the dataset is large, relatively noise-free, and the documents are not overly long, selecting simpler methods like Latent Dirichlet Allocation (LDA) may be the best choice due to their transparency, which allows an analyst to explain how the results were obtained from data. Conversely, when data is noisy, unstructured, and context-rich textual information, LLMs such as BERT, GPT, or BART are often the most suitable due to their ability to capture contextual and semantic nuances.

Effective use of LLMs, however, relies on an accurate definition of the analysis goals, methods, and steps to follow. There are many models already fine-tuned for specific tasks, such as sentiment analysis of Twitter (X) media. Because of its bidirectional architecture, BERT is highly effective for tasks requiring nuanced text understanding, such as sentiment analysis or named entity recognition. Similarly, classification or summarization tasks may benefit from BART's hybrid capabilities, combining text generation and understanding. Fine-tuned models like BERTopic or task-specific LLM variants are particularly advantageous for domain-specific tasks. In addition, it is possible to further improve model performance by providing examples of correctly accomplished tasks.

If the task involves generating coherent text, GPT-based models excel due to their autoregressive nature. A systematic approach to designing the input queries called prompt engineering directs the model to a specific type of analysis, achieving the desired goals. Still, while a GPT-type model can perform a wide range of analytic tasks, it is also outperformed by specialized LLMs fine-tuned to solve a limited range of tasks.

Finally, computational constraints are a critical consideration. Pretrained LLMs and transfer learning alleviate some of these challenges by enabling fine-tuning on smaller datasets, but they still require significant computational resources. Simpler rule-based models may be more practical, even if they are less accurate, offering a practical compromise.

13.6 Abbreviations

This chapter uses multiple abbreviations. The following table provides brief explanations of these abbreviations.

Abbreviation	Explanation
NLP	Natural Language Processing
NER	Named Entity Recognition. Identification of the names of people, places, etc.
POS	Part-of-speech. POS tagging tags each word in a sentence with a corresponding POS.
BoW	Bag-Of-Words representation of a document as a collection ("bag") of words while ignoring grammar, word order, and syntax.
TF-IDF	Term Frequency-Inverse Document Frequency. A measure of how important a word is for understanding a particular document that extends the BoW approach.

PCA	Principal Component Analysis. A popular dimensionality reduction method. PCA was one of the first topic modeling methods. It is typically used with BoW and TF-IDF text representations.
UMAP	Uniform Manifold Approximation and Projection. A dimensionality reduction method excelling in visualizing high-dimensional data such as word space.
HDBSCAN	Hierarchical Density-Based Spatial Clustering of Applications with Noise, an unsupervised clustering algorithm, extension of DBSCAN (▶ Chapter 5).
LDA	Latent Dirichlet Allocation. A popular topic modeling method.
LLM	Large Language Model. An AI model designed to analyze and generate human-like texts.
GPT	Generative Pre-trained Transformer. A family of generative LLMs developed by OpenAI which excels in text generating. The latest version (February 2025) is GPT-4o.
ChatGPT	An AI chatbot application built around a GPT model and optimized for interactive conversation.
BERT	Bidirectional Encoder Representations from Transformers, an LLM developed by Google. BERT is best for language understanding, such as sentiment analysis.
BERTopic	A topic modeling application that uses BERT LLM to capture the semantic meaning of words, followed by dimension reduction and clustering.
BART	Bidirectional and Auto-Regressive Transformer LLM by META. Excels in text understanding and generating tasks such as summarization.
T5	Text-to-Text Transfer Transformer LLM by Google. A typical application is machine translation.
0Shot-TC	Zero-Shot Text Classification. NLP technique that allows an LLM to classify texts into categories without prior training.

13.7 Review Questions

1. What is the main limitation of the bag-of-words approach in text modeling?
2. How does word embedding improve upon the "one-hot" vector representation?
3. In simple words, explain the concept of the attention mechanism.
4. What biological process inspired the attention mechanism?
5. Describe the difference between zero-shot and few-shot learning in LLMs.
6. What distinguishes BERT's training objective from GPT's?

7. List three text analysis tasks that BERT is particularly suited for and explain why.
8. What are the key steps in the BERTopic algorithm for topic modeling?
9. What are the limitations of ChatGPT in performing text analysis tasks compared to specialized LLMs?
10. How does prompt engineering enhance the performance of conversational LLMs like ChatGPT?

13.8 Lab

Objectives

1. Analyze sentiment with BERT Large Language Model (LLM)
2. Classify topics with BART LLM
3. Cluster topics with BERTopic LLM

Data

The data for the first part of this chapter is modified from the Covid-19 Vaccine Tweets with Sentiment Annotation dataset (▶ https://www.kaggle.com/datasets/datasciencetool/covid19-vaccine-tweets-with-sentiment-annotation). The data is stored in a Microsoft Excel file and contains 6,000 records. The data represents tweets about the following widely distributed COVID-19 vaccines: Moderna (USA), Oxford-AstraZeneca (UK), Pfizer-BioNTech (Germany), Sinopharm BIBP (China), Sinovac (China), Covaxin (India), and Sputnik V (Russia).

The data is stored as ▶ https://github.com/andykir/DataMiningLectures/raw/main/covid-19_vaccine_tweets_with_sentiment.xlsx.

Variables in dataset 1.

Variable name	Explanation	Notes
tweet_id	Unique identifier for each tweet	String
label	Manually assigned sentiment	1: negative; 2: neutral; 3: positive

Variable name	Explanation	Notes
tweet_text	Tweet text	String

The data for the second part is abridged from reviews of visitors to Volcano Arenal National Park in Costa Rica. The data is stored in a Microsoft Excel file and contains over 10,242 reviews. The data is stored as ▶ https://github.com/andykir/DataMiningLectures/raw/refs/heads/main/ArenalReviews.xlsx

The Excel file contains a sheet titled "reviews" with the following columns:

Variables in dataset 2.

Variable name	Explanation	Notes
ID	Unique identifier for each interview	Unique number
Title	Review title	Text
Review	Review text	Text

Part 1. Sentiment Analysis

This part aims to extract the sentiment and topics from tweets dealing with COVID-19 vaccination. The overall goal of this assignment is to detect the sentiment expressed towards COVID-19 vaccines. Note that the dataset is labeled, which permits model performance evaluation, but labels may have errors. For instance, tweet #1,382,844,719,848,296,450, "#India at 216,850 new cases of #coronavirus with 1183 deaths in the last 24 h…" is labeled as neutral.

Step 1 (Optional). Install Jupyter Lab on Your Desktop Computer

Large Language Models require significant computational resources. If you use the free Anaconda Cloud distribution, you may exceed your free memory allocation. In this case, you can either upgrade to a paid account or install JupyterLab locally on your computer. Detailed installation instructions are available here: ▶ https://jupyterlab.readthedocs.io/en/4.2.x/getting_started/installation.html

To launch JupyterLab on a Windows computer, open a Command Prompt window and enter the following command:

>jupyter lab.

The Anaconda Cloud interface is identical to JupyterLab's local interface, ensuring a seamless transition between the two.

Step 2. Develop Initial Code for Sentiment Analysis

To accommodate those using the desktop version of the software, the examples assume that we use ChatGPT for code generation.

USER:

Excel file "▶ https://github.com/andykir/DataMiningLectures/raw/refs/heads/main/covid-19_vaccine_tweets_with_sentiment.xlsx" contains tweets on Covid-19 vaccines. The tweets are in the tab "covid-19_vaccine_tweets" in the column "tweet_text". Your goal is to write a Python code to find the sentiment expressed in the tweets. Save results into the file "reviews_with_sentiment_result.xlsx". You have to use BERT LLM.

ChatGPT:

```
import pandas as pd
from transformers import pipeline

# Load the Excel file and the specific sheet
file_path = "https://github.com/andykir/DataMiningLectures/"\
            "raw/refs/heads/main/"\
            "covid-19_vaccine_tweets_with_sentiment.xlsx"
sheet_name = "covid-19_vaccine_tweets"
df = pd.read_excel(file_path, sheet_name=sheet_name)

# Initialize the sentiment analysis pipeline with a BERT model
sentiment_analyzer = pipeline("sentiment-analysis", \
        model="nlptown/bert-base-multilingual-uncased-sentiment")")

# Define a function to map sentiment to numerical values
def get_sentiment_label(sentiment):
    if sentiment == "LABEL_0":  # Extremely negative
        return 1
    elif sentiment == "LABEL_1":  # Negative
        return 1
    elif sentiment == "LABEL_2":  # Neutral
        return 2
    elif sentiment == "LABEL_3":  # Positive
        return 3
    elif sentiment == "LABEL_4":  # Extremely positive
        return 3
    return 2  # Default to neutral

# Apply the sentiment analysis
df["sentiment"] = df["tweet_text"].apply \
    (lambda x: get_sentiment_label(sentiment_analyzer(x)[0]['label']))
# Save the results to a new Excel file
output_path = "reviews_with_sentiment.xlsx"
df.to_excel(output_path, index=False)

print(f"Sentiment analysis completed and saved to {output_path}.")
```

Explanation:

Prior to running the code, ensure you have the transformers, pandas, and torch libraries installed.

- Sentiment Analysis Model: This script uses a multilingual BERT model from Hugging Face, trained specifically for sentiment analysis, which outputs labels ranging from extremely negative to extremely positive.
  Label Mapping: The *get_sentiment_label* function maps model output labels to values of 1 (negative), 2 (neutral), and 3 (positive).
  Saving Results: After calculating sentiment scores, the results are saved to a new Excel file.

Run the code (it should take a few minutes as there are thousands of tweets) and explore the results of sentiment analysis stored in the Excel file:

tweet_id	label	tweet_text	sentiment
1360342002961940000	1	4,000 a day dying from the so called …	2
1382896334886250000	2	Pranam message for today manifested …	2
1375673411846870000	2	Hyderabad-based ?@BharatBiotech? has …	2
1381310901119290000	1	Confirmation that Chinese #vaccines …	2
1362165556091190000	3	Lab studies suggest #Pfizer, #Moderna …	2
…	…	…	…

There appears to be an issue: BERT has evaluated the sentiment of all tweets as 2. To investigate the root cause of this problem, we will begin by closely examining the line of code where BERT is applied to our data. The code snippet applies sentiment analysis to the text data in the *tweet_text* column of a pandas DataFrame named *df*:

```
# Apply the sentiment analysis
df["sentiment"] = df["tweet_text"].apply \
    (lambda x: get_sentiment_label(sentiment_analyzer(x)[0]['label']))
```

There are the following key components of the code.

df["tweet_text"].apply() applies a given function to each element of the *tweet_text* column. In this case, a lambda function is provided as the argument. The lambda function is a one-line function that takes a single parameter *x* (representing each tweet). It processes *x* by passing it to the *sentiment_analyzer* pipeline as an argument and then feeds the resulting label to the *get_sentiment_label* function.

13

Let us first try analyzing the sentiment of a simple text, for instance, "I hate pizza," without calling the get_sentiment_label function.

In a new cell, type:

```
sentiment_analyzer("I hate pizza")
```

Run the cell. Result:
[{'label': '1 star', 'score': 0.7326287031173706}]

Experiment with other texts:

```
sentiment_analyzer("I love pizza")
```

[{'label': '5 stars', 'score': 0.8126732707023621}]

```
sentiment_analyzer("I eat pizza")
```

[{'label': '5 stars', 'score': 0.338176429271698}]

```
sentiment_analyzer("I eat soup")
```

```
sentiment_analyzer("I eat beets")
```

{'label': '3 stars', 'score': 0.2666294276714325}]

The *sentiment_analyzer* function returns an array containing a single element: a dictionary with two keys, 'label' and 'score'. The key 'label' represents the sentiment category, ranging from '1 star' (very negative) to '5 stars' (very positive). The key 'score' represents the confidence level of the prediction. We can now identify the error in the code: the function *get_sentiment_label* incorrectly assumes that the sentiment values returned by *sentiment_analyzer* are "LABEL_0" to "LABEL_4". In reality, the returned values are "1 star" to "5 stars". This mismatch causes the function to behave incorrectly. To resolve the issue, we can make the following adjustments to ensure that the *get_sentiment_label* function maps the returned 'label' values correctly.

```
# Define a function to map sentiment to numerical values
def get_sentiment_label(sentiment):
    if sentiment == "1 star":  # Extremely negative
        return 1
    elif sentiment == "2 stars":  # Negative
        return 1
    elif sentiment == "3 stars":  # Neutral
        return 2
    elif sentiment == "4 stars":  # Positive
        return 3
    elif sentiment == "5 stars":  # Extremely positive
        return 3
    return 2  # Default to neutral
```

Hint: to complete the test computations faster, limit the size of the DataFrame *df* to the first ten lines:

```
df=df[:10]
```

After you finish code debugging, you can remove this line.

Results (in the Excel File):

tweet_id	label	tweet_text	sentiment
1360342002961940000	1	4,000 a day dying from the so called …	1
1382896334886250000	2	Pranam message for today manifested …	3
1375673411846870000	2	Hyderabad-based ?@BharatBiotech? has …	1
1381310901119290000	1	Confirmation that Chinese #vaccines …	1
1362165556091190000	3	Lab studies suggest #Pfizer, #Moderna …	3
…	…	…	…

Step 3. Code Optimization

The code for sentiment analysis may be optimized in several places. First, we can accommodate the sentiment confidence level into model output. For instance, we can change the sentiment of all predictions with confidence below 0.5 to "neutral."

Second, the initial problem with our code is probably caused by the changed format of the "label" key. To make our code more robust, we can modify the function *get_sentiment_label* so that it returns an error message when an unfamiliar sentiment label is encountered.

Finally, the following code snipped can be improved.

```
# Apply the sentiment analysis
df["sentiment"] = df["tweet_text"].apply \
    (lambda x: get_sentiment_label(sentiment_analyzer(x)[0]['label']))
```

As is, the tweets are sent to the sentiment analyzer line by line. Instead, we can send all tweets at the same time. We will also replace the lambda function with a more transparent inline loop.

Improved Code:

```
import pandas as pd
from transformers import pipeline

# Load the Excel file and the specific sheet
file_path = "https://github.com/andykir/DataMiningLectures/"\
            "raw/refs/heads/main/"\
            "covid-19_vaccine_tweets_with_sentiment.xlsx"
sheet_name = "covid-19_vaccine_tweets"
df = pd.read_excel(file_path, sheet_name=sheet_name)

df=df[:10]

# Initialize the sentiment analysis pipeline with a pre-trained BERT model
sentiment_analyzer = pipeline("sentiment-analysis", \
        model="nlptown/bert-base-multilingual-uncased-sentiment")

# Define a function to map sentiment to numerical values
def get_sentiment_label(sentiment, score):
    if score < 0.5:
        return 2
    elif sentiment in ["1 star", "2 stars"] :  # negative
        return 1
    elif sentiment == "3 stars":  # Neutral
        return 2
    elif sentiment in ["4 stars", "5 stars"] :  # positive
        return 3
    else :
        raise Exception("Unknown sentiment value encountered")

# Apply the sentiment analysis
sentiment=sentiment_analyzer(df["tweet_text"].tolist())
df["sentiment"]=[get_sentiment_label(x['label'], x['score']) for x in sentiment]

# Save the results to a new Excel file
output_path = "reviews_with_sentiment.xlsx"
df.to_excel(output_path, index=False)

print(f"Sentiment analysis completed and saved to {output_path}.")
```

Run the code and interpret the output.

Step 4. Improve the Model.

Observe the following line of code:

```
# Initialize the sentiment analysis pipeline with a pretrained BERT model
sentiment_analyzer = pipeline("sentiment-analysis", \
        model="nlptown/bert-base-multilingual-uncased-sentiment")
```

The Hugging Face transformers library's *pipeline* function is a high-level interface that simplifies access to various natural language processing tasks. It allows users to perform tasks like sentiment analysis with just a few lines of code.

The first parameter of the pipeline determines the task it needs to perform. In our example, the task name for sentiment analysis is *sentiment-analysis*. Based on the task name, the pipeline sets up the components necessary for the task's completion.

The second parameter is the name of a particular fine-tuned model to complete the task. In our example, *nlptown/bert-base-multilingual-uncased-sentiment* is a multilingual model fine-tuned for sentiment analysis on over 600,000 product reviews in English, Dutch, German, French, Spanish, and Italian.

Other parameters could be used to define a tokenizer used in text preprocessing and for model performance optimization. Hence, the pipeline makes it easy to load pre-trained models and apply them to common tasks without needing deep knowledge of model architecture, tokenization, or fine-tuning. For instance, a few lines of code are enough to complete sentiment analysis:

```
from transformers import pipeline

# Initialize a sentiment-analysis pipeline
sentiment_analyzer = pipeline("sentiment-analysis")

# Analyze sentiment
result = sentiment_analyzer(["I hate pizza","I love dogs","Paper is white"])
print(result)
```

Result:

```
No model was supplied, defaulted to distilbert/distilbert-base-uncased-finetuned-sst-
2-english and revision 714eb0f (https://huggingface.co/distilbert/distilbert-base-
uncased-finetuned-sst-2-english).
Using a pipeline without specifying a model name and revision in production is not
recommended.
```

```
[{'label': 'NEGATIVE', 'score': 0.9988999366760254},
{'label': 'POSITIVE', 'score': 0.999713122844696},
{'label': 'NEGATIVE', 'score': 0.9991093277931213}]
```

13

In this example, we did not set the name of a model fine-tuned for a specific task; hence, a default "generic" model was used with rather poor results. For instance, the sentiment of the sentence "Paper is white" was determined to be negative. Setting the *model* parameter leads to improved results:

```
from transformers import pipeline

# Initialize a sentiment-analysis pipeline
sentiment_analyzer = pipeline("sentiment-analysis", \
        model="nlptown/bert-base-multilingual-uncased-sentiment")

# Analyze sentiment
result = sentiment_analyzer(["I hate pizza", "I love dogs", "Paper is white"])

print(result)
```

Result:

```
[{'label': '1 star', 'score': 0.7326287031173706},
  {'label': '5 stars', 'score': 0.6776878237724304},
  {'label': '3 stars', 'score': 0.28962281346321106}]
```

The Step 3 code uses a *model fine-tuned* for product review analysis. To further improve sentiment detection, using a model specifically fine-tuned for tweets would be advantageous. One popular sentiment analysis model trained on Twitter data uses an optimized version of BERT called RoBERTa. We will use ChatGPT to change the code to accommodate differences between model outputs.

USER:

I will give you a Python code. Analyze the code.

ChatGPT:

Please provide the Python code you'd like me to analyze, and I'll go through it in detail for you.

USER:

(Copy and paste Python code from Step 3).

ChatGPT:

(Provides analysis of the main blocks of the code).

USER:

Modify the code to be used with cardiffnlp/twitter-roberta-base-sentiment model.

ChatGPT:

```
import pandas as pd
from transformers import pipeline

# Load the Excel file and the specific sheet
file_path = "https://github.com/andykir/DataMiningLectures/"\
            "raw/refs/heads/main/"\
            "covid-19_vaccine_tweets_with_sentiment.xlsx"
sheet_name = "covid-19_vaccine_tweets"

df = pd.read_excel(file_path, sheet_name=sheet_name)

# Initialize the sentiment analysis pipeline
# with cardiffnlp/twitter-roberta-base-sentiment model

sentiment_analyzer = pipeline("sentiment-analysis",
    model="cardiffnlp/twitter-roberta-base-sentiment")

# Define a function to map sentiment to numerical values
def get_sentiment_label(sentiment, score):
    if score < 0.5:
        return 2  # Low confidence, default to Neutral
    elif sentiment == "LABEL_0":  # Negative
        return 1
    elif sentiment == "LABEL_1":  # Neutral
        return 2
    elif sentiment == "LABEL_2":  # Positive
        return 3
    else:
        raise Exception("Unknown sentiment label encountered")

# Apply the sentiment analysis
sentiment = sentiment_analyzer(df["tweet_text"].tolist())
df["sentiment"] = [get_sentiment_label(x['label'], x['score']) for x in sentiment]

# Save the results to a new Excel file
output_path = "reviews_with_sentiment.xlsx"
df.to_excel(output_path, index=False)

print(f"Sentiment analysis completed and saved to {output_path}.")
```

Explanation of Changes:

» Model Change: The model parameter in the pipeline is now set to *cardiffnlp/twitter-roberta-base-sentiment*, a sentiment analysis model specifically fine-tuned for analyzing Twitter data.
Sentiment Labels Update: The *get_sentiment_label* function is modified to handle the new labels:

- "LABEL_0" corresponds to Negative (mapped to 1).
- "LABEL_1" corresponds to Neutral (mapped to 2).
- "LABEL_2" corresponds to Positive (mapped to 3).

If the score (confidence) is below 0.5, it defaults to 2 (Neutral), indicating low confidence in the result.

Note: Running this code will return an error message since some required modules are not installed. For instance, you might need to install the tokenizer developed specifically for LLMs:

```
!pip install tiktoken
```

Additionally, renaming the output file is advisable to facilitate the comparison of the results with those from the previous model.

```
# Save the results to a new Excel file
output_path = "reviews_with_sentiment_twitter_roberta.xlsx"
```

Result (combined with output from Step 3):

tweet_id	label	tweet_text	Sentiment Step 3	Sentiment Step 4
1360342002961940000	1	4,000 a day dying from the so called …	1	1
1382896334886250000	2	Pranam message for today manifested …	3	2
1375673411846870000	2	Hyderabad-based ?@BharatBiotech? …	1	2
1381310901119290000	1	Confirmation that Chinese #vaccines …	1	1
1362165556091190000	3	Lab studies suggest #Pfizer, #Moderna …	3	2
…	…	…	…	…

Part 2. Classification

In this part, we will classify reviews of Arenal Volcano National Park in Costa Rica using pre-assigned topics. The park is known for the views of an active volcano, a picturesque waterfall, a large lake, and abundant wildlife. In addition, there are tourist hotels and lodges. Hence, we include the following categories for the classification analysis: “waterfall”, “volcano”, “wildlife”, “hot springs”, “lake”, and “hotel”. We will use ChatGPT to assist with code development.

USER:

The file “► https://github.com/andykir/DataMiningLectures/raw/refs/heads/main/ArenalReviews.xlsx” contains reviews of a travel destination. The reviews are in the sheet “reviews” in the column “Review”. Your goal is to write a Python code to classify reviews into the following topics: [“waterfall”, “volcano”, “wildlife”, “hot springs”, "lake", "hotel"]. Save results into the file "topics.xlsx". You have to use BART LLM.

ChatGPT:

```
import pandas as pd
from transformers import pipeline

# Load the Excel file and the specific sheet
file_path = \
        "https://github.com/andykir/DataMiningLectures"\
        "/raw/refs/heads/main/ArenalReviews.xlsx"
sheet_name = "reviews"
df = pd.read_excel(file_path, sheet_name=sheet_name)

# Initialize the zero-shot classification pipeline using a BERT model
topic_analyzer = pipeline("zero-shot-classification", \
                 model="facebook/bart-large-mnli")

# Define a set of topics for classification
topics = ["waterfall", "volcano", "wildlife", "hot springs", "lake", "hotel"]

# Apply the topic analysis on the reviews
topic_results = topic_analyzer(df["Review"].tolist(), \
        candidate_labels=topics)

# Extract the highest-confidence topic for each review
df["topic"] = [result["labels"][0] for result in topic_results]
df["topic_score"] = [result["scores"][0] for result in topic_results]

# Save the results to a new Excel file
output_path = "topics.xlsx"
df.to_excel(output_path, index=False)

print(f"Topic classification completed and saved to {output_path}.")
```

Explanation:

The code uses the *zero-shot-classification* pipeline from Hugging Face's transformers library. This approach allows us to classify text without fine-tuning the model for specific topics.

```
df = pd.read_excel(file_path, sheet_name=sheet_name)
```

This loads the ArenalReviews.xlsx file from the specified URL and reads the sheet "reviews" into a DataFrame *df*.

```
topic_analyzer = pipeline("zero-shot-classification", model="facebook/bart-
large-mnli")
```

We initialize the *zero-shot-classification* pipeline with the *facebook/bart-large-mnli* model, a BERT variant that works well for classification tasks, especially in a zero-shot setting.

```
topics = ["waterfall", "volcano", "wildlife", "hot springs", "lake", "hotel"]
```

This list defines the topics we want to classify each review into.

```
topic_results = topic_analyzer(df["Review"].tolist(), candidate_labels=topics)
```

The pipeline processes the reviews and matches them to one of the topics with a confidence score.

```
df["topic"] = [result["labels"][0] for result in topic_results]
df["topic_score"] = [result["scores"][0] for result in topic_results]
```

For each review, we select the topic with the highest confidence score and store it in the topic column along with its *topic_score*.

Finally, the DataFrame *df*, now containing columns for the predicted topic and confidence score, is saved as topics.xlsx:

```
output_path = "topics.xlsx"
df.to_excel(output_path, index=False)
```

Before running the code, it makes sense to limit the size of the data for processing, for instance, to the first 100 rows of reviews:

```
df=df[:100]
```

In addition, notice that only the best topic is saved:

```
# Extract the highest-confidence topic for each review
df["topic"] = [result["labels"][0] for result in topic_results]
df["topic_score"] = [result["scores"][0] for result in topic_results]
```

It would be helpful to save all topics and related scores:

```
# Extract the highest-confidence topic for each review
df["best_topic"] = [result["labels"] for result in topic_results]
df["best_score"] = [result["scores"] for result in topic_results]
df["topic"] = [result["labels"][0] for result in topic_results]
df["topic_score"] = [result["scores"][0] for result in topic_results]
```

The final code:

```
import pandas as pd
from transformers import pipeline

# Load the Excel file and the specific sheet
file_path = \
"https://github.com/andykir/DataMiningLectures/raw/refs/heads/main/ArenalReviews.xlsx"
sheet_name = "reviews"
df = pd.read_excel(file_path, sheet_name=sheet_name)

df=df[:100]

# Initialize the zero-shot classification pipeline
topic_analyzer = pipeline("zero-shot-classification", \
        model="facebook/bart-large-mnli")

# Define a set of potential topics for classification
topics = ["waterfall","volcano","wildlife","hot springs", "lake","hotel"]

# Apply the topic analysis
topic_results = topic_analyzer(df["Review"].tolist(), candidate_labels=topics)

# Extract the topics
df["best_topic"] = [result["labels"][0] for result in topic_results]
df["best_score"] = [result["scores"][0] for result in topic_results]

df["topic"] = [result["labels"] for result in topic_results]
df["topic_score"] = [result["scores"] for result in topic_results]


# Save the results to a new Excel file
output_path = "topics.xlsx"
df.to_excel(output_path, index=False)

print(f"Topic analysis completed and saved to {output_path}.")
```

Note:

There might be multiple warnings; you can run the following code to silence them:

```
import warnings
warnings.filterwarnings('ignore')
```

13

Results (abridged, in the excel file):

ID	Title	Review	best_topic	best_score
1	Be ready to hike	The hike going down is okay with no problem. It gets crowded specially in the morning with all the tour bus from san jose. The water is cold and you can't really swim as it was not that deep. Not allowed to go behind the falls. …	waterfall	0.52
2	Amazing falls crappy steps	The falls are pretty easy to get to you park your car and then walk down to the bottom.. The steps start out fine and then you get to the older steps and they are a bit trickier there is lookouts while you walk down and the views are amazing …	waterfall	0.93

ID	Title	Review	best_topic	best_score
……				
82	Arenal Hanging Bridges	We spent about 2 h visiting the park. Fortunately, at the end of the tour, we saw several monkeys and a snake, but previous to this, there were no animals to be seen…	wildlife	0.87
83	Mixed Emotions but still very nice…	We were lucky as the sky was clear and we were able to enjoy beautiful views of the volcano. There are two zones to visit, Peninsula Zone and Volcano Zone. The ticket is USD 15 per foreigner and you receive Nothing in exchange, just a piece of paper with a mini map…	volcano	0.90
84	Amazing Place	Although this is a real tourist spot, it is amazing. Although there were a lot of people here, it did not seem crowded. We were in pools where there were less than a dozen others. We went in the evening (about 7 pm) so many may have already left. The variety of water temperatures and all the swim up bars were fun. I also went on the fast water slide. Would not recommend this unless you are very experiences. It reaches a speed of 45kph and is all enclosed. The slide is short but insane! Try the others first	hotel	0.24
…				

Notice some misclassified reviews. One example is review #84, which is classified as a hotel, while, in fact, the review is about the hot springs. We can improve results by fine-tuning the model. For that, we can provide examples of correct review classifications.

Part 3. Clustering

In the previous part, we classified destination reviews into pre-existing classes. In this part, we will search for topic clusters based on review semantic similarity using the BERTopic model.

Step 1. Initial Code

USER:

The Excel file "▶ https://github.com/andykir/DataMiningLectures/raw/refs/heads/main/ArenalReviews.xlsx" contains reviews of a travel destination. The reviews are in the sheet "reviews" in the column "Review". Your goal is to write a Python code to cluster reviews into topics. Use the BERTopic model. Save results into the file "clustered_topics.xlsx".

ChatGPT:

The BERTopic model is well-suited for clustering text data into topics, as it uses embeddings to capture the semantic meaning of each review and groups similar reviews together into coherent clusters. To install BERTopic, make sure you have it installed by running pip install bertopic:

```
!pip install bertopic
```

The code:

```
import pandas as pd
from bertopic import BERTopic

# Load the Excel file and specific sheet
file_path =
"https://github.com/andykir/DataMiningLectures/raw/refs/heads/main/ArenalReviews.xlsx"
sheet_name = "reviews"
df = pd.read_excel(file_path, sheet_name=sheet_name)

# Extract the reviews text
reviews = df["Reviews"].tolist()

# Initialize BERTopic model
topic_model = BERTopic()

# Fit the model to the reviews and obtain topics
topics, probabilities = topic_model.fit_transform(reviews)

# Add the clustered topics to the DataFrame
df["topic"] = topics

# Get topic information
topic_info = topic_model.get_topic_info()
topics_overview = topic_model.get_topics()

# Save the results to a new Excel file
output_path = "clustered_topics.xlsx"
with pd.ExcelWriter(output_path) as writer:
    df.to_excel(writer, index=False, sheet_name="Clustered Reviews")
    topic_info.to_excel(writer, index=False, sheet_name="Topic Info")

print(f"Topic clustering completed and saved to {output_path}.")
```


Explanation:

```
topic_model = BERTopic()
topics, probabilities = topic_model.fit_transform(reviews)
```

This code initializes *BERTopic* without additional parameters. For the allowed parameters, see ▶ https://maartengr.github.io/BERTopic/getting_started/parameter%20tuning/parametertuning.html

The function *fit_transform* generates a topic cluster for each review. The results are stored in variables *topics* (topic assignment for each review) and *probabilities* (confidence scores for topic assignment).

```
dtopic_info = topic_model.get_topic_info()
topics_overview = topic_model.get_topics()
```

get_topic_info() retrieves a summary of all topics generated by BERTopic, and *get_topics*() provides a more detailed view of each topic's top words.

Finally, the results are saved to an Excel file.

Results (from the Excel file):
Clustered reviews

ID	Title	Review	topic
1	Be ready to hike	The hike going down is okay with no problem. It gets crowded specially in the morning with all the tour bus from san jose…	0
2	Amazing falls …	The falls are pretty easy to get to you park your car and then walk down to the bottom.. The steps start out fine and then you get to the older steps and they are a bit trickier there is lookouts while you walk down and the views are amazing …	0
…			
7	A lot of fun, beautiful views!	These were a lot of fun! Some of the bridges are very high, so if you don't care for heights then I would pass on it, my husband had much less fun than me and my daughter for this reason:) …	2
…			

Topic info

Topic	Count	Name	Representation	Representative Docs
-1	757	-1_lake_volcano_the_we	['lake', 'volcano', 'the', 'we', 'and', 'of', 'to', 'arenal', 'on', 'was']	One of the highlights of our Costa Rican trip. But do not be too disappointed if you never see the Volcano …
0	3854	0_waterfall_the_down_to	['waterfall', 'the', 'down', 'to', 'it', 'is', 'steps', 'you', 'but', 'and']	It is a beautiful waterfall, but the hike to get to it is very tough. Not an easy hike at all. Going down is one thing, coming back up …

Topic	Count	Name	Representation	Representative Docs
1	2039	1_hot_pools_springs_and	['hot', 'pools', 'springs', 'and', 'the', 'was', 'we', 'dinner', 'were', 'in']	Such a great place to go. The hot springs are in the resort and there are many pools and a small water park...
...				

There are 34 topics numbered from 0 to 33. Topic #-1 represents outliers, that is, the documents that failed to create a large enough cluster.

Step 2. Code Improvement

Read topic representations in the Topic info tab. Notice that the words clustering in topics include common stopwords ("a", "and", and so on). In addition, the word "Costa Rica" is not very useful in topic differentiation. The stopwords not only interfere with topic understanding but also result in an unnecessarily large number of topics (33). We will modify the code to include stop word filtering.

USER:

Modify the code. Remove stopwords. Add "Costa Rica" to the list of stopwords.

ChatGPT:

```
import pandas as pd
from bertopic import BERTopic
import nltk
from nltk.corpus import stopwords
import re

# Download stopwords
nltk.download("stopwords")
stop_words = set(stopwords.words("english"))
# Add custom stopwords
stop_words.update(["costa", "rica"])

# Load the Excel file and specific sheet
file_path =
"https://github.com/andykir/DataMiningLectures/raw/refs/heads/main/ArenalReviews.xlsx"
sheet_name = "reviews"
df = pd.read_excel(file_path, sheet_name=sheet_name)

# Preprocess the reviews to remove stopwords
def preprocess_text(text):
    # Remove non-alphabetic characters and convert to lowercase
    text = re.sub(r"[^a-zA-Z\s]", "", text).lower()
    # Remove stopwords
    filtered_words = [word for word in text.split() if word not in stop_words]
    return " ".join(filtered_words)

# Apply preprocessing to the reviews
df["processed_reviews"] = df["Reviews"].apply(preprocess_text)

# Extract the preprocessed reviews text
processed_reviews = df["processed_reviews"].tolist()

# Initialize BERTopic model
topic_model = BERTopic()

# Fit the model to the processed reviews and obtain topics
topics, probabilities = topic_model.fit_transform(processed_reviews)

# Add the clustered topics to the DataFrame
df["topic"] = topics

# Get topic information
topic_info = topic_model.get_topic_info()

# Save the results to a new Excel file
output_path = "clustered_topics.xlsx"
with pd.ExcelWriter(output_path) as writer:
    df.to_excel(writer, index=False, sheet_name="Clustered Reviews")
    topic_info.to_excel(writer, index=False, sheet_name="Topic Info")

print(f"Topic clustering completed and saved to {output_path}.")
```

Explanation of changes:

To modify the code, we use the *nltk* library to filter out common stopwords in English. This code downloads the English stopwords list using nltk and saves it to stop_words for filtering. It also includes custom stop words "costa" and "rica":

```
nltk.download("stopwords")
stop_words = set(stopwords.words("english"))
# Add custom stopwords
stop_words.update(["costa", "rica"])
```

The function *preprocess_text* removes non-alphabetic characters, converts text to lowercase, and removes stopwords:

```
def preprocess_text(text):
    # Remove non-alphabetic characters and convert to lowercase
    text = re.sub(r"[^a-zA-Z\s]", "", text).lower()
    # Remove stopwords
    filtered_words = [word for word in text.split() if word not in stop_words]
    return " ".join(filtered_words)
```

Then we apply the *preprocess_text* function to each review and save the results in a new column *processed_reviews*:

```
df["processed_reviews"] = df["Reviews"].apply(preprocess_text)
```

Results (from the Excel file):

Clustered reviews

ID	Title	Review	topic
1	Be ready to hike	The hike going down is okay with no problem. It gets crowded specially in the morning with all the tour bus from san jose…	0
2	Amazing falls …	The falls are pretty easy to get to you park your car and then walk down to the bottom.. The steps start out fine and then you get to the older steps and they are a bit trickier there is lookouts while you walk down and the views are amazing …	0
…			
9	So Beautiful	This hot springs was not populated by many people on the day we went. In fact the eight or ten of us ended up being the only ones there. It was very tropical and peaceful…	1
…			

13

Topic info

Topic	Count	Name	Representation	Representative Docs
-1	640	-1_lake_arenal_volcano_views	'lake', 'arenal', 'volcano', 'views', 'boat', 'around', 'beautiful', 'see', 'drive', 'view'	lake arenal man made would never guess drove around lake arenal monteverde pleasant beautiful drive stunning views …
0	4089	0_waterfall_steps_water_worth	'waterfall', 'steps', 'water', 'worth', 'swim', 'back', 'hike', 'stairs', 'beautiful', 'falls'	waterfall one popular waterfalls hike waterfall little tough stairs worth waterfall powerful beautiful swim bottom waterfall river bit look around fish wildlife …
1	2134	1_hot_springs_pools_dinner	'hot', 'springs', 'pools', 'dinner', 'place', 'buffet', 'relaxing', 'food', 'different', 'day'	liked place lot first facilities nice clean lockers showers changing rooms staff nice friendly hot springs nice plenty medium heat water cold pools would preferred hotter water …
…				

Notice that the number of topics is reduced: there are only 18 topics numbered from 0 to 17. Recall that topic # -1 is reserved for outliers.

Step 3. Visualization of the Topic Distance.

The topics can be visualized based on the semantic distance between the topic word representation.

USER:

Write the code to visualize the obtained topics based on cluster distance.

ChatGPT:

```
# Visualize topics based on cluster distance
topic_model.visualize_topics().show()

# Visualize a heatmap of the topic's similarity matrix.
topic_model.visualize_heatmap().show()
```

Explanation:

To visualize the obtained topics based on cluster distance using BERTopic, we use its built-in method *visualize_topics*(). This method creates an interactive visualization of topics and their distances based on embeddings. The visualization shows how closely related different topics are to one another, helping to find overarching themes.

Topic distance visualisation:

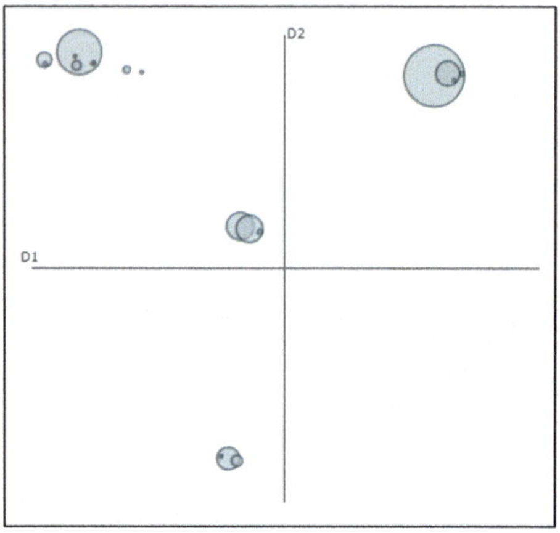

Each circle on this plot corresponds to a separate topic, with the circle size corresponding to the number of documents loading to this topic. Hover the cursor over the topics; notice that the topics that are spatially close are close semantically. We can combine these topics into common themes:

Topic #	Theme
1	Hot springs and dinner
6, 8, 12, 13, 17	Lake
9, 14	Kayaking and fishing on the lake
0, 4, 11, 15	Waterfall La Fortuna and waterfall hike
2, 3, 10	Tours to see animals and hanging bridges
5, 7, 16	Volcano Arenal

We can also use these topic names and representative documents to classify reviews as described in Part 2. Finally, we can change the default parameters in the BERTopic model to accommodate our newly found knowledge on the topics.

Step 4. Fine-Tune the Number and Size of the Topics.

To fine-tune the model, we can use information on hyperparameter tuning found in BERTopic documentation:

▶ https://maartengr.github.io/BERTopic/getting_started/parameter%20tuning/parametertuning.html

Specifically,

- parameter *min_topic_size* defines the minimum number of documents in a topic;
- parameter *min_topic_size* specifies the desirable number of topics.

Code modification

Replace the lines.

```
# Initialize BERTopic model
topic_model = BERTopic()
```

with

```
# Initialize BERTopic model
topic_model = BERTopic(min_topic_size=50)
```

Run the code.

Result:

Topic	Count	Name	Representation	Representative_Docs
0	4112	0_waterfall_steps_water_worth	'waterfall', 'steps', 'water', 'worth', 'swim', 'back', 'beautiful', 'hike', 'get', 'stairs'	great place quite hike well worth waterfall amazing water clean refreshing get little crowded great time …
1	2524	1_lake_volcano_arenal_bridges	'lake', 'volcano', 'arenal', 'bridges', 'see', 'views', 'guide', 'beautiful', 'great', 'tour'	friend took taxiboattaxi la fortuna monteverde booked desafio adventures part tour involves boat ride across lake arenal beautiful offered one best views arenal …
2	2137	2_hot_pools_springs_dinner	'hot', 'pools', 'springs', 'dinner', 'place', 'buffet', 'relaxing', 'great', 'day', 'food'	absolutely loved hot springs tranquil beautiful well cared gorgeous natural feeling settings less touristy beware drinks pricey dinner inclusive ok buffet style go hot springs delicious …

Topic	Count	Name	Representation	Representative_Docs
3	798	3_animals_tour_carlos_guide	'animals', 'tour', 'carlos', 'guide', 'feed', 'great', 'volunteer', 'experience', 'monkeys', 'asis'	took two boys animal rehabilitation center one highlights trip learn many amazing animals learned efforts making teach people taking wild animals pets got prepare food …
4	671	4_fortuna_la_waterfall_hike	'fortuna', 'la', 'waterfall', 'hike', 'worth', 'water', 'back', 'falls', 'steps', 'beautiful'	enjoyed visit la fortuna waterfall many steps deal worth especially went swim …

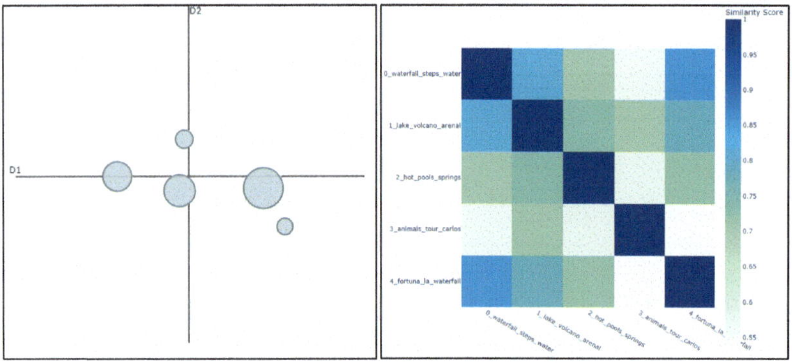

The modified code creates five large clusters of documents:

# cluster	Description
0	Waterfall walking many steps and swimming
1	Tours: lake, volcano, and hanging bridges
2	Hot springs and dinner on-site
3	Animal watching tours
4	La Fortuna waterfall hiking

The heatmap suggests that two clusters (#0 and #4) can be combined into one theme, "waterfall".

Part 4. On Your Own

1. Use the Part 1 semantic analysis example to predict the **emotions** expressed by visitors to the Volcano Arenal National Park.
2. Use Part 2 classification analysis to classify the documents into the topics "Waterfall hiking and swimming", "Tours to the lake, volcano, and hanging bridges", "Hot springs and dinner on-site", and "Animal watching tours"

Build Your Python Skills

Chapter 23 of the free book Learn Python with Jupyter by Serena Bonaretti. ► https://www.learnpythonwithjupyter.com/

Additional Reading

Atkinson-Abutridy, J. (2024). *Large Language Models: Concepts*. CRC Press.

Ekin, S. (2023). Prompt engineering for ChatGPT: a quick guide to techniques, tips, and best practices. Authorea Preprints.

Grootendorst, M. (2022). BERTopic: Neural topic modeling with a class-based TF-IDF procedure. arXiv preprint ► arXiv:2203.05794.

White, J., Fu, Q., Hays, S., Sandborn, M., Olea, C., Gilbert, H., ... & Schmidt, D. C. (2023). A prompt pattern catalog to enhance prompt engineering with ChatGPT. arXiv preprint ► arXiv:2302.11382.

14

Introduction to Social Network Analysis

A. P. Kirilenko, *Practical Data Mining with AI for Social Scientists*,
Springer Texts in Social Sciences, https://doi.org/10.1007/978-3-031-89689-7_14

14.1 Introduction

So far, this book has explored data patterns in records not directly connected to one another. In such datasets, individual records exist independently, and we had to apply a variety of data mining methods to infer connections between them. For instance, in ▶ Chapter 6, we used hospitality industry data to cluster Florida hotels into groups according to their pricing strategy and other characteristics. Notice that data instances in this analysis had no specific fields tying one record to another. Rather, the hotel connection was inferred based on the observed similarities in their features.

Meanwhile, some types of data do have directly connected records. For instance, Facebook users are connected through their friendship links and shared group memberships. Tweets are connected through retweeting links. Scientific writers are connected through joint authorships. Hotels are connected through customer stays during multi-stay trips. Downtown attractions are connected through tourist co-visitations. All these examples have explicit (rather than inferred) links defining connections between dataset records.

Such interlinked datasets can be reimagined as a network of entities, with each record represented by a network *node* and a link between two records represented by a connector (*edge*) between the respective nodes. An analysis of such a network may provide useful insights into the nature of data. For instance, consider a news retweeting network with nodes representing individual users and edges representing message reposting. In this network, densely connected parts may indicate online echo chambers of likely minded users,[1] various *centrality* measures detect influential actors, and the average length of the path connecting different network nodes indicates the efficiency of information flow.

For another example, the graph in ◘ Fig. 14.1 represents a collaborative network of scientists publishing in three leading tourism journals: Annals of Tourism Re-

1 Williams, H. T., McMurray, J. R., Kurz, T., & Lambert, F. H. (2015). Network analysis reveals open forums and echo chambers in social media discussions of climate change. *Global Environmental Change, 32*, 126–138.

search, Tourism Management, and Journal of Travel Research. Each individual point (node) on the graph stands for one scientist and the lines (edges) connecting the nodes mean that these researchers published their results together. The length of the edges in this particular visualization represents the strength of the connection between the nodes with shorter edges connecting authors publishing more frequently within rather than outside their group. Consider the zoomed-in area in ◘ Fig. 14.1, which shows a group of nodes joined in a small network segment. Three authors in the left part of the graph had one joint publication, which included an extra author depicted by a point in the lower right. The edges connect this extra author with another group in the lower right and a single scientist at the top of the plot. We can speculate that this *central* point represents an influential scientist connecting to other investigators in a joint research network.

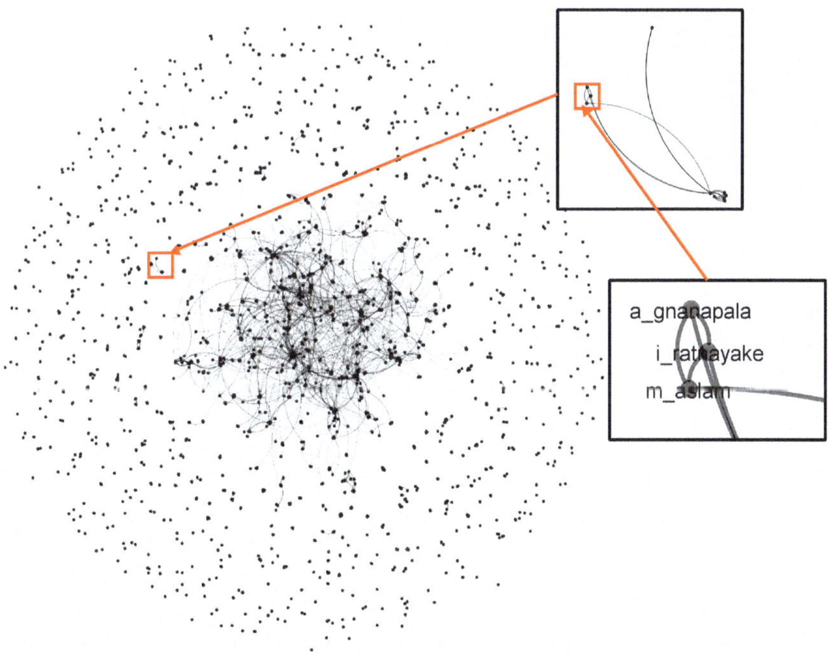

◘ **Fig. 14.1** A network of collaborators in tourism research. The nodes represent individual researchers, and the edges connect researchers with joint publications

Meanwhile, at the center of the plot is a large subnetwork, which consists of authors whose primary research area is travel and tourism. Closer observation reveals several highly interconnected segments representing distinct formal or informal research groups. These groups are linked by relatively few edges, often connecting highly productive and influential scientists who play a pivotal role in advancing the discipline as a whole.

This brief informal examination demonstrates the potential of network analysis methods to provide important insights into the tourism scientific community. Social network analysis comprises analysis and visualization methods that help to understand the patterns and influential actors in a social structure represented by an explicitly connected dataset. This is an interdisciplinary, complex area of study originating in sociology and graph theory. In this chapter, we will touch on this field's basic terminology and methods.

14.2 Terminology and Theoretical Foundations

Key concepts

Node
Edge
Undirected and directed graphs
Path
Connected and disconnected networks
Weighted network
Clique
Bridge
Articulation point
Network size

14.2.1 Key Concepts

The main elements of a network are nodes, also known as vertices. Frequently, the nodes represent individuals (scientists, twitterers, pupils, congressmen, etc.), but they can also represent other entities: companies, hotels, destinations, and so on. The number of vertices defines the size of the network. The publishing network visualized in ◘ Fig. 14.1 includes information on 5048 authors and hence has a size of 5048.

The nodes are connected by edges, also called links. The number of links represents the number of interactions between the entities included in the network. Mathematically, a network of nodes connected by edges is called a *graph*. The graphs can be directed or undirected. The tourism publishing network was represented by an *undirected* graph: if scientist A published with scientist B, then scientist B also published with scientist A; the relationship between the nodes is reciprocal. Other examples of reciprocal relationships creating undirected networks are Facebook friendship, two-way roads connecting populated places, and matches between competing teams. The publishing network contained 8800 undirected edges.

The edges form a *path* through a network. For instance, a "six-degree separated world" hypothesis formulated by psychologist Stanley Milgram states that any two persons are related through a chain of five intermediate social connections at most. In other words, in a social relations ("friendship") network that includes all people, the *shortest path* connecting any two nodes is five or shorter. A network with a path between any two nodes is called *connected.* Contrasting, the publication network is *disconnected* as it includes multiple groups publishing in isolation.

One extreme case is a *fully connected network* in which all nodes are connected to one another. For instance, the graph of phone owners is fully connected: each can directly call another person. Such a network is called a clique. Other examples of a clique include club members, sports teams playing against one another in a round-robin (all-play-all) tournament, and music bands.

Another extreme case is a connected network that would disconnect if a single edge or node were removed. Such an edge that is critical for network connectivity is called a *bridge*, and the "critical" node is called an *articulation point*. For instance, a connection between a business and Congress is frequently made through a single node representing a lobbyist. In a graph formed by the business-congress relationships, the lobbyist is represented by an articulation point.

In the collaboration network example, an edge between two nodes denotes the co-authorship relation between the authors; in this respect, all edges had the same strength. However, it would make sense to give a higher relationship strength to authors frequently co-publishing. The networks that assign weights to the edges to represent the link strength are called *weighted networks*.

14.2.2 Bipartite Networks

The publishing network visualized in ◘ Fig. 14.1 is formed by one type of node: the authors. Meanwhile, relationships can include several node types. Consider a network formed by tourists visiting points of interest (POIs) in a city. This network has two types of nodes: tourists and POIs. The edge on the network represents a tourist visiting an attraction; hence, edges connect tourists with attractions, but not tourists with tourists and not attractions with attractions. This is an ex-

St. Augustine, FL

Key concepts

Unipartite network
Bipartite network
Network projection
Ego (neighborhood) network
Loop

ample of a *bipartite* network. ◘ Figure 14.2(A) shows a bipartite network of tourists and POIs in St. Augustine, FL. St. Augustine is the oldest city in the continental United States; its major attraction is an old Spanish fortress, Castillo de San Marcos National Monument. There are networks formed by an even greater number of distinct disconnected types of nodes. For instance, behavioral ecology may study a tripartite network of herbivores, plants, and plant pollinators.

The network of nodes immediately connected to a selected node of interest is called the "*neighborhood*" or "*ego network*". For instance, ◘ Fig. 14.2B visualizes the neighborhood of one POI node, Castillo de San Marcos National Monument, showing all tourists who wrote reviews about visiting it. Similarly, ◘ Fig. 14.2B visualizes the neighborhood of one tourist and shows all the attractions the tourist visited.

14.2.3 Network Projection

The bipartite network shown in ◘ Fig. 14.2 is complex, preventing visual understanding of the relationship pattern. A frequent method of dealing with this issue is *net-*

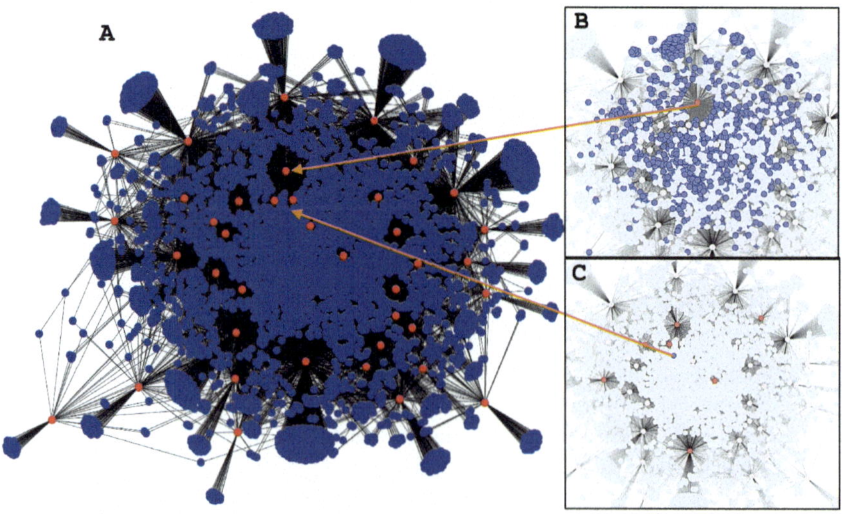

◘ **Fig. 14.2** **A** Bipartite network of tourists (blue dots) and POIs (red dots) in St. Augustine, FL. For clarity, only the major POIs reviewed at least 50 times are included. **B** The neighborhood of one POI node, Castillo de San Marcos, shows tourists visiting and reviewing it. **C** The neighborhood of one tourist showing all attractions the tourist visited and reviewed

work projection. Specifically, a bipartite network can be projected to either set of nodes by connecting two nodes of the same type that share a bipartite connection. For instance, the tourist-POI network can be projected to POIs by connecting two POIs visited by the same tourist. This may greatly simplify network analysis when the primary interest is in the POI co-visitation relationship rather than individual visitors' preferences.

◘ Figure 14.3A displays the *projected* POI network. Distinctly from ◘ Fig. 14.2, this plot focuses solely on the relationships between POIs. Notice the *loops* formed by multiple attraction reviews by the same visitors. To better discriminate between the weaker- and stronger-connected POIs, in ◘ Fig. 14.3B, the links denoting fewer than 50 attraction co-visitations are removed. The POIs in the center of this plot are the most tightly connected ones, co-visited by many tourists. That includes the city's most important attractions: Castillo de San Marcos and the historical St. George Street. The outer nodes are less densely connected with outer attractions. For instance, the Tini Martini bar or a wildlife reserve caters to more niche categories of visitors.

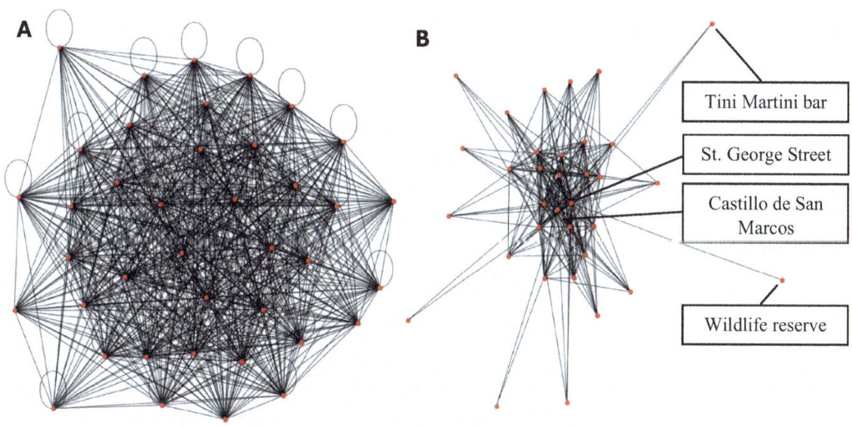

◘ **Fig. 14.3** **A** A unipartite network of POIs in St. Augustine, FL. The nodes represent POIs, and the edges represent reviews of two different POIs written by the same tourist. **B** Simplified network in which the edges representing fewer than 20 POI co-visitations are assigned zero weight

14.3 Measures

Key concepts

Node degree
In- and out-degree
Average degree

14.3.1 Node Degree

The connectivity of a node is represented by its number of edges, called *node degree*. The total number of edges in a network is half the sum of all node degrees, which accounts for counting each edge twice. Note that for a directed network, there are two types of node degrees: the *in-degree* is the sum of edges pointing toward the node, and the *out-degree* points away from the node.

Several important measures related to the statistics of node degrees characterize a network's connectivity as a whole. The mean number of node degrees in a network, called an average degree, defines how densely the network is connected. Note that directed networks have two types of average degrees: the average *in-degree* and the average *out-degree*.

For instance, the scientific publishing network (Fig. 14.1) has an average degree of 3.75. This denotes the mean number of co-authors for a scientist publishing in a high-profile tourism journal. Considering that the theoretical maximum average degree for a network with this number of nodes is 4692, the scientific publishing network is sparsely connected. In comparison, the average degree of the St. Augustine POI unipartite network is 35.6. Considering that the total number of POIs in the network is 39, this is a nearly complete graph: almost every node is directly connected to every other node. The St. Augustine POI network is densely connected.

14.3.2 Degree Distribution

The average node degree is not very informative when a network has few high-degree nodes, as they greatly influence this measure. A better understanding can be derived from the *degree distribution,* which describes how the degrees are distributed across all nodes in the network. The shape of the degree distribution plot provides insights into network properties. Figure 14.4 illustrates the concept with the degree distribution for the filtered POI network illustrated in Fig. 14.4B. The column heights represent the fraction of nodes with the number of connections to other nodes, as indicated on the horizontal axis.

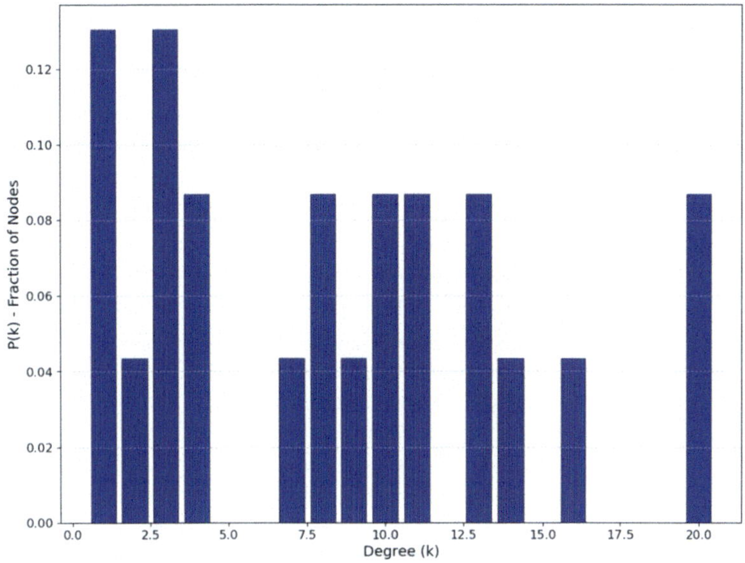

Fig. 14.4 Degree distribution of the POIs network (Fig. 14.3B)

For instance, the right column represents two nodes (0.086 * 23 nodes = 2) with degree k = 20. Those nodes represent the most popular attractions: St. George Street and Castillo de San Marcos.

14.3.3 Centrality Measures

Centrality measures

Degree centrality
Eigenvector centrality
Betweenness centrality
Closeness centrality

Centrality measures estimate the importance of nodes within a network. The term "importance" is ambiguous, which justifies the existence of multiple centrality measures. This section lists the most frequently used ones.

The *degree centrality* is closely related to the degree measure and is calculated as the node degree divided by the number of nodes minus one. The degree centrality values are distributed between 0 (for isolated nodes) and 1 (for fully connected nodes). A node with the highest degree of centrality identifies an entity most connected to other entities. For instance, St. George Street and Castillo de San Marcos have the highest centrality (0.91): these attractions are connected to most other POIs by co-visitations.

The *eigenvector centrality* measures node importance based on the number of its connections and the number of connections of its neighbors. In other words, it considers not only the node degree but also the degree of the nodes it is connected to, assigning higher scores to nodes with many connections to other high-centrality nodes. The centrally located attraction of St. Augustine, St. George Street, has the highest eigenvector centrality (0.33). A very similar measure, Katz Centrality, considers not only direct but also indirect connections.

The *betweenness centrality* measures how often a node lies on the shortest path between two other nodes. Nodes with high betweenness centrality act as "bridges" connecting different network parts. Castillo de San Marcos has the highest betweenness centrality of 0.29.

Closeness centrality measures how close a node is to all other nodes in the network. It measures the distance from a node to all other nodes by considering the shortest paths. If the graph is disconnected, closeness centrality is calculated only for the nodes within the same connected part of the network. In a social network, the news travels through people with high closeness centrality.

The discussed measures prioritize various aspects of node importance; as a result, the node's importance depends on the measure used. For instance, in the co-authorship network, the author with the highest eigenvector centrality is not among the top ten collaborators in terms of degree, betweenness, or closeness centrality.

14.4 Network-Level Metrics

Key concepts

Network density: relative number of edges
Homophily: a measure of "correlation" between connected edges
Modularity: the tendency of a network to partition into distinct communities

Casual observation tells us that the co-authoring network is less cohesive and more sparse than the POI co-visitation network. Network-level measures quantify these differences. The most popular network-level measures are density and modularity.

14.4.1 Density

Density measures the overall connectivity of network nodes. Specifically, it measures the proportion of the actual number of edges in a network E compared to the maximum possible number of edges N:

$$\text{density} = E/N$$

Note that the maximum number of edges for an undirected network with n nodes is

$$N = \frac{n(n-1)}{2}$$

Density is related to the average degree measure: if nodes have a higher average degree, it means they are connected to more nodes, leading to a denser network. Yet, while the average degree gives a general understanding of how connected the network is, the density directly measures it. For instance, the density of the filtered St. Augustine POI network shown in ▣ Fig. 14.3B is 0.38, while the density of the publishing co-authorship network in ▣ Fig. 14.1 is only 0.0008.

14.4.2 Homophily

The *homophily* property is informally called "birds of a feather flock together": it measures the tendency of connected individuals to have similar characteristics. For instance, in a unipartite network of friends, a person tends to relate to someone of the same gender. In an attraction co-visitation network, a POI tends to be connected to POIs of the same type reflecting differences in visitors' interests such as culture, nightlife, or sea and beach.

There are several useful measures of homophily. The *EI Homophily* (External-Internal Homophily) is a simple metric that can be applied to a categorical feature of a node. The EI index calculates homophily based on the number of edges connecting nodes with identical features (internal links, denoted as I) and the number of edges connecting nodes with different features (external links, denoted as E). For example, a link between two male friends would count as an internal link and contribute to the value of I, while a link between a male and a female friend would count as an external link and contribute to the value of E. Then, the homophily would be calculated as

$$EI = \frac{E - I}{E + I}$$

with negative values reflecting a tendency to associate with the same and positive values reflecting a tendency to associate with a different gender. EI homophily close to −1 suggests that nodes connect predominantly with others who share the same characteristics. In contrast, EI homophily close to 1 is observed when the connected nodes tend to have different characteristics.

The *assortativity coefficient* can be used to measure homophily based on the interval or ratio features of a node. This metric is similar to Pearson's correlation. For instance, in the co-authorship network authors, there is a weak tendency of homophily between authors with the same centrality degree: assortativity = 0.26. That is, well-connected authors tend to publish together with other well-connected authors rather than with weakly-connected authors.

14.4.3 Modularity and Community Detection

Modularity Q interpretation

Interpretation of modularity values depend on the context, but values above 0.4 are typically interpreted as strong and values between 0.2 and 0.4 as weak modularity.

Observe the co-authorship network in Fig. 14.1. The nodes on the network tend to form "clumps" or "*communities*" of densely connected authors. These communities define groups of nodes that are densely connected internally but sparsely connected externally. Community detection is a set of algorithms that target discovering communities in a network.

The tendency of a network to partition into communities is called *modularity*. In a highly modular network, most edges connect nodes within the same community, while relatively few edges connect nodes across different communities. One widely used modularity measure is the *modularity Q*, which measures the difference between the *observed* number of edges within communities and the *expected* number of edges if the network is random. Specifically, a "random" network would have the same number of nodes and edges but with edges distributed randomly among the nodes. A higher value of Q indicates a more modular network. The modularity of the co-authorship network is Q = 0.90, indicating a very high tendency of authors to self-organize into tightly knit groups. Compared to that, the modularity of the co-visitation network is only Q = 0.05, indicating an absence of strongly defined communities.

The communities can still be visualized, as illustrated by the St. Augustine attraction network (◘ Fig. 14.5). One may interpret one of these communities as representing the "traditional" tourist attractions, such as the overcrowded with tourists St. George Street and Castillo de St. Marcos, and the other one as "local landmarks": a winery, a distillery, and the Fountain of Youth historical park related to early Spanish search for a magic rejuvenating spring. The observed node clustering into communities is likely driven by business model differences between two tour companies operating in the city: the Red Train Tours, which provides a continuous narrated tour, and the Old Town Tours, which offers hop-on, hop-off flexibility.

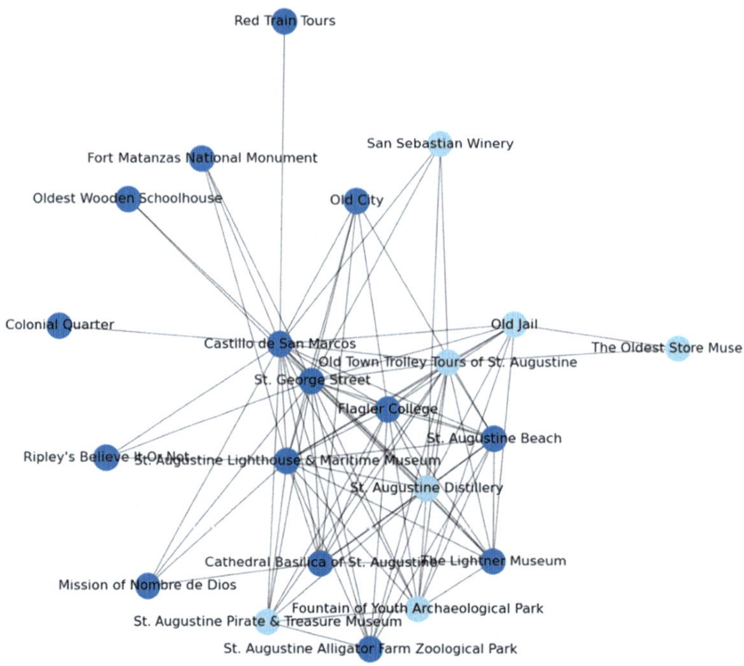

◘ **Fig. 14.5** Two communities of related POIs in St. Augustine, FL

14.5 Data Format

A simple way of network representation is through an *adjacency matrix*. An adjacency matrix is a square matrix with each row and column corresponding to a specific node in the graph. The element at the intersection of row i and column j indicates the presence or absence of an edge between nodes i and j. In its simplest form, the presence of a link between two nodes is coded with 1 and the absence with 0. In a weighted network, matrix elements represent edge weights.

The zoomed-in part of ◻ Fig. 14.1 represents one group of co-authors. For clarity, we re-draw this group in ◻ Fig. 14.6. The respective adjacency matrix (◻ Table 14.1) shows the edges between the nodes as 1 in respective cells. Further, we might propose that the connection strength is determined by node degree centrality: the connection between two authors with a high degree of centrality is stronger than between authors with a low degree of centrality. This would articulate a possible role of Jeremy Buultjens node as an organizer of the collaborative group. ◻ Table 14.2 shows the adjacency matrix for the respective weighted network.

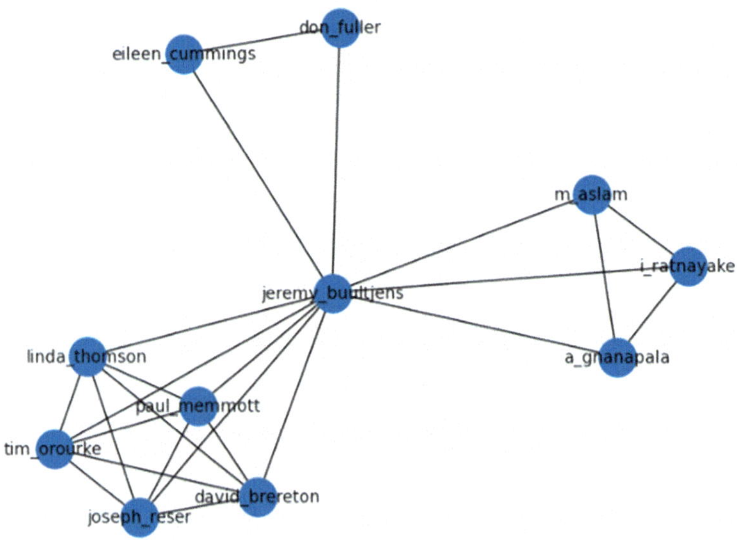

◻ **Fig. 14.6** Zoomed-in part of the co-authorship network (◻ Fig. 14.1)

Note that the adjacency matrices these two tables represent are symmetrical; all undirected networks share this property. Distinctly, directed networks are usually represented by matrices that are not symmetric. In addition, an adjacent matrix is not the only way of representing a network. For instance, a network can be stored as a list

Table 14.1 Adjacency matrix for the co-authorship network (Fig. 14.6)

	Gnanapala	Aslam	Reser	Memmott	Buultjens	Orourke	Thomson	Ratnayake	Brereton	Cummings	Fuller
Gnanapala	0	1	0	0	1	0	0	1	0	0	0
Aslam	1	0	0	0	1	0	0	1	0	0	0
Reser	0	0	0	1	1	1	1	0	1	0	0
Memmott	0	0	1	0	1	1	1	0	1	0	0
Buultjens	1	1	1	1	0	1	1	1	1	1	1
Orourke	0	0	1	1	1	0	1	0	1	0	0
Thomson	0	0	1	1	1	1	0	0	1	0	0
Ratnayake	1	1	0	0	1	0	0	0	0	0	0
Brereton	0	0	1	1	1	1	1	0	0	0	0
Cummings	0	0	0	0	1	0	0	0	0	0	1
Fuller	0	0	0	0	1	0	0	0	0	1	0

Table 14.2 Adjacency matrix for the weighted co-authorship network

	Gnanapala	Aslam	Reser	Memmott	Buultjens	Orourke	Thomson	Ratnayake	Brereton	Cummings	Fuller
Gnanapala	0	0.3	0	0	0.65	0	0	0.3	0	0	0
Aslam	0.3	0	0	0	0.65	0	0	0.3	0	0	0
Reser	0	0	0	0.5	0.75	0.5	0.5	0	0.5	0	0
Memmott	0	0	0.5	0	0.75	0.5	0.5	0	0.5	0	0
Buultjens	0.65	0.65	0.75	0.75	0	0.75	0.75	0.65	0.75	0.6	0.6
Orourke	0	0	0.5	0.5	0.75	0	0.5	0	0.5	0	0
Thomson	0	0	0.5	0.5	0.75	0.5	0	0	0.5	0	0
Ratnayake	0.3	0.3	0	0	0.65	0	0	0	0	0	0
Brereton	0	0	0.5	0.5	0.75	0.5	0.5	0	0	0	0
Cummings	0	0	0	0	0.6	0	0	0	0	0	0.2
Fuller	0	0	0	0	0.6	0	0	0	0	0.2	0

of edges or a list of neighbors for each node. Still, the adjacency matrix has a significant advantage in that it can be directly utilized in mathematical operations.

14.6 Software

Multiple software packages help network analysis. This section lists a few popular applications with a relatively gentle learning curve.

Gephi network visualization example

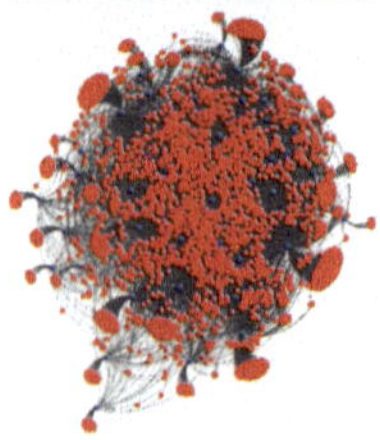

Gephi is a free software package for network visualization and exploratory analysis. Gephi has a graphical user interface (GUI) that allows import, manipulation, and filter networks. It also provides multiple tools for statistical analysis. The network visualization tools are especially impressive. Finally, Gephi is open-source, providing additional benefits in numerous add-ons for deeper analysis. ▶ https://gephi.org/.

Gephi-light is a web-based version of Gephi with a simplified interface fewer analysis options, but easier to use. One can simply drag and drop a file in a supported format to visualize the network. ▶ https://gephi.org/gephi-lite/.

NodeXL visualization example

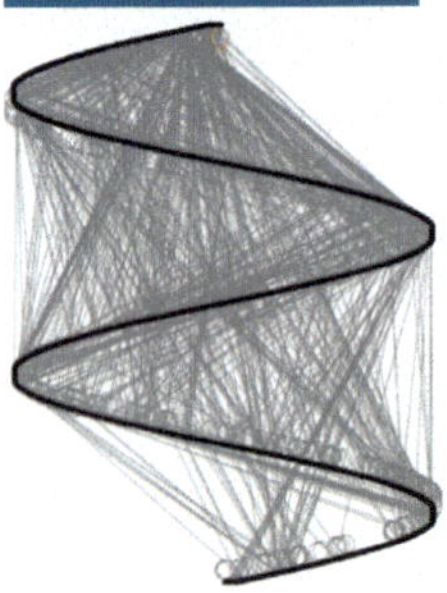

NodeXL is designed as a user-friendly software targeted at social scientists. The tool is integrated as an add-on for Microsoft Excel, which benefits new users by providing a familiar spreadsheet interface. NodeXL permits easy importing, managing, and analyzing network data without requiring programming skills. The basic version of NodeXL is free but quite limited in its analytic capabilities, while the Pro version is paid. Notably, besides the social network analysis NodeXL Pro is capable of text and time series analytics. ▶ https://www.smrfoundation.org/nodexl/.

14.7 Conclusion

This chapter provides only a very basic introduction to social network analysis. Many important topics are omitted in favor of explaining basic concepts. One promising direction for learning is applications of hypothesis testing in network analysis. Hypothesis testing involves evaluating specific propositions about the network properties or structure. This inferential analysis allows a researcher to move beyond descriptive analysis, providing a framework to validate observations about network data statistically.

For instance, this chapter's lab includes community detection. An inferential analysis would allow validation of the identified communities. By comparing the modularity score or other clustering metrics to those in randomized networks, an analyst would understand whether observed communities are statistically significant or merely artifacts of random variation.

Another important area of further learning is dynamic networks. In dynamic networks, the nodes and edges are time-dependent, replicating the evolution of inter-personal or inter-object relationships in the real world. For instance, dynamic networks would permit analysis of retweet cascades following a news event. Dynamic networks allow not only descriptive but also inferential analysis. For example, an analyst might test whether adding new edges follows a preferential attachment mechanism or if the growth rate of a network aligns with theoretical models.

Finally, the potential applications of social network analysis methods go well beyond the social sciences. Its scientific and practical applications include biology, marketing, urban planning, and many other areas.

14.8 Review Questions

1. Describe the key components of a network.
2. What is the difference between the undirected and directed graphs? Provide real-world examples of each.
3. Explain the concept of a "path" in a network. How does it relate to the "six degrees of separation" hypothesis?
4. Define the terms "clique," "bridge," and "articulation point".
5. What are weighted networks, and how do they differ from unweighted networks? Provide a real-world example.
6. What is a bipartite network, and how can it be projected to a unipartite network?
7. Explain the concept of degree centrality. How does it help in identifying influential nodes?
8. Compare eigenvector centrality, betweenness centrality, and closeness centrality. Provide examples of where each measure would be useful.
9. How is network density calculated? What does it indicate about the network?
10. What are network modularity and a community?
11. Be ready to use an adjacency matrix to draw a network.

14.9 Lab

Objectives

1. Build a network of co-authors in leading tourism journals
2. Practice network visualization methods
3. Apply network analysis methods

Data

The data is a list of articles and their authors published in 2000–2018 in three leading tourism journals: Annals of Tourism Research, Tourism Management, and Journal of Travel Research.[2] There are 9673 records in total.

The data is stored in an Excel file TourismJournalAuthors.xlsx, the sheet ► https://github.com/andykir/DataMiningLectures/raw/refs/heads/main/TourismJournalAuthors.xlsx.

Variables

Variable name	Explanation	Notes
doi	Digital Object Identifier (DOI) for the article	Long integer. There may be a matching number of rows with identical DOIs for articles with several authors
Author	Author's name	String in a formal firstName_familyName. When the first name is unavailable, the format is firstInitial_familyName

2 Kirilenko, A. P., & Stepchenkova, S. (2018). Tourism research from its inception to present day: Subject area, geography, and gender distributions. *PLoS One, 13*(11), e0206820.

Part 1. Analysis of Co-publication in Tourism Journals

Step 1. Load Data

USER:

Import data from an Excel file ▶ https://github.com/andykir/DataMiningLectures/raw/refs/heads/main/TourismJournalAuthors.xlsx, sheet "Authors" to Pandas DataFrame df.

Assistant:

```
import pandas as pd

# Define the URL of the Excel file
url =
"https://github.com/andykir/DataMiningLectures/raw/refs/heads/main/TourismJournalAuth
ors.xlsx"

# Read the data from the "Authors" sheet into a DataFrame
df = pd.read_excel(url, sheet_name="Authors")

# Display the first few rows of the DataFrame
print(df.head())
```

Result:

```
           doi        Author
0  78650967235    carlo_aall
1  84920115082    carlo_aall
2  84920138863    carlo_aall
3  33746360240   tham_aaron
4  78650296770 jarle_aarstad
```

Step 2. Create Nodes and Edges; Create the Co-authorship Network

USER:

In DataFrame df, column doi is the publication ID, and column Author is the author's ID. Group authors by publications, create an edge list for co-authorship and create the co-authorship network.

Assistant:

```
import networkx as nx
from itertools import combinations

# Group authors by publications
grouped = df.groupby('DOI')['Author'].apply(list).reset_index()

# Create an edge list for co-authorship
edges = []
for authors in grouped['Author']:
    # Create all combinations of authors for each publication
    edges.extend(combinations(authors, 2))

# Create a co-authorship network
G = nx.Graph()
G.add_edges_from(edges)

# Print the number of nodes and edges in the network
print(f"Number of authors (nodes): {G.number_of_nodes()}")
print(f"Number of co-authorship relationships (edges): {G.number_of_edges()}")
```

Explanation:

Grouping: The *groupby* function groups the DataFrame *df* by the *DOI* column and aggregates the *Author* column into lists.

Creating edges: The *combinations* function from the *itertools* module generates all possible pairs of authors for each publication, which are then added to the *edges* list.

Creating the network: A graph *G* is created using *networkx*, and the edges are added.

Finally, the network's number of nodes (authors) and edges (co-authorship relationships) is printed. Notice how Python's f-strings combine the text with variable values.

Note: The *itertools* library is built-in. To install the *networkx* library, run

```
!pip install networkx
```


Result:

```
Number of authors (nodes): 4693

Number of co-authorship relationships (edges): 8800
```

Step 3. Analyze the Network

USER:

For the network G, compute node degree centrality, betweenness centrality, and closeness centrality. Print the list of 5 top nodes in terms of their degree centrality, betweenness centrality, and closeness centrality. Print average degree centrality, betweenness centrality, and closeness centrality.

Assistant:

```
# Compute centrality measures
degree_centrality = nx.degree_centrality(G)
betweenness_centrality = nx.betweenness_centrality(G)
closeness_centrality = nx.closeness_centrality(G)

# Convert centrality measures to DataFrames for easier manipulation
degree_df = pd.DataFrame(degree_centrality.items(), columns=['Node', 'Degree
Centrality'])
betweenness_df = pd.DataFrame(betweenness_centrality.items(), columns=['Node',
'Betweenness Centrality'])
closeness_df = pd.DataFrame(closeness_centrality.items(), columns=['Node', 'Closeness
Centrality'])

# Get top 10 nodes for each centrality measure
top_degree = degree_df.nlargest(5, 'Degree Centrality')
top_betweenness = betweenness_df.nlargest(5, 'Betweenness Centrality')
top_closeness = closeness_df.nlargest(5, 'Closeness Centrality')

# Print top nodes
print("Top 5 Nodes by Degree Centrality:",top_degree)
print("\nTop 5 Nodes by Betweenness Centrality:",top_betweenness)
print("\nTop 5 Nodes by Closeness Centrality:",top_closeness)

# Calculate and print average centrality measures
avg_degree_centrality = sum(degree_centrality.values()) / len(degree_centrality)
avg_betweenness_centrality = sum(betweenness_centrality.values()) /
len(betweenness_centrality)
avg_closeness_centrality = sum(closeness_centrality.values()) /
len(closeness_centrality)

print(f"\nAverage Degree Centrality: {avg_degree_centrality:.4f}")
print(f"Average Betweenness Centrality: {avg_betweenness_centrality:.4f}")
print(f"Average Closeness Centrality: {avg_closeness_centrality:.4f}")
```

Result:

```
Top 5 Nodes by Degree Centrality:
323   susanne_becken   0.0155
1374    ralf_buckley   0.0155
945    geoffrey_wall   0.0147
215     david_weaver   0.0144
1574    patrick_long   0.0136

Top 5 Nodes by Betweenness Centrality:
572      haiyan_song   0.0264
197     stephen_page   0.0208
560    bob_mckercher   0.0187
1574    patrick_long   0.0160
76     muzaffer_uysal  0.0151

Top 5 Nodes by Closeness Centrality:
1574    patrick_long    0.119
1494        xiang_li    0.119
215     david_weaver    0.119
2279    robin_nunkoo    0.117
414           c_hall    0.117

Average Degree Centrality: 0.0008
Average Betweenness Centrality: 0.0002
Average Closeness Centrality: 0.0385
```

Explanation:

The authors on the high degree centrality list are influential in terms of their collaborative links and collaborate with many other authors.
Authors with high betweenness centrality act as an intermediary between different groups of authors who might not otherwise collaborate. They bridge fragmented parts of the academic network.
Authors with high closeness centrality have diverse collaborative links, quickly interact with other authors, and are efficient regarding collaborative reach across the network.

Overall, the low value of the centrality measure points to a highly fragmented network.

Step 4. Visualize the Network

USER:
Visualize network G.

Assistant:

```
import matplotlib.pyplot as plt

# Set the size of the plot
plt.figure(figsize=(12, 12))

# Draw the network
pos = nx.spring_layout(G)  # positions for all nodes
nx.draw_networkx_nodes(G, pos, node_size=300, node_color='skyblue', alpha=0.7)
nx.draw_networkx_edges(G, pos, width=1.0, alpha=0.5)
nx.draw_networkx_labels(G, pos, font_size=10)

# Set the title and display the plot
plt.title("Co-authorship Network")
plt.axis('off')  # Turn off the axis
plt.show()
```


Result:

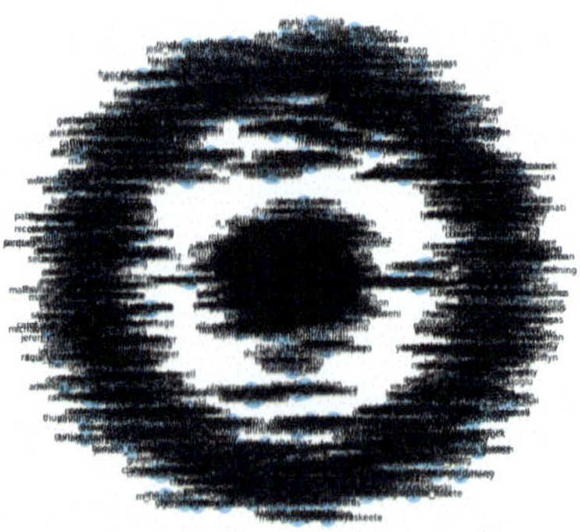

This network visualization is hard to understand since it includes the names of many hundreds of authors. The first step in improving it is to remove node labels, reduce the size of nodes, and change the colors to more contrastive ones.

Improved Code:

```
# Set the size of the plot
plt.figure(figsize=(12, 12))

# Draw the network
pos = nx.spring_layout(G)  # positions for all nodes
nx.draw_networkx_nodes(G, pos, node_size=5, node_color='red', alpha=0.7)
nx.draw_networkx_edges(G, pos, width=1.0, edge_color='black', alpha=0.5)

# Set the title and display the plot
plt.title("Co-authorship Network")
plt.axis('off')  # Turn off the axis
plt.show()
```

Result:

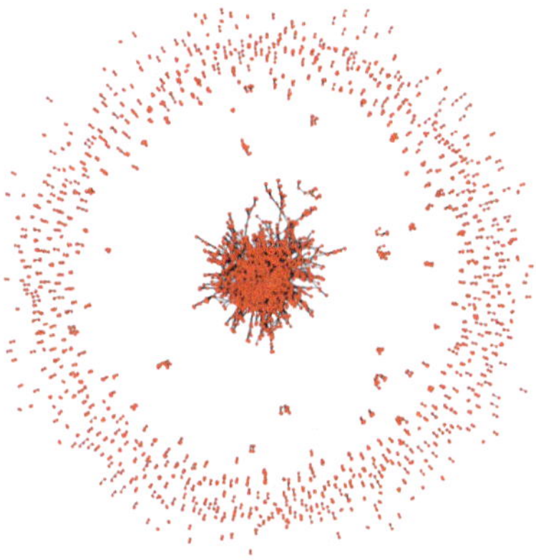

Evidently, the co-authorship network has a well-connected subnet of authors central to the field and multiple small groups with few publications in the field. It makes sense to concentrate our analysis on the central group. For that, we will extract the largest subnetwork.

Part 2. Analysis of the Core Group of Authors

Step 1. Extract the Central Part of the Network

USER:

Use the network G. Extract a list of connected components. Identify the largest component. Extract the subgraph corresponding to the largest component.

Assistant:

```
# Get the list of connected components
connected_components = list(nx.connected_components(G))

# Identify the largest component
largest_component = max(connected_components, key=len)

# Extract the subgraph corresponding to the largest component
largest_component_subgraph = G.subgraph(largest_component)

# Print the size of the largest component and the number of nodes and edges in the
subgraph
print(f"Size of the largest component: {len(largest_component)}")
print(f"Number of nodes in the largest component subgraph:
{largest_component_subgraph.number_of_nodes()}")
print(f"Number of edges in the largest component subgraph:
{largest_component_subgraph.number_of_edges()}")
```

Result:

```
Size of the largest component: 2205
Number of nodes in the largest component
subgraph: 2205

Number of edges in the largest component
subgraph: 5974
```

Explanation:

Connected Components: The *nx.connected_components(G)* function returns a generator of sets, where each set contains the nodes of a connected component. This is converted to a list for easier manipulation.

Identifying the Largest Component: The *max* function is used with the *key=len* argument to find the component with the maximum size.

Extracting the Subgraph: The subgraph method is called on *G* with the largest component to create a subgraph that includes only the nodes and edges of that component.

Step 2. Analyze and Visualize the New Network

To analyze and visualize the new network, we will reuse the code of Part 1, Step 3.

Code:

```
# Compute centrality measures
degree_centrality = nx.degree_centrality(largest_component_subgraph)
betweenness_centrality = nx.betweenness_centrality(largest_component_subgraph)
closeness_centrality = nx.closeness_centrality(largest_component_subgraph)

# Calculate and print average centrality measures
avg_degree_centrality = sum(degree_centrality.values()) / len(degree_centrality)
avg_betweenness_centrality = sum(betweenness_centrality.values()) /
len(betweenness_centrality)
avg_closeness_centrality = sum(closeness_centrality.values()) /
len(closeness_centrality)

print(f"\nAverage Degree Centrality: {avg_degree_centrality:.4f}")
print(f"Average Betweenness Centrality: {avg_betweenness_centrality:.4f}")
print(f"Average Closeness Centrality: {avg_closeness_centrality:.4f}")
```

Result:

```
Average Degree Centrality: 0.0025
Average Betweenness Centrality: 0.0022

Average Closeness Centrality: 0.1729
```

Notice that all centrality measures have increased by a factor of 10, hinting at the authors' significantly higher connectivity. We can reuse the code from Step 4 to visualize the network.

Code:

```
# Set the size of the plot
plt.figure(figsize=(12, 12))

# Draw the network
pos = nx.spring_layout(largest_component_subgraph)  # positions for all nodes
nx.draw_networkx_nodes(largest_component_subgraph, pos, node_size=5,
node_color='red', alpha=0.7)
nx.draw_networkx_edges(largest_component_subgraph, pos, width=1.0,
edge_color='black', alpha=0.5)

# Set the title and display the plot
plt.title("Co-authorship Network")
plt.axis('off')  # Turn off the axis
plt.show()
```

Result:

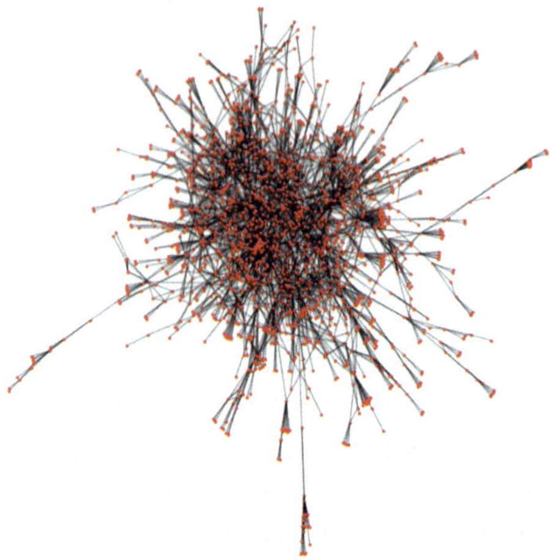

Step 3. Calculate Network Density and Modularity. Extract Communities

USER:

Compute density and modularity of network largest_component_subgraph.

Assistant:

```
# Compute the density of the largest component subgraph
density = nx.density(largest_component_subgraph)

# Compute the communities using greedy modularity
communities =
nx.algorithms.community.greedy_modularity_communities(largest_component_subgraph)

# Calculate modularity
modularity = nx.algorithms.community.modularity(largest_component_subgraph,
communities)

# Print the results
print(f"Density of the largest component subgraph: {density:.4f}")
print(f"Modularity of the largest component subgraph: {modularity:.4f}")
```


Explanation:

Density Calculation: The *nx.density* function computes the density of the *largest_component_subgraph*.

Community Detection: The *greedy_modularity_communities* function identifies communities in the graph.

Modularity Calculation: The *nx.algorithms.community.modularity* function computes the graph's modularity based on the identified communities.

Result:

```
Density of the largest component subgraph:
0.0025

Modularity of the largest component sub-
graph: 0.7958
```

The network modularity of 0.8 is very high. It indicates that many authors frequently publish within their group but have limited collaborations outside the group.

Comparatively, the network density of 0.0025 is very low, suggesting that the network is sparse: generally, authors collaborate only with a few peers out of a much larger academic community. Low network density frequently coincides with high network modularity.

Step 4. Visualize the Communities

USER:

Visualize the communities.

Assistant (the code is edited for clarity):

```
import random

# Create a color map for the communities
color_map = {}
for i, community in enumerate(communities):
    for node in community:
        color_map[node] = i  # Assign a unique color index to each node based on its
community

# Set the size of the plot
plt.figure(figsize=(12, 12))

# Draw the largest component subgraph with community colors
pos = nx.spring_layout(largest_component_subgraph)  # positions for all nodes
node_colors = [color_map[node] for node in largest_component_subgraph.nodes()]  # Get
colors for nodes

nx.draw_networkx_nodes(largest_component_subgraph, pos, node_size=5,
node_color=node_colors, cmap=plt.cm.tab20, alpha=1)

# Set the title and display the plot
plt.title("Community Visualization of the Largest Component Subgraph")
plt.axis('off')  # Turn off the axis
plt.show()
```

Explanation:

Community Detection: The *greedy_modularity_communities* function finds the communities in the *largest_component_subgraph*.

Color Mapping: A dictionary *color_map* is created to assign a unique color index to each node based on its community membership.

Result:

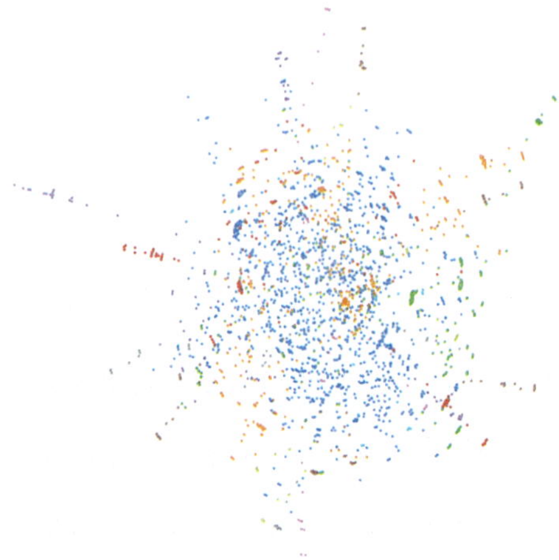

This visualization of authors' communities still needs improvement. Even though the colors show the presence of multiple communities, it is hard to differentiate between them. Instead, we will create a new network of authors' *communities*.

USER:

Use the identified communities to make a new network. Use the the number of nodes within the community as a node weight. Use the number of connections between members of different communities as edge weights. Print the number of nodes and the number of edges.

Assistant:


```
# Create a new graph for the community network
community_network = nx.Graph()

# Add nodes to the community network with weights
for community in communities:
    community_size = len(community)  # Number of nodes in the community
    community_network.add_node(f"Community_{communities.index(community)}",
weight=community_size)

# Add edges between communities based on connections
for i, community_a in enumerate(communities):
    for j, community_b in enumerate(communities):
        if i < j:  # To avoid double counting
            # Count the number of edges between members of community_a and
community_b
            inter_edges = sum(1 for node_a in community_a for node_b in community_b
if G.has_edge(node_a, node_b))
            if inter_edges > 0:
                community_network.add_edge(f"Community_{i}", f"Community_{j}",
weight=inter_edges)

# Print the number of nodes and edges in the new community network
print(f"Number of nodes in the community network:
{community_network.number_of_nodes()}")
print(f"Number of edges in the community network:
{community_network.number_of_edges()}")
```

Explanation:

New Graph Creation: A new graph *community_network* is created to represent the communities.
Adding Nodes: Each community is added as a node in the new graph, with the node's weight set to the number of nodes in that community.
Adding Edges: The code iterates over pairs of communities to count the number of edges between them in the original graph *G*. If there are connections, an edge is added between the corresponding community nodes in the new graph, with the weight set to the number of inter-community edges.

Result:

```
Number of nodes in the community network: 53

Number of edges in the community network: 124
```

We have identified 53 communities of authors active in tourism research. We can reuse the code from Step 3 to visualize their network.

USER:

Visualize the weighted network community_network. Use normalized node weight for node color using a viridis colormap. Extract the numerical part of the community label and use it to label the nodes.

Assistant:

```
import numpy as np

# Normalize the node weights for coloring
node_weights = np.array([community_network.nodes[node]['weight'] for node in
community_network.nodes()])
normalized_weights = (node_weights - node_weights.min()) / (node_weights.max() -
node_weights.min())

# Set the size of the plot
plt.figure(figsize=(10, 10))

# Draw the community network with normalized node weights
pos = nx.spring_layout(community_network)  # positions for all nodes
node_colors = plt.cm.viridis(normalized_weights)  # Use viridis colormap for node
colors

# Extract numerical part of the community label for node labels
node_labels = {node: node.split('_')[1] for node in community_network.nodes()}

# Draw nodes and edges
nx.draw_networkx_nodes(community_network, pos, node_size=500, node_color=node_colors,
alpha=0.7)
nx.draw_networkx_edges(community_network, pos, width=1.0, alpha=0.5)
nx.draw_networkx_labels(community_network, pos, labels=node_labels, font_size=12)

# Set the title and display the plot
plt.title("Weighted Community Network Visualization")
plt.axis('off')  # Turn off the axis
plt.show()
```

Explanation:

We used *viridis* colormap for node weight. This colormap maps low weights to cool colors and high values to warm colors.

See other matplotlib colormaps at ► https://matplotlib.org/stable/users/explain/colors/colormaps.html.

The communities are labeled as community_0 … community_53. For clearer visualization, we use only the numerical part to label the nodes on the plot.

To visualize the network, we used the "spring" layout:

```
pos = nx.spring_layout(community_network)
```

There are other layouts worth experimenting with, for instance Kamada-Kawai, circular, and spectral.

Result:

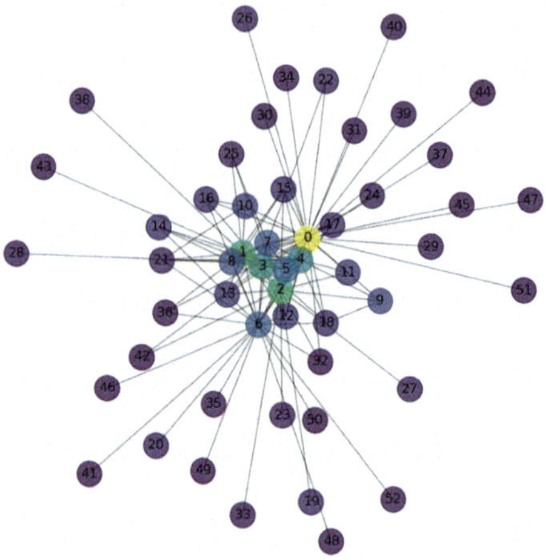

Step 5. Find the Authors from the Most Influential Community

Notice that among the network of core groups of authors, there is a handful of highly influential groups numbered 0–4 and perhaps also 5–8. To confirm that, we can reuse Part 1 (Step 3) code to compute the node centrality measures.

```
# Compute centrality measures
degree_centrality = nx.degree_centrality(community_network)
betweenness_centrality = nx.betweenness_centrality(community_network)
closeness_centrality = nx.closeness_centrality(community_network)

# Convert centrality measures to DataFrames for easier manipulation
degree_df = pd.DataFrame(degree_centrality.items(), columns=['Node', 'Degree
Centrality'])
betweenness_df = pd.DataFrame(betweenness_centrality.items(), columns=['Node',
'Betweenness Centrality'])
closeness_df = pd.DataFrame(closeness_centrality.items(), columns=['Node', 'Closeness
Centrality'])

# Get top 10 nodes for each centrality measure
top_degree = degree_df.nlargest(5, 'Degree Centrality')
top_betweenness = betweenness_df.nlargest(5, 'Betweenness Centrality')
top_closeness = closeness_df.nlargest(5, 'Closeness Centrality')

# Print top nodes
print("Top 5 Nodes by Degree Centrality:",top_degree)
print("\nTop 5 Nodes by Betweenness Centrality:",top_betweenness)
print("\nTop 5 Nodes by Closeness Centrality:",top_closeness)
```

Result:

```
Top 5 Nodes by Degree Centrality:
0  Community_0  0.634615
6  Community_6  0.442308
2  Community_2  0.403846
1  Community_1  0.365385
4  Community_4  0.288462

Top 5 Nodes by Betweenness Centrality:
0  Community_0  0.509175
6  Community_6  0.309856
2  Community_2  0.200803
1  Community_1  0.121709
8  Community_8  0.051995

Top 5 Nodes by Closeness Centrality:
0  Community_0  0.732394
6  Community_6  0.641975
2  Community_2  0.626506
1  Community_1  0.611765
4  Community_4  0.584270
```

Evidently, Community 0 is the most influential. We can list members of this community.

USER:

List the nodes from community community_0. Print the number of nodes.

Assistant:

```
community_0_nodes = communities[0]  # Access the first community

# Convert the set of nodes to a list for easier handling
community_0_nodes_list = list(community_0_nodes)

# Print the nodes in community_0
print("Nodes in community community_0:")
print(community_0_nodes_list)

# Print the number of nodes in community_0
print(f"Number of nodes in community community_0: {len(community_0_nodes_list)}")
```

Result:

<The result is a list of 333 authors.>

Part 3. On Your Own

Analyze and visualize the social network formed by St. Augustine POI visitors. Note that this is a bipartite network with two types of nodes: visitors and points of interest (POI).

Data

The data lists visitors who left a review of St. Augustine, FL, attractions.[3] In total, there are 13,075 records. The data is stored in the sheet reviews in the Excel spreadsheet: ▶ https://github.com/andykir/DataMiningLectures/raw/refs/heads/main/StAugustinePOI.xlsx.

Variables

Variable name	Explanation	Notes
ID	Record ID	Integer, categorical
reviewerID	Unique reviewer's ID	Integer, categorical
POIid	Unique POI's ID	Integer, categorical
POI	Name of the point of interest	String

For simplicity, you can use the following prompt to project the network:

USER:

In the DataFrame df, the column "reviewer" contains visitor IDs; the column "POI" contains visited points of interest (POI). The goal is to write Python code for social

3 Kirilenko, A. P., Stepchenkova, S. O., & Hernandez, J. M. (2019). Comparative clustering of destination attractions for different origin markets with network and spatial analyses of online reviews. *Tourism Management, 72*, 400–410.

network analysis. The result should be a unipartite network, with POI representing the nodes. The same tourist visiting two POIs forms an edge connecting these POIs.

Build Your Python Skills

Chapters 24 and 25 of the free book Learn Python with Jupyter by Serena Bonaretti. ► https://www.learnpythonwithjupyter.com/.

Additional Reading

Aggrawal, N., & Anand, A. (2022). *Social networks: Modelling and analysis*. CRC Press.

Camacho, D., Panizo-LLedot, A., Bello-Orgaz, G., Gonzalez-Pardo, A., & Cambria, E. (2020). The four dimensions of social network analysis: An overview of research methods, applications, and software tools. *Information Fusion, 63*, 88–120.

Understanding Data Storage and Databases

A. P. Kirilenko, *Practical Data Mining with AI for Social Scientists*,
Springer Texts in Social Sciences, https://doi.org/10.1007/978-3-031-89689-7_15

15.1 Introduction to Data Storage

Database

A *database* is a digital collection of data, organized in a way that allows efficient search, retrieval, modification, and deletion of records. The manipulations, such as the search and extraction of records, are provided by a specialized software called a *database management system (DBMS)*.

Data storage

Data storage can be organized on-premises, cloud, or hybrid. Cloud data storage is typically more cost-effective but may raise concerns about data security and privacy.

15

An information management system that has been well thought out is necessary for every project that deals with large amounts of data. The goal is to ensure that amounts of data are processed, stored, and retrieved in a logical and effective manner. A well-designed data management system is especially important in industries dealing with people, which have access to numerous sources of noisy, unstructured conversational data that include texts and videos in addition to numerical data. For example, hotels and travel firms handle large datasets, including transactional data, client booking information, reputation management data, and many more. This data must be kept in a secure manner that abides by data protection laws and permits quick and easy processing and retrieval. Neglecting the data management component of a data mining project might result in data processing errors and inconsistencies, as well as possible data loss.

Numerous functions, including data storage, retrieval, manipulation, backup, and security, are integrated into a data management system. The main focus of this chapter is on the data management systems' data storage, retrieval, and modification features. A *database*, a particular kind of data management system, provides data storage and query functionality. The database can be compared to a "virtual cabinet" that holds the gathered data. A database management system (DBMS) is a specialized software that performs manipulations, including searching and extracting data.

Structured data, usually arranged in tables, is a common feature of databases. In this sense, databases and spreadsheets—like those found in Microsoft Excel—are comparable. This is why the lab for this chapter is focused on typical table manipulation tasks. But a table isn't all that a database is. Databases facilitate procedures that link fields from various tables and uphold data integrity standards. Additionally, databases are designed to facilitate simultaneous analysis by several analysts, retrieve data quickly, and include strong security features like access control, authentication, and encryption to safeguard sensitive data.

The data storage options can be arranged for on-premises, cloud, or hybrid storage based on their physical location. Computers housed on the organization's premises are used for data storage in on-premises storage systems. Although this data storage solution gives you a lot of control over the data, it can be costly and not very scalable. Conversely, cloud data storage keeps information "in the cloud" or on distant servers. Although both Google Drive and Microsoft OneDrive are instances of cloud data storage, they do not have database features; for example, they lack the ability to perform complex queries, join operations, or enforce consistency across datasets. Contrasting, Google Cloud, Microsoft Azure, and Amazon Web Services (AWS) offer cloud database services. Although cloud data storage is usually more affordable, there are additional privacy and data security risks. Lastly, businesses can keep critical data onsite while taking advantage of cloud storage for less sensitive data thanks to hybrid data storage, which incorporates aspects of both on-premises and cloud storage.

15.2 Understanding Databases

A database is an organized collection of data stored and accessed electronically from a computer system. Usually, a database stores data organized in tables. Every row of a table represents a single record or an example in the data mining terms—for example, one comment published on social media. Each column of a table denotes a specific field (in database terms) or an attribute (in data mining terms) of the records, such as the comment author's name, social media platform ID, date of posting, and so on. The way the databases are designed emphasizes the efficiency of managing, querying, and manipulating the data they store. Since data mining is the process of extracting patterns, trends, and insights from extensive data, having databases as a secure and efficient data storage medium is very important.

SQL

Structured query language (SQL) is a programming language for storing and processing information in a relational database.

By consolidating data in one place, databases make it easier to analyze relationships and patterns across different datasets. Before data mining algorithms can be applied, the data often needs to be preprocessed to clean, transform, and reduce noise. Databases provide tools and functionalities for preprocessing tasks such as data cleaning, normalization, and aggregation, improving data quality. The querying mechanism allows for retrieving subsets of data while meeting specific criteria (for example, social media comments published during the weekends only). Finally, databases often include built-in tools that can be used to perform basic data analysis tasks directly within the database environment.

The particular needs of a data mining task will determine which of the several database types should be used. As of December 2023, 72% of databases are relational. Relational databases store data in tables where rows correspond to individual records or specific instances of the data, and columns reflect the attributes of the data that is being stored.

Relational databases that are widely used include MySQL, PostgreSQL, Oracle, and numerous others. Furthermore, many databases, like Microsoft Access, have graphical front-end interfaces, which make learning database operations easier. These front-ends allow intuitive data creation, modification, and querying. To extract information from a relational database, a special programming language called SQL (Structured query language) is frequently used.

Not all types of data are well-suited to relational databases, particularly unstructured data like text and images. Relational databases rely on a structured schema, with data organized into tables. However, text data is inherently unstructured, lacking a predefined format, and often comes with significant variability in size, content, and context. Representing such data in a rigid relational schema is inefficient, requiring complex transformations or additional layers to parse and store the information meaningfully. Non-relational databases, on the other hand, are specifically designed to handle large volumes of unstructured or semi-structured data. They offer greater flexibility in storing, indexing, and querying such data, making them better suited for use cases involving text and images. This adaptability is especially important for applications like natural language processing, image recognition, or any scenario where data does not conform to a traditional tabular format.

15.3 Basic Database Operations

CRUD

Basic database operations are Create, Read, Update, and Delete, commonly known as CRUD.

The basic operations performed on data in a database are often categorized into four types, collectively known as CRUD operations: Create, Read, Update, and Delete. Each of these operations can be completed using an SQL statement, which can be executed from the database software as well as from Python code.

Operation	Explanation	SQL
Create	Create new data records in a database	INSERT
Read	Retrieve data records from the database for processing	SELECT
Update	Modify existing data records in the database	UPDATE
Delete	Remove existing data records from the database	DELETE

Other important database operations include indexing (creating unique indexes for each record in database tables), greatly improving data retrieval speed, especially when dealing with large datasets. Transactions allow the execution of multiple database operations as one unit. Concurrency control permits multiple users to work with a database at the same time while protecting data consistency.

15.4 Merging the Tables

15.4.1 The Keys

The relational databases are optimized for storing data in interconnected tables. The interconnectivity is provided by matching keys in different tables. Keys, which are distinct identifiers that connect similar data across tables, establish the linkage between tables and allow for combining related data from two or more tables. In the illustration on the right, the linkage between two tables, one with the social media comments and the other with information on the registered users, is provided by the key "Author" present in both tables. The field "Author" in the first table is called the primary key, and the field "Author" in the second table is called the foreign key. A primary key is a unique identifier for each record in a database table. It ensures that no two records have the same value in the primary key column or columns. A foreign key is a column or a set of columns in one table that references the primary key of another table. The linkage functionality allows merging data from multiple tables into a single table containing required data in an easy-to-understand and process format.

Relations between tables in a database

ID	Author	TEXT
1	Andy51	I lav catz
2	lov4me	Holiday!
3	lov4me	My car

Author	Followers
Andy51	15
lov4me	48967

15.4.2 Table Join

The four main types of table join operations are inner join, left join (also called a left outer join), right join (also called a right outer join), and full outer join. Consider a database of social media messages. The database contains two tables. The first table has information about the individual bloggers: their nicknames and the number of followers. The second table contains the messages written by the bloggers. Our goal is to find out whether the popularity of bloggers is related to the topics of their messages. For this analysis, we need to merge two tables to have both the messages' texts and the followership in one place. There are several ways to do that.

The most intuitive way would be to use the *inner join*. An inner join returns records with matching values in both tables. Notice that the join result does not include blogger lov4me since he/she is absent from the right table. It also does not include the blogger named Travel3 since he/she is absent from the left table.

Outer Join returns all records from both tables and fills in special "NULL" values for missing matches on either side. *The left outer join* keeps all records in the left table. From the join result, we can easily see that there is a very popular blogger *lov4me* for whom we have no information in the other table.

Author	N followers
Andy51	15
Super	1005
lov4me	48967

ID	Author1	Text
1	Andy51	I lav cats
2	Andy51	I lav birds
3	Travel3	Hawaii mode
4	Super	New look

Author	N followers	ID	Author1	Text
Andy51	15	1	Andy51	I lav cats
Andy51	15	2	Andy51	I lav birds
Super	1005	3	Super	New look
lov4me	48967	NULL	NULL	NULL

Similarly, the *Right join* keeps all records in the right table:

Author	N followers
Andy51	15
Super	1005
lov4me	48967

ID	Author1	Text
1	Andy51	I lav cats
2	Andy51	I lav birds
3	Travel3	Hawaii mode
4	Super	New look

Author	N followers	ID	Author1	Text
Andy51	15	1	Andy51	I lav cats
Andy51	15	2	Andy51	I lav birds
NULL	NULL	3	Travel3	Hawaii mode
NULL	NULL	4	Super	New look

In these examples, we used two tables, but one can merge records from multiple tables (in SQLite database, up to 64 tables) based on their related columns. See the chapter lab for a specific illustration of how it can be done.

Notice that in the previous examples, the inner and left outer joins resulted in duplicated records from the left table. The reason was that a single record in the left table was associated with multiple records in the right table: the author Andy51 published two messages. This type of association is called a "*one-to-many association.*" There could also be a *one-to-one association* that does not allow record duplication and a *many-to-many* association that allows for duplicate rows in both tables.

15.4.3 Key Matching

Recall that we were using the word "match" when discussing key matching in join operations. In our example, we were using the *exact match*: that is, the records from two tables were matched one to the other when the values in the respective *Author* fields were exactly the same. A *partial match* occurs when two strings are not *entirely* identical so that only a portion of one string matches the other. For example, partial matching can be useful when the goal is to match information from different pages of the same website.

Fuzzy match: similar, but not identical

Finally, a *fuzzy match* occurs when two strings have only an approximate matching. Fuzzy matching algorithms are designed to account for variations such as misspellings, typos, abbreviations, and other forms of variation or inconsistency in the data. For example, the author of this book may spell his name as Andrei Kirilenko or Andrei P. Kirilenko. Fuzzy matching is very useful for matching personal names and the names of businesses coming from different sources.

Fuzzy matching typically includes a similarity score. This allows for more flexibility in matching strings that are similar but not exact. For example, the following table shows fuzzy matches between business names in a global business database acquired from Kaggle.com and the mentions encountered in Instagram posts of top tourism influencers. This analysis aimed to detect the scope of potential business sponsorship of social media influencers. Notice how the fuzzy matching score was used to make a matching decision.

Business name in global business database	Matching Instagram @ mention	Matching Score	Accept? (score < 0.8)
Zuvi Collective	collective	0.71	No
London, Travel + Lifestyle	lifestyle pa london	0.75	No
Zipp Speed Weaponry	zipp speed weaponry	1.00	Yes
Zipcar	zipcar	1.00	Yes
The Zen House	house of zen	0.85	Yes
ZEITZ Foundation	on foundation	0.66	No
Zeffer Cider	cider	0.66	No
SAINT LAURENT	saint laurent	0.99	Yes
YouTube India	youtube	0.84	Yes

15

Look at the record with the attribute “Business name in global business database” = “The Zen House”. The potentially matching Instagram account name found by the fuzzy matching algorithm is the “house of zen.” These two strings have no exact nor partial match due to the shuffled words. The matching score is, however, above the acceptance threshold, indicating a potential match. To summarize, exact matches require matching fields to be identical, partial matches look up shared elements, and fuzzy matches use complex algorithms to calculate matching scores. Each matching type serves different purposes depending on the context and requirements of the task at hand.

15.5 Data Security and Privacy

One of the key advantages of databases is an integrated data security mechanism. This is particularly important in industries such as tourism and hospitality, where vast amounts of sensitive customer data are collected, stored, and processed daily. The main goals of database security are to (1) secure sensitive data and maintain customers' confidentiality, (2) guarantee data integrity, and (3) ensure timely data availability for analysis. These requirements not only protect data from unauthorized use but also help to build trust with customers, comply with legal and regulatory requirements, and prevent financial losses in case of a data breach.

Data security and privacy are governed by a variety of best practices and administrative regulations. The most demanding system of regulations has probably been adopted by the European Union in the General Data Protection Regulation (GDPR) law (► http://GDPR.eu). The GDPR penalties reach up to 20 million euros for serious data privacy violations. The United States currently lacks a single comprehensive federal data protection law. Instead, multiple federal and state laws cover specific sectors (The US Privacy Act of 1974, HIPAA for medical data, COPPA for data on minors, the California Consumer Privacy Act (CCPA), and others). Similar laws exist in many other countries. These regulations mandate that businesses implement robust data protection measures, respect individuals' data rights, and promptly report data breaches.

Best practices that businesses employ to protect their data include:

- Strong data encryption, including encryption of data storage and data in transit.
- Strong authentication of users attempting database connections. A good practice is to employ two-factor

identification, which combines two separate, distinctive forms of identification, such as a password and a user's fingerprint.
- Anonymizing collected data, replacing customers' names with unique IDs, and deleting collected information not needed for data analysis, such as home addresses.
- Data auditing for sensitive information, which requires special protection.
- Continuous access control and auditing for authorized data use. Tracking and recording data access and modification activities for accountability and forensic purposes.
- Timely revoking rights from those who don't need database access. A frequent problem in database management is the "privilege creep": not revoking database access from its former users. In the university environment, this includes the graduating students, colleagues departing for the project, and participants in the completed projects.
- Physical protection of database servers.
- Training personnel on data protection principles and procedures.

A well-protected database system that follows legal and internal administrative regulations not only supports continuity of business data storage and access but also provides a competitive advantage in today's data-driven world.

15.6 Review Questions

1. What is the importance of secure data storage?
2. Compare On-Premises, Cloud, and Hybrid data storage.
3. What are the databases?
4. Explain the difference between Relational and NoSQL databases.
5. Explain the basic database operations: Create, Read, Update, Delete (CRUD).
6. Explain the different types of join operations: Inner Join, Outer Join, Left Join, and Right Join.
7. Explain different types of merge operations: One-to-One, Many-to-One, and Many-to-Many.

8. Explain the following match operations: Exact Match, Partial Match, Fuzzy Match.
9. Discuss the importance of data security and privacy in databases.
10. Explain the techniques for ensuring data security and privacy: Encryption, Access Control, and Auditing.

15.7 Lab

Objectives

1. Get familiar with combining data coming from different datasets.
2. Learn the basics of using Python libraries for database data storage and retrieval.
3. Practice controlling the quality of crowdsourced data.

In this lab, we practice typical database operations without using a "real" database. Python distributions typically include SQLite, a lightweight, file-based database that ensures data can be easily shared and transferred across different environments.

Data

The data comes from tweets on climate change.[1] The specific goal of data analysis was to differentiate between tweets sent by those concerned about climate change ("activists") and those who expressed a denial or skepticism ("skeptics"). The following methodology was suggested:

1. Human raters classify a sample of tweets into the "activists," "skeptics," and "other/unsure" classes.
2. The classified tweets are used to train a machine learning model.
3. The trained model is applied to classify the rest of the tweets.

---

1 Kirilenko, A. P., Molodtsova, T., & Stepchenkova, S. O. (2015). People as sensors: Mass media and local temperature influence climate change discussion on Twitter. *Global Environmental Change, 30*, 92–100.

Implementation of the first step of this plan was outsourced to a crowdsourcing platform, Amazon Mturks. To deal with accidental or malicious misclassifications of tweets by human raters, the same tweets were analyzed by multiple raters.

Your goal is to combine data from multiple tables, describe data, and analyze the work quality of paid workers.

Content

The data was collected using Twitter API. The actual text of the tweets is removed from the data. The dataset list is as follows:

tweetsCSClassif.xlsx: The attitude towards climate change in tweet processing by human raters.

Variables

Variable	Explanation	Code explanation
ID	Record ID	Unique ID
idTweet	The ID of a tweet	Unique ID
idRater	The ID of a rater who analyzed the tweet	Unique ID
classif_attitude	The attitude towards the climate change issue expressed in the tweet as classified by the rater.	-2 (very skeptical) … 0 (neutral) … 2 (very supportive)

raterCountry.xlsx: The country of the human rater.
Variables

Variable	Explanation	Code explanation
idRater	The ID of a rater	Unique ID
CountryCode	ISO country code of the rater	Unique ID

CountryISOcode.xlsx: The correspondence between country code and country name.

Variables

Variable	Explanation	Code explanation
ID	Record id	Unique ID
CountryID	Unique ID of a country according to the International Organization for Standardization	Unique ID
CountryName	Country names as registered by the International Organization for Standardization	Unique ID

Part 1. Understand the Data

Step 1. Read the Data

We will write the code using the Anaconda AI assistant. Start a new empty notebook. Give the notebook a meaningful name. Enter the following text to the Anaconda Assistant window and start the chat:

USER:

Read the data from ► https://github.com/andykir/DataMiningLectures/raw/main/tweetsCSClassif.xlsx to pandas DataFrame tweetsCSClassif.

Read the data from ► https://github.com/andykir/DataMiningLectures/raw/main/CountryISOcode.xlsx to pandas DataFrame CountryISOcode.

Read the data from ► https://github.com/andykir/DataMiningLectures/raw/main/raterCountry.xlsx to pandas DataFrame raterCountry.

Assistant:

```
import pandas as pd

# Read the data into pandas DataFrames
tweetsCSClassif =
pd.read_excel('https://github.com/andykir/DataMiningLectures/raw/main/
tweetsCSClassif.xlsx')
CountryISOcode =
pd.read_excel('https://github.com/andykir/DataMiningLectures/raw/main/
CountryISOcode.xlsx')
raterCountry =
pd.read_excel('https://github.com/andykir/DataMiningLectures/raw/main/
raterCountry.xlsx')
```

Explanation:

In this code, we use the *pd.read_excel* function to read the Excel files from the provided URLs into pandas DataFrames *tweetsCSClassif*, *CountryISOcode*, and *raterCountry*.

Step 2. Describe the Data

Recall that previously, to describe data, we used the methods *DataFrame.info( )*, *DataFrame.head( )*, and *DataFrame.describe( )* from the pandas library. Instead of relying on the Assistant, we can type the code directly into a new cell of the notebook. This time, we are not interested in descriptive statistics of the data; hence, the *DataFrame.describe( )* method will not be used.

Code:

```
for data in [tweetsCSClassif, CountryISOcode, raterCountry]:
    #print the summary of the DataFrame structure
    print(data.info(), '\n')

    #print the top rows of the DataFrame and skip two lines
    print(data.head(), '\n\n')
```

For better readability, we use a loop variable *data* to cycle through a list of three DataFrames. We also skip two lines after each iteration by printing the “new line” symbol ‘\n’ twice. For comparison, this is a code that does not utilize loops:

```
print(tweetsCSClassif.info(), '\n')
print(tweetsCSClassif.head(), '\n\n')
print(CountryISOcode.info(), '\n')
print(CountryISOcode.head(), '\n\n')
print(raterCountry.info(), '\n')
print(raterCountry.head(), '\n\n')
```

15

Result:

```
Data columns (total 4 columns):
 #   Column           Non-Null Count  Dtype
---  ------           --------------  -----
 0   ID               8198 non-null   int64
 1   idTweet          8198 non-null   int64
 2   idRater          8198 non-null   int64
 3   classif_attitude 8198 non-null   int64

   ID  idTweet  idRater  classif_attitude
0  1        92      352                 0
1  2        92   107970                 0
2  3        92    84968                -1
3  4        92    20308                -1
4  5        92      838                 0
```

The dataset *tweetsCSClassif* contains 8198 records with four attributes (*ID, idTweet*, *idRater*, and *classif_attitude*). All values are stored as integer values; there are no missing values.

```
Data columns (total 3 columns):
 #   Column       Non-Null Count  Dtype
---  ------       --------------  -----
 0   ID           30 non-null     int64
 1   CountryID    30 non-null     object
 2   CountryName  30 non-null     object

   ID CountryID                          CountryName
0  1         US  United States
1  2         GB  United Kingdom
2  3         AU  Australia
3  4         CA  Canada
4  5         IN  India
```

The dataset *CountryISOcode* contains 30 records with 3 attributes (*ID*, *CountryID*, and *CountryName*). There is no missing data. *idRater* is stored as integer values. Note that the *CountryID* and *CountryName* attribute types are shown as Object.

```
<class 'pandas.core.frame.DataFrame'>
RangeIndex: 127 entries, 0 to 126
Data columns (total 2 columns):
 #   Column          Non-Null Count  Dtype
---  ------          --------------  -----
 0   idRater         127 non-null    int64
 1   FirstOfcountry  127 non-null    object
dtypes: int64(1), object(1)
memory usage: 2.1+ KB
None
   idRater FirstOfcountry
0        1             BR
1       24             US
2       54             US
3       96             US
4      200             US
```

The dataset *raterCountry* contains 127 records with two attributes (*idRater* and *FirstOfCountry*). There is no missing data. *idRater* is stored as an integer, and the *FirstOfcountry* type is Object.

Data type Object can be a string, integer, float, or another type. It makes sense to convert all attributes with an Object data type to a string type. We can use the following code:

```
CountryISOcode['CountryID'] =
CountryISOcode['CountryID'].astype('string')
CountryISOcode['CountryName'] =
CountryISOcode['CountryName'].astype('string')
print("Information of DataFrame CountryISOcode:", '\n')
print(CountryISOcode.info(),'\n\n')

raterCountry['FirstOfcountry'] =
raterCountry['FirstOfcountry'].astype('string')
print("Information of DataFrame raterCountry:", '\n')
print(raterCountry.info())
```

15

Result:

```
Information of DataFrame CountryISOcode:
<class 'pandas.core.frame.DataFrame'>
RangeIndex: 30 entries, 0 to 29
Data columns (total 3 columns):
 #   Column       Non-Null Count  Dtype
---  ------       --------------  -----
 0   ID           30 non-null     int64
 1   CountryID    30 non-null     string
 2   CountryName  30 non-null     string
dtypes: int64(1), string(2)
memory usage: 852.0 bytes
None

Information of DataFrame raterCountry:
<class 'pandas.core.frame.DataFrame'>
RangeIndex: 127 entries, 0 to 126
Data columns (total 2 columns):
 #   Column          Non-Null Count  Dtype
---  ------          --------------  -----
 0   idRater         127 non-null    int64
 1   FirstOfcountry  127 non-null    string
dtypes: int64(1), string(1)
memory usage: 2.1 KB
None
```

Part 2. Table Query

To search (query) either of the three tables we have imported into the computer memory, we have to specify which values we are interested in from the table. For example, to print a list of IDs for raters coming from the UK (note that the 2-letter ISO code for the UK is GB, derived from the full name of the country: United Kingdom of Great Britain and Northern Ireland):

USER:

Use the data in DataFrame "raterCountry". Print the rows for which the column FirstOfcountry = 'GB'

Assistant:

```
import pandas as pd

# Filter the rows where FirstOfcountry is equal to 'GB'
filtered_data = raterCountry[raterCountry['FirstOfcountry'] == 'GB']

# Print the filtered data
print(filtered_data)
```

Explanation:

In this code, we use the pandas to filter the rows where the *FirstOfcountry* column equals 'BR' and then print the filtered data.

Result:

```
        idRater FirstOfcountry
15          806             GB
28         1362             GB
45        78873             GB
63        94094             GB
78       105893             GB
87       109668             GB
99       124213             GB
101      125572             GB
111      127784             GB
115      129357             GB
121      130687             GB
122      130712             GB
125      132657             GB
```

This code can be customized. For example, to print a list of IDs for raters coming from the UK and Brazil (ISO codes UK and BR), we can modify the code as follows:

```
# Filter the rows where FirstOfcountry is equal to 'GB' or 'BR'
filtered_data =
raterCountry[raterCountry['FirstOfcountry'].isin(['GB', 'BR'])]
```

Explanation:

In this code, we select the rows of *raterCountry*, for which the values in column *FirstOfcountry* are in the list ['GB', 'BR'].

An alternative code:

15

```
# Filter the rows where FirstOfcountry is equal to 'GB' or 'BR'
filtered_data = raterCountry[(raterCountry['FirstOfcountry']=='GB') |
(raterCountry['FirstOfcountry']=='BR')]
```

Explanation:

In this code, we select the rows from *raterCountry* where the values in the column *FirstOfcountry* are equal to 'GB' or 'BR'. Note the use of the symbol | to represent the logical OR operation.

Part 3. Combining the Data from Different Tables

Step 1. Draw the Relationships Between the Tables

In the previous section, we queried a single table. However, the three tables are logically related to each other through the fields they share in common. These relationships can be visualized as follows:

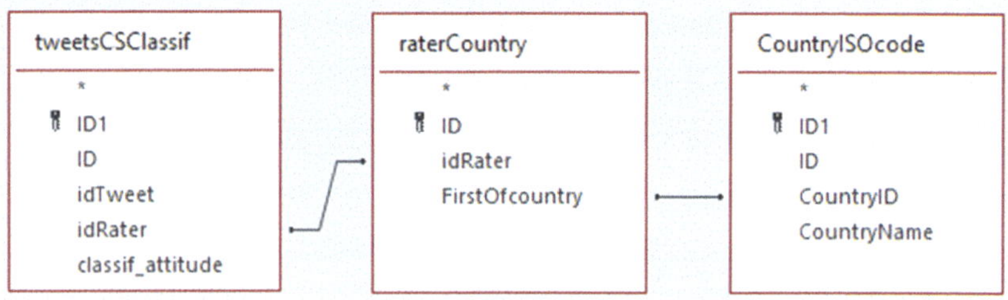

The field *idRater* contains information about the worker's ID, which can be used to match rows related to the same workers in the tables *tweetCSClassif* and *raterCountry*. These fields, used for matching rows across tables, are referred to as *keys*.

In this example, the matching keys in the tables *tweetCSClassif* and *raterCountry* share the same name; however, this is not a requirement. For instance, the key *FirstOfCountry* in the table *raterCountry* matches the key *CountryID* in the table *CountryISOcode*. Matching keys allow you to extend the information in one table by incorporating data from related tables. This process can be achieved through *joining* and *merging* operations.

Step 2. Combine the Tables

Let us create a query that would expand the *tweetsCSClassif* table data with information on the human rater's country from the table *raterCountry*.

USER:

Join the DataFrames "tweetsCSClassif" and "raterCountry" on the key "idRater".

Assistant:

```
merged_df = tweetsCSClassif.merge(raterCountry, on='idRater')
```

Explanation:

To join the DataFrames *tweetsCSClassif* and *raterCountry* on the key *idRater*, we use the *merge*() function. The resulting DataFrame, *merged_df*, will contain the combined data from both DataFrames. Note that there is a similar, but not identical, function called *join*().

To evaluate the merge result, we can use the *DataFrame.info*() and *DataFrame.head*() methods. Open a new cell and enter the following code:

```
print(merged_df.info(),'\n')
print(merged_df.head())
```

Result:

```
<class 'pandas.core.frame.DataFrame'>
RangeIndex: 8198 entries, 0 to 8197
Data columns (total 5 columns):
 #   Column           Non-Null Count  Dtype
---  ------           --------------  -----
 0   ID               8198 non-null   int64
 1   idTweet          8198 non-null   int64
 2   idRater          8198 non-null   int64
 3   classif_attitude 8198 non-null   int64
 4   FirstOfcountry   8198 non-null   string
dtypes: int64(4), string(1)
memory usage: 320.4 KB
None

   ID  idTweet  idRater  classif_attitude FirstOfcountry
0   1       92      352                 0             US
1  12       98      352                 1             US
2  20      113      352                 1             US
3  34      140      352                 1             US
4  45      161      352                 1             US
```

15

Notice that the merged table contains columns from the tables *tweetsCSClassif* and *raterCountry*.

Additionally, we can join tables where the matching keys have different names:

USER:

Join the DataFrames "raterCountry" and "CountryISOcode" on the key "FirstOfcountry" in DataFrame "raterCountry" and the key "CountryID" in DataFrame "CountryISOcode".

Assistant:

```
merged_df = raterCountry.merge(CountryISOcode,
left_on='FirstOfcountry', right_on='CountryID')
```

Explanation:

The parameters *left_on* and *right_on* specify the join keys from the left and right tables, respectively.

To evaluate the result of joining the tables, we can use the *DataFrame.info*() and *DataFrame.head*() methods:

```
print(merged_df.info(),'\n')
print(merged_df.head())
```

Result:

```
<class 'pandas.core.frame.DataFrame'>
RangeIndex: 127 entries, 0 to 126
Data columns (total 5 columns):
 #   Column          Non-Null Count  Dtype
---  ------          --------------  -----
 0   idRater         127 non-null    int64
 1   FirstOfcountry  127 non-null    string
 2   ID              127 non-null    int64
 3   CountryID       127 non-null    string
 4   CountryName     127 non-null    string
dtypes: int64(2), string(3)
memory usage: 5.1 KB
None

   idRater FirstOfcountry  ID CountryID                            CountryName
0        1             BR  10        BR                                 Brazil
1       24             US   1        US  United States
2       54             US   1        US  United States
3       96             US   1        US  United States
4      200             US   1        US  United States
```

To join multiple tables, the *merge* (or *join*) functions can be applied consequently:

```
merged_df = tweetsCSClassif.merge(raterCountry,
on='idRater').merge(CountryISOcode, left_on='FirstOfcountry',
right_on='CountryID')

print(merged_df.info(),'\n')
print(merged_df.head())
```

Result:

```
RangeIndex: 8198 entries, 0 to 8197
Data columns (total 8 columns):
 #   Column             Non-Null Count  Dtype
---  ------             --------------  -----
 0   ID_x               8198 non-null   int64
 1   idTweet            8198 non-null   int64
 2   idRater            8198 non-null   int64
 3   classif_attitude   8198 non-null   int64
 4   FirstOfcountry     8198 non-null   string
 5   ID_y               8198 non-null   int64
 6   CountryID          8198 non-null   string
 7   CountryName        8198 non-null   string
dtypes: int64(5), string(3)
memory usage: 512.5 KB
None

   ID_x  idTweet  idRater  classif_attitude FirstOfcountry  ID_y CountryID  \
0     1       92      352                 0             US     1        US
1    12       98      352                 1             US     1        US
2    20      113      352                 1             US     1        US
3    34      140      352                 1             US     1        US
4    45      161      352                 1             US     1        US

   CountryName
0  United States
1  United States
2  United States
3  United States
4  United States
```

Notice that the attributes with identical names (*ID*) were renamed in the combined table to *ID_x* and *ID_y*, respectively.

Part 4. Inner, Left, and Right Join

In the previous example, every key in the left table had a corresponding match in the right table. Consequently, the number of records in the resulting table was the same as the number of keys in the initial table where the join began. However, this is not always the case. For example, consider the following Python code:

15

```
print("raterCountry")
print(raterCountry.info(),'\n')

print("CountryISOcode")
print(CountryISOcode.info(),'\n')

merged_df = raterCountry.merge(CountryISOcode,
left_on='FirstOfcountry', right_on='CountryID')

print("merged_df")
print(merged_df.info(),'\n')
```

Result:

```
raterCountry
<class 'pandas.core.frame.DataFrame'>
RangeIndex: 127 entries, 0 to 126
Data columns (total 2 columns):
 #   Column          Non-Null Count  Dtype
---  ------          --------------  -----
 0   idRater         127 non-null    int64
 1   FirstOfcountry  127 non-null    string
dtypes: int64(1), string(1)
memory usage: 2.1 KB
None

CountryISOcode
<class 'pandas.core.frame.DataFrame'>
RangeIndex: 30 entries, 0 to 29
Data columns (total 3 columns):
 #   Column       Non-Null Count  Dtype
---  ------       --------------  -----
 0   ID           30 non-null     int64
 1   CountryID    30 non-null     string
 2   CountryName  30 non-null     string
dtypes: int64(1), string(2)
memory usage: 852.0 bytes
None

merged_df
<class 'pandas.core.frame.DataFrame'>
RangeIndex: 127 entries, 0 to 126
Data columns (total 5 columns):
 #   Column          Non-Null Count  Dtype
---  ------          --------------  -----
 0   idRater         127 non-null    int64
 1   FirstOfcountry  127 non-null    string
 2   ID              127 non-null    int64
 3   CountryID       127 non-null    string
 4   CountryName     127 non-null    string
dtypes: int64(2), string(3)
memory usage: 5.1 KB
None
```

The number of records in the merged table matches the number of records in *raterCountry* because each key in the *FirstOfcountry* column has a corresponding key in the *CountryID* column. However, this is not always guaranteed. For example, we can remove the code for the UK from the *CountryISOcode* table.

USER:

Use the DataFrame CountryISOcode. Remove the rows for which CountryISOcode.CountryID='GB'. Save the result in DataFrame CountryISOcode_reduced.

Assistant:

```
CountryISOcode_reduced = CountryISOcode[CountryISOcode['CountryID'] !=
'GB']
```

USER:

Merge the DataFrames raterCountry and CountryISOcode_reduced.

Assistant:

```
merged_df = raterCountry.merge(CountryISOcode_reduced,
left_on='FirstOfcountry', right_on='CountryID')

print("merged_df")
print(merged_df.info(),'\n')
```

Result:

```
merged_df
<class 'pandas.core.frame.DataFrame'>
RangeIndex: 114 entries, 0 to 113
Data columns (total 5 columns):
 #   Column          Non-Null Count  Dtype
---  ------          --------------  -----
 0   idRater         114 non-null    int64
 1   FirstOfcountry  114 non-null    object
 2   ID              114 non-null    int64
 3   CountryID       114 non-null    object
 4   CountryName     114 non-null    object
dtypes: int64(2), object(3)
memory usage: 4.6+ KB
None
```


The merged data contains 114 rows, compared to 127 in the previous example. Rows from the *raterCountry* table with key values that do not match any key in the *CountryISOcode_reduced* table are excluded from the resulting table. This type of operation is known as an *inner join*. An inner join retrieves only the records with matching key values in both tables. If no match exists, the corresponding rows from both tables are omitted from the result. This is the default type of join.

Other types of join operations include the *left join* (or left outer join) and *right join*. A left join returns all records from the left table and the matched records from the right table. If there is no match, the result includes NULL for columns from the right table. For example, you can perform a left join on the *raterCountry* and *CountryISOcode_reduced* DataFrames as follows:

```
merged_df = raterCountry.merge(CountryISOcode_reduced, how='left',
left_on='FirstOfcountry', right_on='CountryID')

print("merged_df")
print(merged_df.info(),'\n')
```

Result:

```
merged_df
<class 'pandas.core.frame.DataFrame'>
RangeIndex: 127 entries, 0 to 126
Data columns (total 5 columns):
 #   Column          Non-Null Count  Dtype
---  ------          --------------  -----
 0   idRater         127 non-null    int64
 1   FirstOfcountry  127 non-null    object
 2   ID              114 non-null    float64
 3   CountryID       114 non-null    object
 4   CountryName     114 non-null    object
dtypes: float64(1), int64(1), object(3)
memory usage: 5.1+ KB
None
```

The merged table has 127 rows, but the *ID*, *CountryID*, and *CountryName* columns have empty cells corresponding to the UK-based raters:

```
print(merged_df[merged_df.FirstOfcountry == 'GB'].head())
```

Result:

```
    idRater FirstOfcountry  ID CountryID CountryName
15      806             GB NaN       NaN         NaN
28     1362             GB NaN       NaN         NaN
45    78873             GB NaN       NaN         NaN
63    94094             GB NaN       NaN         NaN
78   105893             GB NaN       NaN         NaN
```

Contrasting, the cells corresponding to raters from other countries are filled in:

```
print(merged_df[merged_df.FirstOfcountry != 'GB'].head())
```

Result:

```
   idRater FirstOfcountry    ID  CountryID    CountryName
0        1             BR  10.0         BR         Brazil
1       24             US   1.0         US  United States
2       54             US   1.0         US  United States
3       96             US   1.0         US  United States
4      200             US   1.0         US  United States
```

Similar to a left join, a right join retrieves all records from the right table and the matching records from the left table. If no match is found, the resulting dataset will include NULL values on the side of the left table. By contrast, a full outer join (or full join) includes all rows from both tables, with NULL values in places where the join condition is not satisfied. In the *merge*() method, the type of join is specified using the *how* parameter.

```
merged_df = raterCountry.merge(CountryISOcode_reduced) #inner join
merged_df = raterCountry.merge(CountryISOcode_reduced, how='left')
#left join
merged_df = raterCountry.merge(CountryISOcode_reduced, how='right')
#right join
merged_df = raterCountry.merge(CountryISOcode_reduced, how='outer')
#outer join
```

Part 5. Data Selection, Aggregation, Grouping, and Ordering

Data Selection

Data selection selects specific columns from the tables. Start with merging all three tables:

```
merged_df = tweetsCSClassif.merge(raterCountry,
on='idRater').merge(CountryISOcode, left_on='FirstOfcountry',
right_on='CountryID')

print(merged_df.head(),'\n')
```

15

Result:

```
   ID_x  idTweet  idRater  classif_attitude FirstOfcountry  ID_y CountryID  \
0     1       92      352                 0             US     1        US
1    12       98      352                 1             US     1        US
2    20      113      352                 1             US     1        US
3    34      140      352                 1             US     1        US
4    45      161      352                 1             US     1        US

   CountryName
0  United States
1  United States
2  United States
3  United States
4  United States
```

To select only the essential columns with information on the tweet, rater, the attitude, and rater's country, we can use the following code:

```
relevantData=merged_df[['idTweet','idRater','classif_attitude',
'CountryName']]
print(relevantData.head(),'\n')
```

Note how the double square brackets are used in column selection.

Result:

```
   idTweet  idRater  classif_attitude  CountryName
0       92      352                 0  United States
1       98      352                 1  United States
2      113      352                 1  United States
3      140      352                 1  United States
4      161      352                 1  United States
```

Data Aggregation

The *describe*() method offers data aggregation operations (see Part 1), while the *include* parameter enables the selection of specific aggregation methods.

```
print(relevantData.describe(include='all'))
```

Result:

```
            idTweet        idRater  classif_attitude    CountryName
count   8198.000000    8198.000000       8198.000000           8198
unique          NaN            NaN               NaN              9
top             NaN            NaN               NaN  United States
freq            NaN            NaN               NaN           7555
mean   10190.491705   55134.223591          0.251769            NaN
std     5719.336105   50071.743068          0.920690            NaN
min       92.000000       1.000000         -2.000000            NaN
25%     5355.000000     913.000000          0.000000            NaN
50%    10072.000000   75478.000000          0.000000            NaN
75%    14951.000000   92706.000000          1.000000            NaN
max    20074.000000  133168.000000          2.000000            NaN
```

Other data aggregations can be obtained using the methods *count*(), *mean*(), *median*(), *mode*(), *std*(), *min*(), *max*(), *sum*(), and others.

```
print(relevantData.count())
```

Result:

```
idTweet              8198
idRater              8198
classif_attitude     8198
CountryID            8198
dtype: int64
```

```
print(relevantData['classif_attitude'].mode())
```

Result:

```
0    0
```

From this output, we can conclude that the most frequent code for the attitude towards climate change is 0 (neutral).

Data Grouping by Clause

In data analysis, we often focus on more advanced forms of data aggregation. The *group by clause* organizes rows into groups based on the values in one or more columns. This is especially powerful when used alongside aggregate functions like *count*(), *sum*(), *mean*(), *max*(), and others, which calculate metrics for each group rather than individual rows. For instance, we might be curious whether raters from different countries perceive people's attitudes on climate change in tweets differently. To explore this, we can group the *classif_attitude* column by the *CountryName* column and compute the mean statistics for each group.

```
print(relevantData[['CountryName','classif_attitude']].groupby('Countr
yName').mean())
```

Result:

```
CountryName                          classif_attitude
Australia                                    0.156863
Brazil                                       0.300000
Canada                                       0.064516
Columbia                                     0.000000
Germany                                     -0.250000
Indonesia                                    0.222222
Ireland                                      0.260870
United Kingdom                               0.130233
United States                                0.257710
```


To handle more complex groupings, we can use the *agg*() method. For example, in the previous case, we might be interested in calculating the mean value of people's attitudes, but only when there is a sufficiently large number of records for a specific country. The updated code is shown below:

```
relevantData_grouped=relevantData[['CountryName','classif_attitude']].
groupby('CountryName').agg(['count','mean'])
print(relevantData_grouped)
```

Result:

```
                     classif_attitude
CountryName                     count      mean
Australia                          51  0.156863
Brazil                             10  0.300000
Canada                             31  0.064516
Columbia                            1  0.000000
Germany                             8 -0.250000
Indonesia                         189  0.222222
Ireland                           138  0.260870
United Kingdom                    215  0.130233
United States                    7555  0.257710
```

To limit results to the countries with at least 50 records, we can use a data selection query:

```
print(relevantData_grouped[relevantData_grouped['classif_attitude']['c
ount']>=50])
```

Result:

```
                     classif_attitude
CountryName                     count      mean
Australia                          51  0.156863
Indonesia                         189  0.222222
Ireland                           138  0.260870
United Kingdom                    215  0.130233
United States                    7555  0.257710
```

This output demonstrates significant differences in results returned by Mturk raters from different countries processing identical data.

Data Ordering

The tables can be ordered using the *sort_values* method. For example, to order the table *relevantData* according to the country of the rater, we can use the following code:

```
print(relevantData.sort_values(by='CountryName'))
```

Result:

```
      idTweet  idRater  classif_attitude CountryName
8009    19855   131530                 0   Australia
7989     3143   120911                 0   Australia
7988    19529   128526                 0   Australia
7987    16996   128526                 1   Australia
7986    16251   128526                 0   Australia
...       ...      ...               ...         ...
2724     4772    75478                 1 United States
2723     4728    75478                 1 United States
2722     4720    75478                 0 United States
3072    16209    75478                 0 United States
4098    10257   129610                 0 United States
```

We can also sort by multiple columns:

```
print(relevantData.sort_values(by=['CountryName','idRater']))
```

Result:

```
      idTweet  idRater  classif_attitude  CountryName
7959      295      863                 0    Australia
7960      817      863                 1    Australia
7961     3116      863                 1    Australia
7962     4323      863                -1    Australia
7963     4795      863                 0    Australia
...       ...      ...               ...          ...
7467    12372   133168                -1 United States
7468    15043   133168                 0 United States
7469    18158   133168                 0 United States
7470    19444   133168                -1 United States
7471    19476   133168                 0 United States
```

Part 6. On Your Own

Each tweet in the dataset was evaluated by multiple raters. Answer the following questions:

1. Provide descriptive statistics on how many raters evaluated a single tweet. Compute the minimum, maximum, mean, and medium number of raters per tweet.
2. The standard deviation of the ratings for a tweet (attribute *classif_attitude*) reflects the difficulty of evaluating it. Tweets that are easy to understand tend to receive similar ratings, resulting in a small standard deviation. Conversely, tweets that are hard to understand often receive diverse evaluations, leading to a higher standard deviation. Identify and print the IDs (attribute *idTweet*) of the **five tweets with the highest standard deviations**, as these were the most challenging to evaluate.

15

3. Some human raters may attempt to complete their tasks quickly, which can result in low-quality evaluations that need to be discarded. To identify such raters, find the five raters whose ratings deviate the most from the mean ratings of the same tweets. There are multiple approaches to solving this problem.

Improve Python skills

Chapters 26 and 27 of the free book Learn Python with Jupyter by Serena Bonaretti.
▶ https://www.learnpythonwithjupyter.com/.

Additional Reading

McFadyen, R. (2021). *Relational Databases and Microsoft Access 365-Version 4.0*. Free online: ▶ https://ecampusontario.pressbooks.pub/relationaldatabasesandmicrosoftaccess365/

Ethics and Explainable AI

A. P. Kirilenko, *Practical Data Mining with AI for Social Scientists*,
Springer Texts in Social Sciences, https://doi.org/10.1007/978-3-031-89689-7_16

16.1 Privacy Concerns and Data Security

Data mining

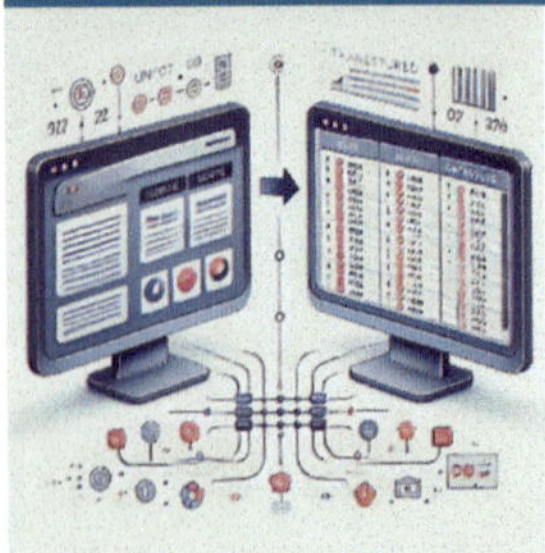

Hundreds of scientific papers and thousands of popular articles have explored ethical considerations of using data mining, extending to web data scraping, machine learning, AI, and other closely related fields. Some ideas found in this literature have eventually percolated into laws and regulations. For instance, the European Union General Data Protection Regulation (GDPR) provides a comprehensive legal framework aimed at protecting personal data, including the digital footprint left on websites. In comparison, the US federal regulations are mostly limited to medical and financial personal information. Yet, some states have their own online privacy laws, such as the California Consumer Privacy Act (CCPA).

Data mining relies on the availability of large datasets, which inevitably leads to data ownership questions. The resulting network of regulations is complex, patchy, and spatially heterogeneous. In the US, the common understanding is that data scrapping is legal as long as it does not interfere with website functionality.[1] In 2003, X Corp (Twitter) attempted to sue Bright Data Ltd. over Twitter data scraping. Yet, the lawsuit was dismissed with the court ruling that scraping publicly available data is not necessarily illegal. Yet it is a good practice to read the Terms of Service (TOS), limit the amount of scraped data to a minimum, limit the number of data requests so that data scrapping does not interfere with web service functionality, refrain from publishing raw data for public access, and avoid interfering with commercial activity of web platform.

Some sources to consult prior to scraping

European Union General Data Protection Regulation (GDPR)

State laws such as California Consumer Privacy Act (CCPA)

Terms of Service (TOS)

robots.txt web bot instructions, for instance https://www.tripadvisor.com/robots.txt

This section aims to introduce general *ethical* considerations of data collection, processing, and using data mining results in decision-making. Even in the case of social media, when users publish their content publicly, the frequent expectation is of a limited audience, a known context, and a specific purpose for which the content is shared. Indeed, multiple incidents of a single tweet, stripped of its original meaning or intent, can become a target for ridicule, outrage, or condemnation. This phenomenon is amplified by the viral nature of social media, where negative and sensational content tends to spread

16

1 Stringam, B., Gerdes, J. H., & Anderson, C. K. (2023). Legal and ethical issues of collecting and using online hospitality data. *Cornell Hospitality Quarterly*, *64*(1), 54–62.

rapidly to audiences far removed from the original poster's intended circle.Contributing to the potential of harm are:

- loss of nuances such as local knowledge, cultural references, humor, or sarcasm;
- selective editing of content with malicious goals;
- echo chambers, reinforcing the original negative narrative and provoking widespread public outrage; and
- anonymity and presumed lack of accountability for cyberbullies.

Combined with the practice of doxing (malicious publishing of victims' personal data), misinterpretations of social media posts result in victims losing their jobs, being ostracized from their communities, and reputational damage.

Aside from these extremes, users may not comprehend how their data is repurposed for marketing, surveillance, or manipulation beyond the scope of their original intent. One wide-publicized example was the 2018 Cambridge Analytica scandal,[2] in which a consulting firm used the personal data of tens of millions of Facebook users to assist political campaigns by building profiles of voters to target political advertisements. Notably, an integral part of this data breach was through a Facebook personality quiz, "This is your digital life", developed by a University of Cambridge data scientist and originally intended to be taken voluntarily.

In social science research, data often pertains to behavior patterns, preferences, and personal information. Consequently, the potential for causing harm is high. This is why one important principle in using social data for research is anonymization, even the data that is presumably already anonymous—for instance, social media posts that use handles (usernames). People frequently have identical or similar usernames across various networks, making matching data from different sources easy. This can lead to accidental de-anonymization and expose information that is supposed to be private. University research dealing with human subjects should be approved through an Institutional Review Board (IRB), tasked with ensuring study participants' safety and rights. Counterintuitively, even non-interventional social media research that uses publicly available data is subjected to

IRB approval

Even when a university research study uses publicly available social media data, it is a good idea (and frequently required) to obtain approval from an Institutional Review Board (IRB). Extending research requires a new approval!

2 ▸ https://www.bbc.com/news/technology-64075067.

the IRB approval process. Note, however, that this protection is not foolproof. After all, the University of Cambridge IRB properly approved the study, which led to the Cambridge Analytica scandal, yet it deviated from the declared purpose and extent.

Responsible data mining

Following the law

Respect for public rights and interests

Informed consent, if applicable

Transparent data collection, use, and results communication policies

Proper data generalization should help with privacy but not with the intended use issue. Scholars suggest that when generalized data is associated with specific groups, it tends to be used as individual characteristics of people belonging to them. The proposed solution was to limit data analysis to specific purposes, explicitly disclosed to each person prior to data collection.[3] This is clearly unrealistic for data mining, where data patterns resulting from analysis are not apriori known.

Preventing unintended harmful consequences of a study requires transparent data usage policies, ensuring informed consent where applicable, and following legal frameworks to protect individuals while balancing the benefits of data-driven insights. We may generalize the main messages of this section into two principles: accountability and transparency. Accountability refers to analysts'responsibility to ensure that their practices are ethical, legal, and respectful of public rights and interests. The transparency principle requires clarity and openness in how data mining is conducted and how its results are communicated. However, transparency requires that the methodologies, algorithms, and decision-making processes involved in data mining are made explicit and understandable to all relevant stakeholders, which is typically impossible with black-box machine learning algorithms. This is why making AI applications not only effective but also explainable is critically important.

16.2 Explainable AI

16.2.1 Explainability and Interpretability

Machine learning algorithms are frequently opaque "black boxes," with their internal workings obscured. This lack of transparency makes it difficult to explain

3 Van Wel, L., & Royakkers, L. (2004). Ethical issues in web data mining. *Ethics and Information Technology*, *6*(2), 129–140.

how a certain result or a decision recommendation based on this result was made. Consequently, it leads to a lack of stakeholders' trust in data mining. Applying black box algorithms for decision-making may also be illegal; for instance, GDPR in Europe protects an individual's right to an explanation of a decision made after an algorithmic assessment.

Explainable AI (XAI) is a set of methods aiming to alleviate the transparency problems introduced by black box algorithms. Obviously, it is next to impossible to explain the reasoning process underlying results obtained with machine learning, especially to a non-specialist. Instead, explainable AI targets a simplified explanation based on the main factors leading to its results. In a way, explainable AI relates to machine learning algorithms in the same way popular science relates to science: it bridges the gap between complex, technical, scientific knowledge and the public's understanding, making science and technology more approachable and relevant to everyday life.

Explainable AI

Explainability: the ability to justify and explain decisions based on the model's output in a user-friendly format

Interpretability: the ability to explain the model's mechanism.

Two sides of explainable AI are explainability and interpretability. Interpretability is the ability to explain how the model is working. Explainability relates to justifying decisions based on the model's output in a user-friendly format. For instance, decision trees are inherently highly interpretable: one may follow the branches of a tree. In turn, this high interpretability makes a decision based on the results of a decision tree algorithm easy to explain. For instance, if a customer were denied a loan based on a decision tree algorithm, this decision would be significantly more transparent (and easy to dispute should an error be noticed) than an artificial neural network decision. Such highly interpretable data mining methods are sometimes called "white box" or "glass box" models in contrast to opaque "black box" models such as artificial neural networks.

Since explainability and interpretability share the same concept, they are frequently used interchangeably. In this section, we will use the term "explainability" to refer to either.

16.2.2 Algorithms

Certain data mining algorithms, like decision trees, are inherently explainable due to their design. Similarly, many statistical models traditionally taught in courses are also straightforward to interpret. For example, conclusions from linear regression can be understood by analyzing the normalized coefficients. However, even slightly more complex models, such as ridge regression, significantly complicate the explanation process. As a result, the majority of methods require specialized tools to produce clear explanations of their outputs.

Popular XAI algorithms

SHAP: Frequently updated; rich in methods and visualization features. Model agnostic; provides both local and global explanations. May be hard for beginners

LIME: creates interpretable surrogate models; provides local explanations

ELI5: supports scikit-learn classification algorithms; provides both local and global explanations

InterpretML: easier to use for beginners. Focuses on inherently interpretable models; provides both local and global explanations

Examples of popular tools include SHAP (SHapley Additive Explanations), LIME (Local Interpretable Model-agnostic Explanations), ELI5 (Explain Like I am Five), and InterpretML. In addition, some LLMs, such as GPT, can provide reasoning for their analyses. The Additional Reading section includes a highly advisable book by Pradeepta Mishra, which lists multiple approaches and methods and includes Python code examples.

Unfortunately, the code of many of the XAI libraries is rarely updated. Given the reliance on other Python packages, finding the code calling on a depreciated function is not infrequent. This problem is easy to deal with: one should simply downgrade the version of the Python package. In this section, we describe the XAI library SHAP, with our selection partially driven by the fact that this library is frequently updated.

SHAP library is able to provide explanations to an output of a very wide class of machine learning tools. Mathematically, SHAP uses a result from the game theory, which describes a "fair allocation" of a reward (or a loss) between multiple players in a coalition. Applied to data mining, the players are represented by features in a dataset and the game reward is represented by the model prediction. For instance, ▸ Chapter 7 used the Naïve Bayes algorithm to classify newspaper publications into two classes: those that discuss climate change's impact on the tourism industry and those that do not. Its topics represented the features ("players") of a particular publication, and the prediction ("award") was represented by the result of publication classification. The SHAP algorithm would then quantify each topic's "fair" contribution to the model's prediction.

SHAP is a *model-agnostic* XAI method: it can be applied to various machine learning models, including decision tree-based, artificial neural networks, regression models, LLMs, and many others. For instance, applied to sentiment analysis with BERT LLM described in ▸ Chapter 14, SHAP can quantify each word's contribution to a document's predicted sentiment. Feature contri-

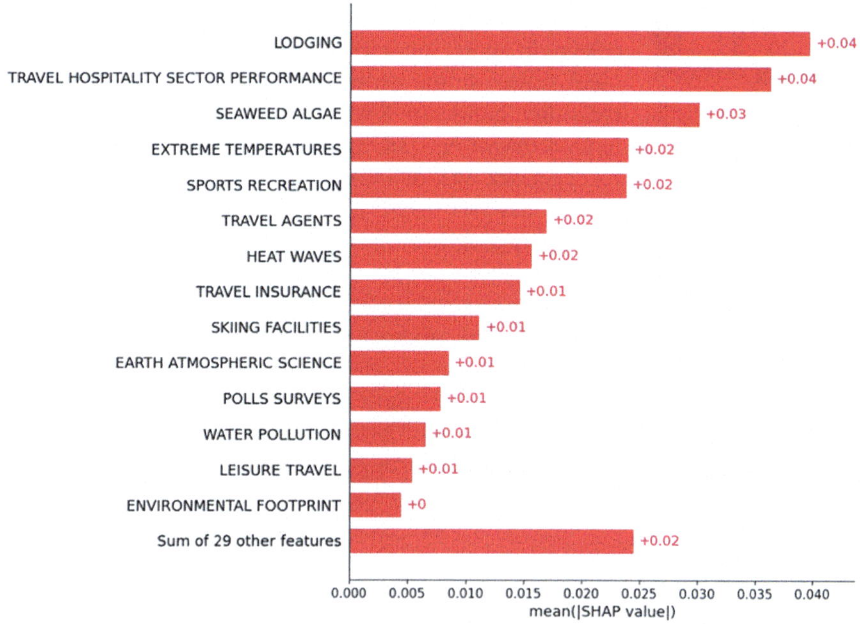

Fig. 16.1 Explaining predictions of a Naïve Bayes model (▸ Chapter 7). The SHAP chart indicates the most important predictors that a newspaper article discusses climate change's impact on the tourism industry

bution to model results is visualized with multiple plots, helping interpretation (Fig. 16.1).

SHAP provides an explanatory analysis of model behavior at two levels: *global* and *local*. At a global level, an XAI provides insights into the *overall* behavior of a black box model. For instance, the global explanation could identify the most important features contributing to model results, capture feature interaction effects, and summarize model behavior at different ranges of features. At a local level, XAI explains a specific prediction, for instance, the classification of a specific newspaper publication. Local explanation helps in answering the question: Why did the model predict this output in this particular case?

Consider a bank loan approval scenario in which the bank reviews applicants' data to determine loan eligibility. The bank uses a machine learning model trained on data from past applicants to compute the probability of new applicants defaulting on a loan. The global XAI aims to explain how different parts of the loan application package impact eligibility decisions in general.

Global and local explanations

Global level: explaining the overall behavior of a model, for instance, the most important features *on average*.

Local level: explaining a particular model decision, for instance, the most important features in one concrete decision.

One would expect the leading factors to be the customer's credit score, income, loan repayment history, employment status, home ownership, and similar measures. On the other hand, an XAI analysis may uncover an unintended illegal bias leading to making decisions based on race, gender, age, or location of an applicant. Meanwhile, a local analysis would help explain why a particular customer's application was rejected.

In another example, SHAP analysis in ◻ Fig. 16.1 provides a generic explanation of what factors determine that a particular newspaper article was classified as related to climate change's impact on the tourism industry. Notice that there are no single most important factors: climate change is a multifaceted problem, and its impact on the tourism industry may be through intensified algae blooming, reef bleaching, heat waves, shortened skiing season, and so on.

Meanwhile, ◻ Fig. 16.2 gives a local explanation of why a particular article titled "Reef damage threatens tourism jobs" was classified as related to climate change's impact on the tourism industry. The major predictors for this article were the presence of topics of Leisure Travel, Water Pollution, Polls Surveys, and Island Reefs. In comparison, a local explanation for the decision in the article "Global warming may overheat Med beaches" relies on a very different list of predictors: Travel hospitality sector performance, Heatwaves, and Extreme temperatures (◻ Fig. 16.3).

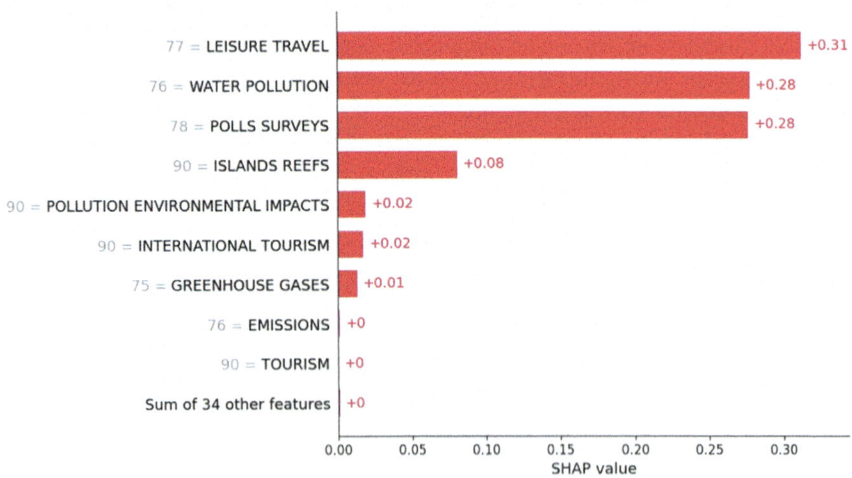

◻ **Fig. 16.2** Local explanation. Explaining a decision to classify the article "Reef damage threatens tourism jobs" as related to climate change's impact on tourism

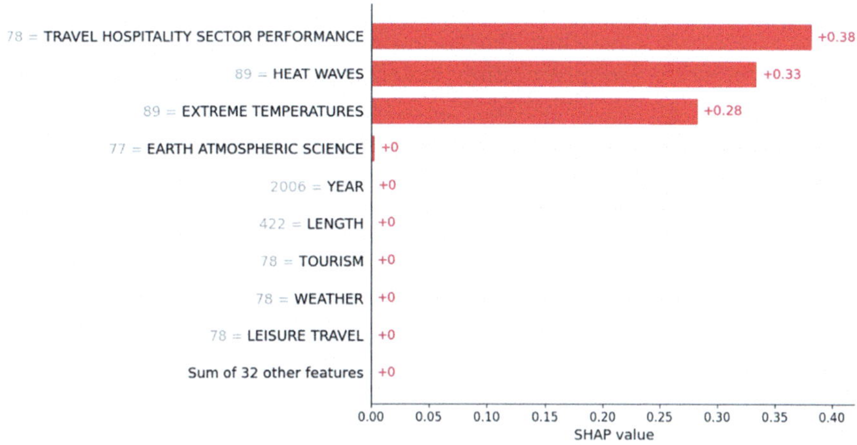

Fig. 16.3 Local explanation. Explaining a decision to classify the article "Global warming may overheat Med beaches" as related to climate change's impact on tourism

SHAP provides many other tools for explaining model output. Overall, model-agnostic XAIs are powerful tools for explaining black box model decisions, but they should not be trusted blindly. While XAIs do provide a post-hoc analysis of model performance, they do not reflect the actual model mechanism. The simple surrogate imitation provided by an XAI does not capture the nuances of the original black-box model. This may create a false overreliance on XAI, leading to unwarranted acceptance of faulty models, in turn damaging public trust in AI. This issue of trust is discussed in the next section.

16.3 Trust in AI

Evidently, there is a lack of trust in machine learning methods and artificial intelligence algorithms in general. This lack of trust is sometimes called a "Black Box Problem." An article by Brożek et al.[4] deconstructed the black box problem into four issues:

4 Brożek, B., Furman, M., Jakubiec, M., Kucharzyk, B.: The black box problem revisited. Real and imaginary challenges for automated legal decision making. Artif. Intell. Law. 1–14 (2023).

- opacity: a limited understanding of the process by which the conclusions were derived;
- justification, inability to explain a reason for a specific outcome;
- unpredictability, arising from humans' aversion to "surprises"; and.
- strangeness, arising from humans' aversion to "soulless" algorithms.

Black box problem

The black box problem in AI and machine learning is a lack of trust in decisions made by models too complex to understand. The underlying issues are:

model opacity;

lack of reasoning justification;

unpredictability of model results, especially with new data;

"strangeness" of the decision-making process fundamentally different from human reasoning

Notice how the issues escalate from rational concerns like opaque model mechanisms to irrational and emotional ones. The opacity challenge may be resolved with a simplified description of the basis for the model decision. In this way, even the challenge of learning LLM ideas, such as encapsulation, transformers, or attention mechanisms, can be replaced with a superficial explanation: "An LLM generates answers by predicting the most likely sequence of words based on patterns learned from vast amounts of text data during training".

The previous section on XAI demonstrates how to deal with the justification of model prediction at both global and local levels. The unpredictability problem is related. Once a person can "understand " how a black-box model functions and how the model has arrived at a particular solution, justification of accepting model results becomes much easier.

According to Brożek et al. (2024) the most important problem in accepting AI as a decision-maker is the *strangeness* of the very fact of accepting the decision made by an AI as opposed to a human. Authors explain the term "strangeness" as the unwillingness of humans to accept the very way decisions are made. For instance, in the LDA topic classification (Volcano Arenal National Park, ▶ Chapter 12), one can easily accept the decision to interpret a string of tokens/probabilities "pool (0.026); hot (0.02); water (0.018); spring (0.017); time (0.011)" as "hot spring visit" since the final decision on topic assignment is made by a human expert, relegating the AI to an advisory role. At the same time, an identical decision made by an AI algorithm, is not well acceptable due to the irresolvable difference in how this decision is made.

Regarding this strangeness effect, it is important to remember that the final decision is always made by a human. Even when an "opaque" and "unpredictable" data mining process aids in research or business analysis, the final decision is the responsibility of a human researcher, analyst, or decision-maker. We hope this book helps to build the fundamentals for that.

16.3.1 Review Questions

1. Explain the ethical considerations of data scraping and provide examples of best practices to minimize harm.
2. What are the potential consequences of misinterpreting content shared on social media platforms?
3. How did the Cambridge Analytica scandal illustrate the misuse of personal data?
4. Why is anonymization important in social science research, and what are the challenges in ensuring true anonymization?
5. Define "explainability" and "interpretability" in the context of Explainable AI. How are they related?
6. How do inherently interpretable models like decision trees differ from models like neural networks regarding explainability?
7. What is the SHAP library, and how does it use game theory to provide explanations for model predictions?
8. Describe the difference between global and local explanations in Explainable AI, providing an example of each.
9. What is the "Black Box Problem" in AI, and what are the four key issues identified by Brożek et al.?
10. How can Explainable AI help address the challenges of opacity and justification in AI decision-making?
11. Why might people find decisions made by AI algorithms less acceptable than those made by humans, even if the outcomes are the same?
12. How does the concept of "strangeness" affect the acceptance of AI-generated decisions, and how can this be mitigated?

16.4 Lab

Objectives

1. Provide a global and a local explanation for Naïve Bayes newspaper article classifier decisions.
2. Provide a local explanation for the BERT LLM sentiment analyzer.

Data

Part 1 Data

The data for the first part of the chapter is the same as in ▶ Chapter 7. It comes from the LexisNexis Newspapers and Wires database of global news media (▶ https://www.lexisnexis.com/ap/academic/form_news_wires.asp). The data is stored in a Microsoft Excel file with two sheets: the LabeledData sheet contains 324 labeled records, and the UnlabeledData sheet contains 2141 unlabeled records.

The goal is to:

(1) Using the labeled dataset, train a classifier to group the publications into two classes: (1) those that are relevant for discussion of climate and tourism interaction and (2) others.

(2) Explain model decision-making by finding the most important attributes generally contributing to the prediction.

(3) Explain the model's classification decision in a particular instance.

The data is stored as ▶ https://github.com/andykir/DataMiningLectures/raw/main/Newspapers.xlsx.

Variable	Explanation	Notes
LABEL	Label. YES: relevant: NO: irrelevant	Binominal. Only for labeled data
YEAR	Year of the publication	Ratio
PUBLICATION	The outlet	String variable
TITLE	Publication title	String variable
LENGTH	Number of words	Ratio
AIRLINES … WETLANDS	41 attributes showing the presence/absence of a topic in the publication. The attribute name defines the topic	Ratio. Values: 0 (lesser than 50% chance of topic presence); 50–100 (probability of the topic presence)

Part 2 Data

The data for the second part is the same as in ► Chapter 13 and comes from the Covid-19 Vaccine Tweets with Sentiment Annotation dataset (► https://www.kaggle.com/datasets/datasciencetool/covid19-vaccine-tweets-with-sentiment-annotation). The data represents tweets about the following widely distributed COVID-19 vaccines: Moderna (USA), Oxford-AstraZeneca (UK), Pfizer-BioNTech (Germany), Sinopharm BIBP (China), Sinovac (China), Covaxin (India), and Sputnik V (Russia) and is stored as ► https://github.com/andykir/DataMiningLectures/raw/main/covid-19_vaccine_tweets_with_sentiment.xlsx.

Variable name	Explanation	Notes
tweet_id	Unique identifier for each tweet	String
label	Manually assigned sentiment	1: negative; 2: neutral; 3: positive
tweet_text	Tweet text	String

The goal of the second part is to explain predictions of tweet sentiment.

Part 1. Provide a Global and Local Explanation for a Naïve Bayes Classifier

We will reuse and simplify the code from ► Chapter 7 for this part.

Step 1. Read and Preprocess Data

Code:

```
import sklearn
import shap

import pandas as pd

# Read the Excel file from the URL
url = "https://github.com/andykir/DataMiningLectures/raw/main/Newspapers.xlsx"
labeledData = pd.read_excel(url, sheet_name="LabeledData") #we need only the labeled
data

# Print information about the labeledData DataFrame
print(labeledData.info())
labeledData.iloc[0:5, 0:6]

#split data into features and labels
X=labeledData.drop(['LABEL'], axis=1)
y=labeledData['LABEL'].map({'YES': True, 'NO': False}) #remap labels to True/False

from sklearn.naive_bayes import GaussianNB
from sklearn.model_selection import train_test_split

# Split the data into training and testing sets. Oversample "YES" class to ensure
balance data.
X_train, X_test, y_train, y_test = train_test_split(X, y, test_size=0.4,
random_state=42,stratify=y)  #"stratify = y" ensure balanced class representation

# Ensure the preprocess data is as intended
print(X_train)
print(y_train)
```

Note the key "stratify=y":

```
X_train, X_test, y_train, y_test = train_test_split(X, y, test_size=0.4,
random_state=42,stratify=y)
```

We use this key to ensure that in our sample used for model training and testing, each label is represented equally.

Result:

```
#    Column                                        Non-Null Count   Dtype
---  ------                                         --------------   -----
 0   LABEL                                          324 non-null     object
 1   YEAR                                           324 non-null     int64
 2   PUBLICATION                                    324 non-null     object
 3   TITLE                                          324 non-null     object
 4   LENGTH                                         324 non-null     int64
 5   AIRLINES                                       324 non-null     int64
.....
321      True
322      True
323      True
```

Step 2. Train Naïve Bayes Classifier

Code:

```
#model train
nb_classifier = GaussianNB()

#we will not use features 'PUBLICATION','TITLE','LENGTH' in the model
X_tr=X_train.drop(['PUBLICATION','TITLE','LENGTH'], axis=1)
X_ts=X_test.drop(['PUBLICATION','TITLE','LENGTH'], axis=1)

nb_classifier.fit(X_tr, y_train)
```

Step 3. Make Predictions

Code:

```
#make predictions
y_pred = nb_classifier.predict(X_ts)
y_pred
```

Result:

```
array([1, 0, 0, 1, 1, 1, 0,… 0, 0])
```

Naïve Bayes classifier result is an array. Each element of the array represents a prediction for one instance; 1 means that the related publication is about climate change in tourism, and 0 – that it is not. To know the corresponding publications, we can print their titles:

```
#print titles of respective publications
X_test['TITLE']
```

Result:

```
170                      Reef damage threatens tourism jobs
43      Mandela of the Maldives forced out by police c...
92      Biodiversity in art Children to show they care...
300                   Global warming may overheat Med beaches
267               Killer heat floods and storms the forecast
                                ...
```

Notice that publications with ID 170, 300, and 267 were correctly classified as "1," while publications 43 and 92 were classified as "0", which is also correct.

Step 4. Evaluate Model Performance on Test Data

Code:

```
# model validation on independent data
print("Naive Bayes score: ", nb_classifier.score(X_ts, y_test))

from sklearn.metrics import accuracy_score, precision_score, recall_score, f1_score

# Calculate evaluation metrics
accuracy = accuracy_score(y_test, y_pred)
precision = precision_score(y_test, y_pred, average='weighted')
recall = recall_score(y_test, y_pred, average='weighted')
f1 = f1_score(y_test, y_pred, average='weighted')

# Print the evaluation metrics
print("Accuracy:", accuracy)
print("Precision:", precision)
print("Recall:", recall)
print("F1-score:", f1)
```

Result:

```
Naive Bayes score:  0.7230769230769231
Accuracy: 0.7230769230769231
Precision: 0.7957767722473605
Recall: 0.7230769230769231
F1-score: 0.7064615384615385
```

Step 4. The global explanation for model output.

We will use ChatGPT for code generation. Start with installing the SHAP library:

```
!pip install shap
```

User:

Nb_classifier is a trained Naive Bayes classifier. X_tr and X_ts are training and testing features; y_train and y_test are training and testing labels. Your goal is to write Python code that uses SHAP to make a global explanation for model results.

ChatGPT code (after correcting initial errors and dropping unused features)

```
# shap model explain

def f(x):
    return nb_classifier.predict_proba(x)[:, 1]

masker = shap.maskers.Independent(data = X_tr)

explainer = shap.Explainer(f, masker)
shap_values = explainer(X_ts)

shap.plots.bar(shap_values, max_display=15)
```

16

Explanation:

```
def f(x):
    return nb_classifier.predict_proba(x)[:, 1]
```

We define a function named *f* that would return Naïve Bayes probabilities computed by the Naïve Bayes classifier. This function's only purpose is to reformat the Naïve Bayes output to the format the SHAP explainer requires.

```
masker = shap.maskers.Independent(data = X_tr)
```

The masker provides data to "train" the SHAP explainer against.

```
explainer = shap.Explainer(f, masker)
shap_values = explainer(X_ts)
```

shap.Explainer uses the function *f* to provide class probabilities and calculates SHAP values for each feature based on its contribution to the model's predictions.

```
shap.plots.bar(shap_values, max_display=15)
```

This snippet draws a bar global explanation plot based on SHAP values for the entire dataset. The key max_display defines how many features to include.

Result:

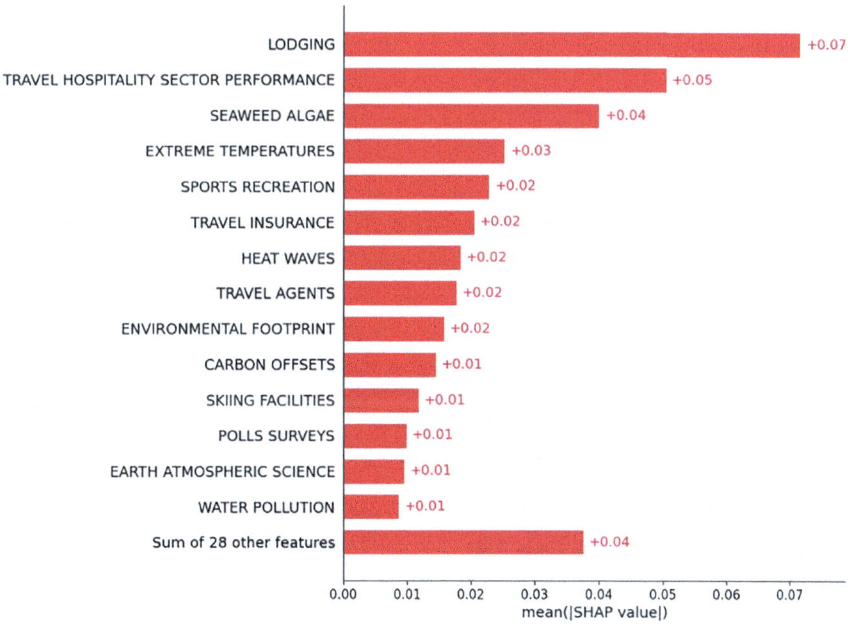

This plot lists feature contributions to model projections in order of importance.

Step 5. Local Explanations

To provide local explanations for a specific instance, we can call the respective visualization function. For instance, the fourth instance in the X_*testing* dataset is "Global warming may overheat Med beaches". To understand which features of this instance have contributed to the model's prediction to classify it as related to climate change and tourism, we can use the following code:

```
shap.plots.bar(shap_values[3])
```

Recall that indexing in Python starts with zero, so the fourth instance's index is 3.

Result:

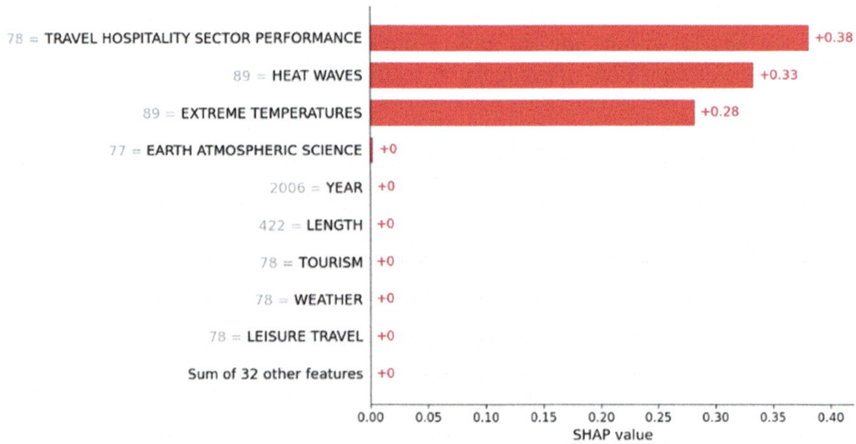

Part 2. Provide Local Explanations for Tweet Sentiment Analysis

Step 1. Reuse Sentiment Analysis Code from Lab 13

Copy the final code from the BERT sentiment analysis lab. Run the code and review the results.

```
import pandas as pd
from transformers import pipeline

# Load the Excel file and the specific sheet
file_path = "https://github.com/andykir/DataMiningLectures/raw/refs/heads/main/covid-
19_vaccine_tweets_with_sentiment.xlsx"
sheet_name = "covid-19_vaccine_tweets"
df = pd.read_excel(file_path, sheet_name=sheet_name)

df=df[:10]

# Initialize the sentiment analysis pipeline with cardiffnlp/twitter-roberta-base-
sentiment model
sentiment_analyzer = pipeline("sentiment-analysis", model="cardiffnlp/twitter-
roberta-base-sentiment")

# Define a function to map sentiment to numerical values
def get_sentiment_label(sentiment, score):
    if score < 0.5:
        return 2  # Low confidence, default to Neutral
    elif sentiment == "LABEL_0":  # Negative
        return 1
    elif sentiment == "LABEL_1":  # Neutral
        return 2
    elif sentiment == "LABEL_2":  # Positive
        return 3
    else:
        raise Exception("Unknown sentiment label encountered")

# Apply the sentiment analysis
sentiment = sentiment_analyzer(df["tweet_text"].tolist())
df["sentiment"] = [get_sentiment_label(x['label'], x['score']) for x in sentiment]

# Save the results to a new Excel file
output_path = "reviews_with_sentiment_twitter_roberta.xlsx"
df.to_excel(output_path, index=False)

print(f"Sentiment analysis completed and saved to {output_path}.")
```

Run the code.

Step 2. Use SHAP to Explain Sentiment Prediction for the First Tweet

We will use ChatGPT for code generation.

USER:

I will give a Python code. Analyze it.
<insert the BERT sentiment analysis code>

ChatGPT:

<returns code analysis>

User:

Your goal is to explain sentiment classification for the first tweet. Write Python code that uses SHAP.

ChatGPT code (after correction):

```
# … reused BERT sentiment classifier code is not included …

import shap

text= [df["tweet_text"][0]]

# Initialize the sentiment analysis pipeline with return_all_scores=True key

sentiment_analyzer = pipeline("sentiment-analysis", model="cardiffnlp/twitter-roberta-base-sentiment", return_all_scores=True )

prediction = sentiment_analyzer(text)
print(prediction[0])

explainer = shap.Explainer(sentiment_analyzer)
shap_values = explainer(text)

print(text, "sentiment: ", df["sentiment"][0])
shap.plots.text(shap_values)
```

Explanation:

```
text = [df["tweet_text"][0]]
```

Extracts the first tweet (df["tweet_text"][0]) from the dataset and stores it in a list. The reason is that the sentiment analysis pipeline expects the input in list format for tokenization and processing.

```
sentiment_analyzer = pipeline("sentiment-analysis", model="cardiffnlp/twitter-roberta-base-sentiment", return_all_scores=True )
```

Without the key *return_all_scores = True*, the pipeline outputs just the prediction with the highest confidence. Now, it outputs probabilities for all sentiment labels (LABEL_0, LABEL_1, and LABEL_2).

```
prediction = sentiment_analyzer(text)
print(prediction[0])
```

The sentiment_analyzer pipeline processes the text to predict the sentiment label and confidence scores.

Result:

[{'label': 'LABEL_0', 'score': 0.8618995547294617},

{'label': 'LABEL_1', 'score': 0.12867186963558197},

{'label': 'LABEL_2', 'score': 0.009428661316633224}]

16

Label_0 has the highest probability, indicating a negative sentiment prediction.

```
explainer = shap.Explainer(sentiment_analyzer)
shap_values = explainer(text)
```

SHAP calculates the importance of each token in the input text for the prediction. With return_all_scores = True, SHAP can explain the contributions to all three sentiment classes, not just the most probable one.

```
print(text, "sentiment: ", df["sentiment"][0])
shap.plots.text(shap_values)
```

Print the text and the most probable sentiment. Then show the tokens contributing positively or negatively to the predicted sentiment probabilities.

Result:

```
['4,000 a day dying from the so called Covid-19 "vaccine" @DailyBeast
reports. #vaccine #PfizerVaccine #Moderna \n\nhttps://t.co/p1nQWWZpk4']
sentiment:  1
```

Explanation for Label_0 (negative sentiment):

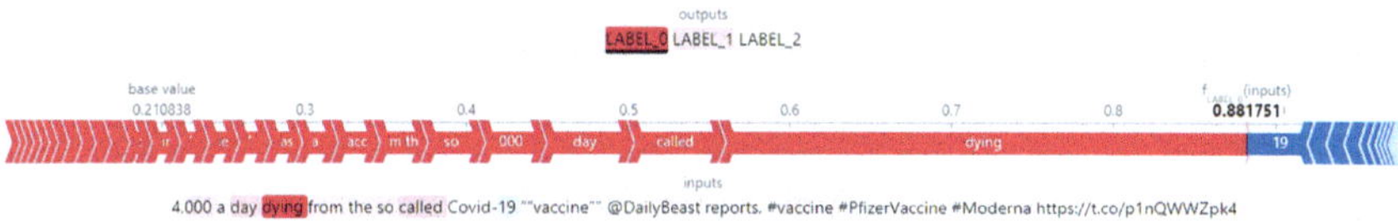

The text in the bottom part of the plot shows the importance of each token contributing to prediction LABEL_0 (negative sentiment). The red color indicates tokens that increase the probability of the label, while the blue-shaded tokens decrease the probability. Notice that "dying" is the most important word in prediction justification.

The force plot in the upper part of the visualization shows how the tokens' influence combined, with the red part increasing the confidence in the LABEL_0 prediction and the blue part reducing it.

Now press the label selector LABEL_1 at the very top of the plot to observe tokens influencing the decision for and against Label_1 prediction (neutral sentiment):

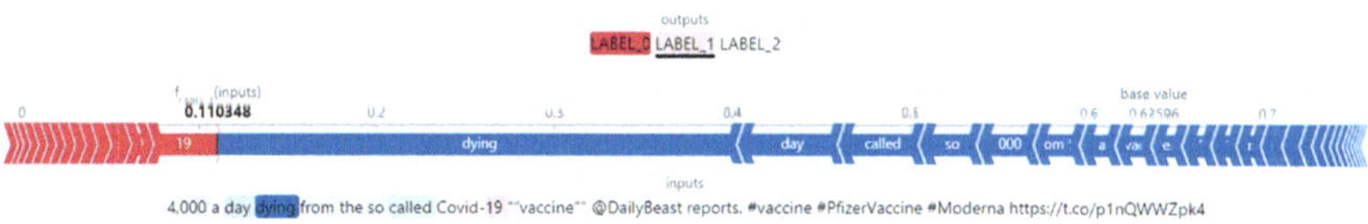

Notice that now the token "dying" reduces the confidence in the LABEL_1 prediction.

Finally, press the label selector LABEL_2 to understand tokens contributing towards and against Label_2 prediction (positive sentiment):

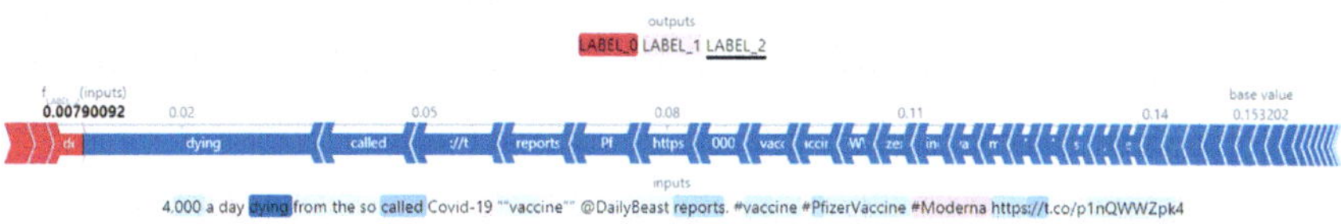

There are very few red tokens indicating no reason to classify this tweet as "positive".

Part 3. On Your Own

Read the SHAP overview and examples: ▶ https://shap.readthedocs.io/en/latest/overviews.html

Continue Part 1, exploring other SHAP functions, such as:

- *shap.plots.beeswarm( )*
- *shap.plots.heatmap( )*
- *shap.summary_plot( ).*

Build Your Python Skills

Chapter 28 of the free book Learn Python with Jupyter by Serena Bonaretti. ▶ https://www.learnpythonwithjupyter.com/

Additional Reading

Michael Galarnyk publishes a highly recommended YouTube podcast on explainable AI. ▶ https://www.youtube.com/@adataodyssey.

Mishra, P. (2022). *Practical explainable AI using Python: Artificial intelligence model explanations using Python-based libraries, extensions, and frameworks*. Apress.

Van Wel, L., & Royakkers, L. (2004). Ethical issues in web data mining. *Ethics and Information Technology, 6*(2), 129–140.